Lecture Notes in Mathematics

Edited by A. Dold and B. Eckmann

496

Topics in K-Theory

L. Hodgkin

The Equivariant Künneth Theorem in K-Theory

V. Snaith

Dyer-Lashof Operations in K-Theory

Springer-Verlag
Berlin · Heidelberg · New York 1975

Authors
Dr. Luke H. Hodgkin
Department of Mathematics
Kings College
London WC2R 2LS/Great Britain

Dr. Victor P. Snaith
Department of Mathematics
Purdue University
W. Lafayette, Indiana 47907
USA

Library of Congress Cataloging in Publication Data
Main entry under title:

Topics in K-theory.

(Lecture notes in mathematics ; 496)
Includes bibliographies and index.
CONTENTS: Hodgkin, L. The equivariant Künneth theorem in K-theory.--Snaith, V. Dyer-Lashof operations in K-theory.
1. K-theory. 2. Spectral sequences (Mathematics) 3. Algebra, Homological. I. Hodgkin, Luke Howard, 1938- The equivariant Künneth theorem in K-theory. 1975. II. Snaith, Victor Percy, 1944- Dyer-Lashof operations in K-theory. 1975. III. Series: Lecture notes in mathematics (Berlin); 496.

QA3.L28 no. 496 [QA612.33] 510'.8s [512'.55]
75-41435

AMS Subject Classifications (1970): 22E15, 55B15, 55B20, 55D35, 55G25, 55G50, 55H05, 55H20, 55J25, 57F35

ISBN 3-540-07536-4 Springer-Verlag Berlin · Heidelberg · New York
ISBN 0-387-07536-4 Springer-Verlag New York · Heidelberg · Berlin

Printed in Germany
Offsetdruck: Julius Beltz, Hemsbach/Bergstr.

These two papers were submitted independently of one another for publication in the Lecture Notes in Mathematics. However, the editors felt that the topics were sufficiently closely related to warrant their publication as one volume.

Each paper contains its own detailed table of contents.

THE EQUIVARIANT KÜNNETH THEOREM IN K-THEOREM

Luke Hodgkin

Table of Contents

Introduction

This book is an improved and much expanded version of a long unavailable University of Warwick preprint (An Equivariant Künneth Formula in K-theory, Warwick 1967; referred to as [0] in this introduction). The aims are to prove the existence of the equivariant Künneth formula; to situate the proof in a general theory of Eilenberg-Moore sequences; and to use the formula to obtain interesting applications. The main difference between the present treatment and [0] are:

1. A general treatment of 'negative filtration' spectral sequences and geometric resolutions, applicable to a wide range of categories of spaces and cohomology theories (§§1 - 5).
2. An investigation of generalized difference elements in the K-theory of homogeneous spaces (§10).
3. As a specific example, the use of the Künneth formula to compute the K-theory of non-simply connected simple groups (§12).

The fundamental theorems, however, are the same as in [0]. In this introduction I shall outline the theorems and the structure of the book; I shall also give an account of what work has been done on the subject since this text (dating from 1970-2) was written.

The main theorems

Let G be a compact Lie group and let X,Y be G-spaces. (We shall suppose all G-spaces locally contractible and of finite covering dimension.) Then following Segal [18] we can define the equivariant K-theory $K_G^*(X)$, which takes values in the category of $\mathbb{Z}_2$-graded modules over the representation ring $R(G) = K_G^*(\text{point})$.[1] We hope to find a formula expressing $K_G^*(X \times Y)$ in terms of $K_G^*(X)$, $K_G^*(Y)$, where G acts on X × Y by the diagonal action. Such a formula, by analogy with the case G = 1 (ordinary K-theory, where one exists, see [4]), would be called a Künneth formula.

---

[1] Strictly, $K_G^0(\text{point}) = R(G)$, $K_G^1(\text{point}) = 0$.

The standard Künneth formula of [4] is a short exact sequence; reference to the proof shows that this is related to the fact that $K^*(\text{point}) = \mathbb{Z}$ has global dimension 1. It would be natural to suppose - compare the situation with other (non-equivariant) cohomology theories [1] - that the appropriate generalization to a coefficient ring like R(G) of arbitrary global dimension is a spectral sequence involving the derived functors $\text{Tor}^i_{R(G)}$. This is in fact what we find. The basic process (§5) constructs a spectral sequence which has the right E_2 term and converges to <u>some</u> definite limit; it is then a non-trivial result (and not always true) that the limit of this natural spectral sequence is indeed $K^*_G(X \times Y)$.

Specifically, the result is the following (a synthesis of Theorem 5.1, Proposition 7.1 and Theorem 8.1.(iii) see §7 for the necessary deduction):

Theorem 1. *Given* G,X,Y *as above, there is a strongly convergent spectral sequence* $\{E_r(X,Y)\}$, *graded by* $\mathbb{Z} \oplus \mathbb{Z}_2$, *such that*

(i) $E_2(X,Y) = \text{Tor}_{R(G)}(K^*_G(X), K^*_G(Y))$.

(ii) *The limit* H(X,Y) *of* $\{E_r(X,Y)\}$ *is a functor of the* G-*spaces* X,Y.

(iii) *There is a natural homomorphism*

$$\Phi: H(X,Y) \to K^*_G(X \times Y)$$

which is an isomorphism when X *or* Y *is a trivial* G-*space*.

The 'equivariant Künneth formula' is the spectral sequence defined by theorem 1. It is, however, only useful when Φ is an isomorphism; some conditions for this are given in the other main theorem.

Theorem 2. *Let* G *be a connected group such that* $\pi_1(G)$ *is torsion-free. Then the* I(G)-*adic completion* $\Phi^\wedge: H(X,Y)^\wedge \to K^*_G(X \times Y)^\wedge$ *is an isomorphism; and* Φ *is an isomorphism provided that* X *or* Y *is a free* G-*space*.

This is proved below as theorem 8.1. Theorems 1 and 2 together correspond to theorem 2.1 of [0]. Note that the restriction that $\pi_1(G)$ should be torsion-free is natural - see §11 for the way in which it is used. The presence of the completion in theorem 2 is much less desirable; I shall comment on the problem below.

By successive specializations from theorem 2 we get

Corollary 1. Consider G as a (free) G-space by left-translation. Then $K_G^*(G) = \mathbb{Z}$ *and* $K_G^*(X \times G) \cong K^*(X)$. (Segal, [18, 2.1.]). *Hence we have a strongly convergent spectral sequence*

$$\{E_r(X,G)\} : \mathrm{Tor}_{R(G)}\,(K_G^*(X), \mathbb{Z}) \Rightarrow K^*(X),$$

when G is connected and $\pi_1(G)$ *is torsion-free.*

Corollary 2. For G as above, let $H \subset G$ *be a closed subgroup; G/H is a G-space in the usual way. Then* $K_G^*(G/H) \cong R(H)$ *and the above sequence becomes*

$$\{E_r(G/H,G)\} : \mathrm{Tor}_{R(G)}(R(H), \mathbb{Z}) \Rightarrow K^*(G/H).$$

(See §9 for these two corollaries). This last very attractive - and useful - special case was the one which first led me to look for the existence of the spectral sequence. It provides the most group-theoretic method available for studying the K-theory of homogeneous spaces, and allows us to gather together a variety of different results in a common framework - see §§9, 11, 12. As I have said before, all of the above results were already proved in [0].

The idea of an Eilenberg-Moore sequence

Let

$$\begin{array}{ccc} F & \longrightarrow & E \\ \downarrow & & \downarrow \\ X & \longrightarrow & B \end{array}$$

be a fibre square of topological spaces and let h^* be a multiplicative cohomology theory. By an 'Eilenberg-Moore' spectral sequence, I mean a spectral sequence

(1) $$\mathrm{Tor}^{p,q}_{h^*(B)}\;(h^*(E), h^*(X)) \Rightarrow h^*(F).$$

Taking $E = \Lambda B$, X = point, this includes the important special case of a sequence which leads from $h^*(B)$ to $h^*(\Omega B)$. The whole problem has been studied from several viewpoints in ordinary cohomology and homology ([10], [19], [16], [2] for example); and it has been pointed out that one way of understanding (1) is to view it as a Künneth formula on the category of spaces over B, i.e., maps with B as target. In this category, $F \to B$ is the product of $X \to B$ and $E \to B$.

The Künneth formula of theorem 1 is not an Eilenberg-Moore sequence, but it is closely related to one. If X is a G-space, then $X_G = X \times_G EG$ is naturally a space over BG. We obtain a functor $(X \longmapsto X_G)$ from G-spaces to spaces over BG which preserves products.[1] And, by a theorem of Atiyah and Segal [7], $K^*(X_G)$ is the I(G)-

---

[1] This functor has been studied by various people including Jon Beck [8].

adic completion $K_G^*(X)^\wedge$.

Hence, applying completion (a very well-behaved exact functor for the size of modules which interest us), we obtain from theorem 1 a spectral sequence

$$(2) \qquad \mathrm{Tor}_{K^*(BG)}(K^*(X_G), K^*(Y_G)) \Rightarrow K^*((X \times Y)_G).$$

If G is connected and $\pi_1(G)$ is torsion-free the spectral sequence converges without further restrictions on X,Y, precisely because of the presence of completion already in theorem 2; and it is an Eilenberg-Moore sequence in the sense defined above. On this basis, with a few other examples, it occurred to me that the right context for the machinery introduced to prove theorem 1 was a general treatment of Künneth formulas, which would include Eilenberg-Moore sequences in general.

Hence the generality of Part I (§§1 - 5), which I hope will be useful to anyone who plans to follow up this programme. However, the more I have pursued my own attempts (which were to form the unwritten §6), the more I have become convinced that there are very serious difficulties at least when $g\ell.\dim.h^0(B) = \infty$.

To take a concrete case, which ought to be very simple, let h^* be K-theory and let us choose $B = S^n$, $E = \Lambda S^n$, X = point, so that as above $F = \Omega S^n$. If we use the machinery of §5 and the cobar resolution (i.e., the construction outlined in I p.18 below), we have a spectral sequence with

$$(3) \qquad E_2 = \mathrm{Tor}_{K^*(S^n)}(\mathbb{Z},\mathbb{Z}) \Rightarrow H = H(\Lambda S^n, \text{point};\ K^*_{S^n})$$

in the notation of §§3, 5. The theory of §§1 - 5 implies that if we use S-maps

$$\Omega S^n \overset{f_k}{\rightsquigarrow} S^{n-1} \vee S^{2(n-1)} \vee \dots \vee S^{k(n-1)} = Y_k \qquad (n > 1)$$

defining the stable decomposition of ΩS^n (James), then in (3) $H = \varinjlim_k K^*(Y_k)$ and the canonical map Φ defined as in theorem 1 above is

$$\Phi = \varinjlim f_k^* : \varinjlim K^*(Y_k) \to K^*(\Omega S^n).$$

On the face of it, this is quite a good result. For example, if n is odd, $H = \varinjlim K^*(Y_k)$ with the appropriate product is a divided polynomial algebra; $K^*(\Omega S^n)$ is its completion, a divided power series algebra. We might hope that the relation between the two could be naturally expressed as in [6], and that this result would give a guide to what can be obtained in general.

What raises the question to a different level of difficulty from the convergence of the Atiyah-Hirzebruch spectral sequence? Simply this: that in order to complete the image $\Phi(H) \subset K^*(\Omega S^n)$ we need a topology which has nothing to do with the spectral sequence. The natural topology on $K^*(\Omega S^n)$, defined by skeleta following Atiyah-Hirzebruch, restricts to a topology on $\Phi(H)$; but this topology cannot be derived from the groups $K^*(Y_k)$, themselves naturally discrete, by any easy passage to the direct limit. Some new methods, new algebraic categories etc., are needed.

I have spent some time on this example - it should be easy for the reader to work out the details, as an exercise in the application of part I. Before leaving it, I should point out (i) that even if we use finite coefficients (K-theory mod p) the same problem arises, and (ii) that we cannot escape the difficulty, as is sometimes possible, by passing to homology theories; in the corresponding K_* version precisely the dual problem is present, with $K_*(\Omega S^n)$ a polynomial algebra and the spectral sequence converging to a power series algebra. And an Eilenberg-Moore sequence which fails to give a result for the case of ΩS^n is not much use.

As one more point about the difficulties to be faced in establishing such sequences, let me mention the following. I have sometimes stated in the past that convergence of an Eilenberg-Moore spectral sequence for h^* in the test case (X = point, $E = \Lambda B$, $F = \Omega B$) was sufficient to establish it for any pair of spaces over B (X,E). This was based on generalizing Proposition 8.1 below, which does prove precisely what I claimed in the case $B = BG$, for G as in Theorem 2. However the proof in Proposition 8.1 that Γ is an additive cohomology theory uses the finiteness of $g\ell$. dim. $K^*(BG)$. Hence again, the proof does not work in cases like $B = S^n$.

To look on the positive side, aside from theorems 1 and 2 one rather mysterious fact does suggest to me that a convergence result for the Eilenberg-Moore sequence can be established in K-theory under special conditions. Let us take h^* to be K-theory mod p, and let $B = K(\pi, n+1)$ where π is finitely generated abelian. Then we have the spectral sequence (as above for ΩS^n)

$$\mathrm{Tor}_{K^*(K(\pi,n+1);\mathbb{Z}_p)}(\mathbb{Z}_p,\mathbb{Z}_p) \Rightarrow H \xrightarrow{\Phi} K^*(K(\pi,n);\mathbb{Z}_p). \tag{4}$$

By [2a], either of the conditions (a) $n \geq 2$ or (b) π finite and $n \geq 1$ ensure that $K^*(K(\pi,n+1);\mathbb{Z}_p) = \mathbb{Z}_p$, and so immediately $H = \mathbb{Z}_p$. We find that Φ is an isomorphism if $n > 2$ or if π is finite and $n = 2$; but that taking $n = 2$, $\pi = \mathbb{Z}$ or $n = 1$, π a p-group give examples of cases where Φ is not an isomorphism. (Compare below, p. 30). Finally, if $n = 1$ and π is free, $K(\pi,2)$ is the classifying space of a torus so that Φ is an isomorphism by an analogue of Theorem 2. There appears to be some link between these conditions for convergence of (4) and the condition

(Tors $\pi_1(G)$ =) Tors $\pi_2(BG) = 0$ required for theorem 2.

I offer the above remarks as an apology for §6, and as indications for the reader who may be interested in developing a satisfactory theory.

Plan of the work

Part I is straightforward, and has as its aim the definition of a general Künneth formula spectral sequence. §1 introduces a family of categories of spaces which include the two of interest to us (G-spaces and spaces over B). Next we deal with the 'negatively filtered spaces' which are required to describe the sequence. A major part of the difficulty in Eilenberg-Moore theory comes from the fact that we are filtering not the space F itself but successively higher suspensions of it, and the necessary modifications to ordinary filtration theory (§2) and to the associated spectral sequence theory (§3) are important for what follows. The form in which it seems to me natural to express the results, using the Cartan-Eilenberg model of H(p,q) systems, provides us with a spectral sequence which converges almost trivially to a certain group H; and an obstruction Γ to the group H being the one we are looking for.

The main requirement for a Künneth formula spectral sequence, introduced in §4, is a geometric resolution of a space X by 'Künneth spaces'. This idea is a direct extension of Atiyah's method of proving the formula in [4] - compare also the 'displays' in Larry Smith's version [19]. Because we know nothing about the global dimension of the ground ring, and for other reasons, we have to allow resolutions to be infinite in length. Simplifying somewhat, a geometric resolution determines naturally a negative filtration; and the product of resolutions of X,Y gives a negative filtration of $X \times Y$, hence a spectral sequence applying h^*. In §5 this is shown to be the required Künneth formula spectral sequence

$$\mathrm{Tor}_{\bar{h}}(h^*(X),h^*(Y)) \Rightarrow H \xrightarrow{\Phi} h^*(X \times Y) \quad .$$

Here h^* is a cohomology theory on our category, with coefficient ring $\bar{h}$; and $X \times Y$ is the product in the category.

The spectral sequence is defined in theorem 5.1. For it to exist, given a category and a cohomology theory h^*, there need to be enough Künneth spaces[1] to construct resolutions. For it to converge to $h^*(X \times Y)$ we need to prove that Φ is an isomorphism. These are the fundamental specific problems for any Eilenberg-Moore sequence.

---

[1] In a sense which corresponds to the usual 'enough projectives'.

In part II I deal with these specific problems for the category of G-spaces and the theory K_G^*; identify geometrically the primary and secondary edge homomorphisms in the spectral sequence; and apply it in a particular case. §7 is a proof of the existence of enough Künneth spaces. The spaces used are equivariant Grassmannians (generalizing [4] again) and the proof, though formally simple, depends on elliptic operators via [5], as do so many interesting results in equivariant K-theory. §8 does most of the work for the second problem, reducing the proof of theorem 2 (as mentioned above) to the single case X = Y = G. I also exhibit two examples where the conditions on G are not satisfied and theorem 2 fails. Before dealing with the remaining test case it is perfectly possible to define and identify edge homomorphisms (composed with Φ if necessary)

$$R(H) \underset{R(G)}{\otimes} \mathbb{Z} = \mathrm{Tor}^{0}_{R(G)} (K_G^*(G/H), K_G^*(G)) \xrightarrow{\bar{\alpha}} K^*(G/H)$$

and

$$\mathrm{Tor}^{1}_{R(G)} (R(H), \mathbb{Z}) \xrightarrow{\bar{\beta}} K^*(G/H)/\mathrm{Im}(\bar{\alpha}) \quad .$$

This is done for $\bar{\alpha}$ in §9; here it turns out that we are dealing essentially with Atiyah and Hirzebruch's α [6]. $\bar{\beta}$ follows in §10, and here we have quite a large generalization of the β used in [11], of independent interest as it defines a wide class of difference elements including some well known special cases.

Having connected β with the spectral sequence, we are in a good position to prove the basic special case:

$$\{E_r(G,G)\}: \mathrm{Tor}_{R(G)}(\mathbb{Z},\mathbb{Z}) \Rightarrow K^*(G)$$

since by the main result of [11, 3], $K^*(G)$ is generated by the image of β when G is connected and $\pi_1(G)$ is torsion-free. The link is completed in §11, with the necessary work on the homological algebra of R(G). This completes the work of establishing theorems 1 and 2 (and the corollaries mentioned above).

Finally in §12, the results are applied to the particular case of $\{E_r(G/\Gamma,G)\}$ when G is simple and simply-connected and Γ is a subgroup of its centre. The sequence converges to $K^*(G/\Gamma)$, which I calculate without much trouble when Γ is of prime order; and with more trouble and some special arguments when Γ is the centre of Spin (2n). This leaves the projective unitary groups PU(n) for n composite and their coverings; here the spectral sequence seems to become much more difficult, and even if it is known to collapse (see below) the extension problems are not, so far as I can see, easily disposed of.

Parallel and subsequent work

The original idea of defining the Eilenberg-Moore spectral sequence via some sort of geometric resolution seems to have occurred to several people around the same time. The best source for such sequences is still the Lecture Notes volume by Larry Smith [19] whose first part deals with the theme of this book's Part I, concentrating on ordinary homology and cohomology where the convergence results work. Further interesting work on Eilenberg-Moore sequences, again in the cohomology case but possibly available for generalization, is to be found in Rector's very interesting paper [16]. But since these there appears to have been little further developed of the general theory.

With regard to the particular case of the equivariant Künneth formula, Haruo Minami proved independently in 1969 the following short exact sequence:

$$(5) \qquad 0 \to K_G^*(X) \otimes K_H^*(Y) \to K_{G\times H}^*(X \times Y) \to \mathrm{Tor}(K_G^*(X), K_H^*(Y)) \to 0$$

[14]. Here X is a G-space, Y is an H-space, and $G \times H$ acts on $X \times Y$ in the obvious way. The sequence, which Minami proves in the Atiyah way using a geometric resolution of X, is in fact a special case of theorem 1, with the help of a generalized argument of the type found at the beginning of §8. We take $G \times H$ to be the group; set up the spectral sequence for X,Y as $(G \times H)$-spaces where H acts trivially on X and G on Y; and use

$$K_{G\times H}^*(X) = K_G^*(X) \otimes R(H); \quad K_{G\times H}^*(Y) = R(G) \otimes K_H^*(X)$$

The relationship of (5) with the spectral sequence now follows from (a) identifying the $\mathrm{Tor}^p_{R(G\times H)}$ for the two modules, where we know $R(G \times H) = R(G) \otimes R(H)$; (b) if X (resp. Y) is any homogeneous space of G (resp. H) then Φ is an isomorphism - in fact, $K_G^*(X) \otimes K_H^*(Y) \cong K_{G\times H}^*(X \times Y)$.

Since the appearance of [0] there has been a certain amount of work published in connection with the equivariant Künneth formula. I shall summarise the work I know of, with apologies to anyone who may have been omitted.

The most interesting field in which improvements have been made to the results of Theorems 1 and 2, is the removal of the annoying completion in the statement of theorem 2. Here the first step was taken by Vic Snaith in [20]; he proved that when G is a torus T, Φ is an isomorphism for all X,Y (without completion). The basic idea of the proof is the following: on the principle of p. it is sufficient to show that Φ is an isomorphism when X,Y are homogeneous spaces of T, and these can be precisely described. Snaith then computes the E_2 term of the spectral sequence explicitly and compares it with $K_T^*(X \times Y)$. The spectral sequence collapses.

In §3 of [20], Snaith asks the wider question: can one now prove the same result for a general G with $\pi_1(G)$ torsion-free by using the above result in the case of a maximal torus $T \subset G$? The answer is that the problem can be reduced to

<u>Conjecture</u> $\phi : R(T) \otimes_{R(G)} R(T) \to K_T^*(G/T)$ is an isomorphism. Here ϕ is the Cartesian product map

$$K_G^*(G/T) \otimes_{R(G)} K_G^*(G/T) \to K_G^*(G/T \times G/T) = K_T^*(G/T) .$$

And, by a result of Pittie (see below), R(T) is a free, hence flat, R(G)-module, so that the K_G^* Künneth formula spectral sequence of $G/T \times G/T$ collapses, and Φ as defined in theorem 1 reduces to the Cartesian product, i.e., to ϕ. The conjecture is therefore a special case of what we want to prove; Snaith shows that it implies the general case.

A note on the reduction in [20] would perhaps be interesting here, since it is rather condensed and involves a result which is of independent interest. The important statement is made (p.176) that a G-resolution

$$X \to Z_o \to Z_1 \to Z_2 \to \qquad \cdots \tag{6}$$

for a G-space X is also a T-resolution; this is done first in the case $X = G/_T$ and then for X a general G-space. In each case enough has been proved to ensure that the sequence

$$0 \leftarrow K_T^*(X) \leftarrow K_T^*(Z_o) \leftarrow K_T^*(Z_1) \leftarrow \qquad \cdots$$

is exact, being obtained from the K_G^* sequence by applying $R(T) \otimes_{R(G)} -$. It remains to ensure that Z_i is actually a Künneth space for K_T^*, given that it is one for K_G^*. But here we can use the fact that $K_G^*(Z_i)$ is finitely generated and projective over R(G). (Definition 4.1) and hence $K_T^*(Z_i) = R(T) \otimes_{R(G)} K_G^*(Z_i)$ is finitely generated and projective over R(T). Now we have the following corollary of Theorem 1.

<u>Corollary 3</u>. <u>Suppose Φ has been proved to be an isomorphism for all G-spaces X,Y. Then Z is a Künneth space for G if and only if $K_G^*(Z)$ is finitely generated and projective over R(G).</u>

<u>Proof</u>. In this case any spectral sequence $\{E_r(Z,Y)\}$ collapses and reduces to the Cartesian product morphism

$$K_G^*(Z) \otimes_{R(G)} K_G^*(Y) \to K_G^*(Z \times Y)$$

(see §9), which must therefore be an isomorphism.

Applying the corollary in the case where G is T (and we know that Φ is an isomorphism), we can deduce that the spaces Z_i are Künneth spaces for T.

Recently, John McLeod has announced that the above conjecture is true, and hence Φ is an isomorphism without completion for all G. [13] .

Further important work on the equivariant Künneth formula has been done by Snaith in several other papers [21, 22, 23]. In particular he identifies the differentials in the spectral sequence (of which no mention has been made here) as generalized Massey products, [21]; and shows that in the particular sequence for a homogeneous space G/H (Cor. 2) under slight restrictions on H all these differentials vanish, [22], thus proving a parallel to a theorem of Peter May [12] for ordinary cohomology.

Apart from these general results, there have been two successful applications (to my knowledge) of the spectral sequence to the K-theory of homogeneous spaces. The first is that of Harsh Pittie [15] for $K^*(G/U)$ when U is of maximal rank in G - I refer to this in more detail below, §9. The second is A. Roux's work [17] on the K-theory of Stiefel manifolds - here the covering group has to be taken to be the spinor group and the results generalize what we find in §12 on the K-theory of SO(n).

I am sure that a great deal more could be done with the G/H spectral sequence, drawing if necessary on the collapsing theorem of Snaith referred to above. Apart from this, as I have said already, the main problem raised by this work is the proper formulation and proof of convergence for the Eilenberg-Moore spectral sequence. The two problems might come together in an investigation of what the spectral sequences told us about the K-theory of the iterated loopspaces of BO, in particular (see [9]) spaces like $U/O = \Omega^6 O$, etc. I hope these various challenges will attract some readers.

I must thank a large number of people for helpful conversations during this work at various stages; in particular J.F. Adams, M.F. Atiyah, Jon Beck, David Epstein, Rolph Schwarzenberger, Graeme Segal and Vic Snaith; also Professor B. Eckmann and the ETH Zürich for providing me with time and place to begin the work during a month in Zürich (four years ago!). Finally, Mrs. Joan Bunn has typed and retyped the manuscript under increasingly demanding conditions; I am especially grateful to her.

Introduction - bibliography

[0] L. Hodgkin, An Equivariant Künneth formula for K-theory, preprint, University of Warwick, 1967.

[1] J.F. Adams, Lectures on Generalized Cohomology, in 'Category Theory Homology Theory and Applications', Lecture Notes in Math. 99, Springer, 1969.

[2] J.F. Adams, On the Cobar construction, Proc. Nat. Acad. Sci. U.S.A. 42 (1956), 409-412.

[2a] D.W. Anderson and L.H. Hodgkin, The K-theory of Eilenberg-MacLane complexes, Topology 7 (1968), 317-329.

[3] S. Araki, Hopf Structures attached to K-theory: Hodgkin's Theorem, Ann. Math. 85 (1967), 508-525.

[4] M.F. Atiyah, Vector bundles and the Künneth formula, Topology. 1(1962), 245-248.

[5] ————, Bott periodicity and the index of elliptic operators, Quart.J.Math.19(1968),113-140

[6] ————, Characters and cohomology of finite groups, Publ. Math. IHES 9 (1961), 23-64.

[7] M.F. Atiyah and F. Hirzebruch, Vector bundles and homogeneous spaces, Proc. Sympos. Pure Math. AMS 3 (1961), 7-38.

[7a] M.F. Atiyah and G. Segal, Equivariant K-theory and completion, J. Diff. Geom. 3 (1969). 1 - 19.

[8] J. Beck, On H-spaces and infinite loop spaces, in 'Category Theory, Homology Theory and Applications, Lecture Notes in Math. 99, Springer, 1969.

[9] R. Bott, The Stable Homotopy of the Classical Groups, Ann. Math. 70 (1959), 313-337.

[10] S. Eilenberg and J.C. Moore, Homology and Fibrations I, Comm. Math. Helv. 40 (1966), 199-236.

[11] L Hodgkin, The K-theory of Lie groups, Topology 6 (1967), 1-36.

[12] J.P. May, The cohomology of principal bundles, homogeneous spaces and 2-stage Postnikov systems, Bull.A.M.S. 74 (1968), 334-339.

[13] J. McLeod (to appear).

[14] H. Minami, A Künneth formula for equivariant K-theory, Osaka J. Math. 6 (1969), 143-6.

[15] H. Pittie, Homogeneous vector bundles on homogeneous spaces, Topology 11 (1972), 199-204.

[16] D.L. Rector, Steenrod operations in the Eilenberg-Moore spectral sequence, Comm. Math. Helv. 45 (1970), 540-552.

[17] A. Roux, Application de la suite spectrale d'Hodgkin aux variétés de Stiefel, Bull. Soc. Math. France 99 (1971), 345-368.

[18] G. Segal, Equivariant K-theory, Publ. Math. IHES 34 (1968), 129-151.

[19] L. Smith, Lectures on the Eilenberg-Moore spectral sequence, Lecture Notes in Math. no. 134 (1970), Springer.

[20] V.P. Snaith, On the Künneth formula spectral sequence in equivariant K-theory, Proc. Camb. Phil. Soc. 72 (1972), 167-177.

[21] ————, Massey products in K-theory, Proc. Camb. Phil. Soc. 68 (1970), 303-320.

[22] ————, Massy products in K-theory II, Proc. Camb. Phil. Soc. 69 (1971), 259-289.

[23] V.P. Snaith, On the K-theory of homogeneous spaces and conjugate bundles of Lie groups, Proc. L.M.S. (III) 22 (1971), 562-584.

Part I

Künneth Formula Spectral Sequences

§1. Categories of spaces

In what follows 'topological space' is to be taken as meaning 'compactly generated space having the homotopy type of a CW complex' (see [17, 22] for these ideas). The category of these spaces, and arbitrary continuous mappings between them, will be called Top. The product in Top is the compactly generated product [22]. I shall be dealing mainly in the sequel with two families of topological categories with extra structure, which generalize Top.

1. Categories of G-spaces, where G is a compact Lie group*; i.e., full subcategories of the category G-Top whose objects are G-spaces and morphisms G-maps (equivariant maps).

2. Categories of 'spaces over B' where B is in Top; i.e., full subcategories of the category Top/B whose objects are continuous mappings $X \xrightarrow{f_X} B$, and whose morphisms $(X,f_X) \to (Y,f_Y)$ are mappings $g: X \to Y$ such that the triangle

(1)

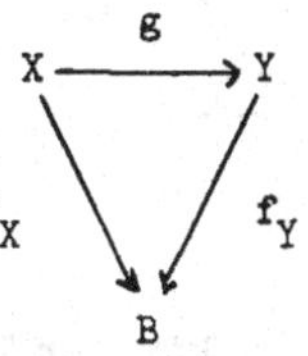

is commutative.

(I shall write an object in Top/B as a pair (X,f_X), shortening it to X when I can get

---

* This restriction on G could be relaxed in much of what follows.

away with it. Many formulations of these ideas exist in more general settings - see for example [15]).

We note the following properties shared by G-Top and Top/B, together with their reasonable subcategories.

A.1. Both have a terminal object θ ; i.e., an object such that for any X there is only one morphism $X \to \theta$. For G-Top, θ = point (as a G-space in the unique, trivial way). For Top/B,

$$\theta = (B \xrightarrow{1_B} B).$$

A.2. Both are closed under the formation of (finite) sums and products. 'Sum' in both cases is just disjoint union with appropriate structure. The categorical product of G-spaces X,Y is the product $X \times Y$ in Top given the diagonal action of G:

$$g \cdot (x,y) = (g \cdot x, g \cdot y)$$

The categorical product of (X,f_X) and (Y,f_Y) in Top/B is the fibred product

$$X \underset{B}{\Pi} Y = \{(x,y) \in X \times Y : f_X(x) = f_Y(y)\} \quad .$$

Again, these are standard observations.

A.3. Top 'operates' on the two categories via functors

$$\text{G - Top} \times \text{Top} \to \text{G-Top}$$

$$\text{Top/B} \times \text{Top} \to \text{Top/B}$$

which I shall write $(X,U) \longmapsto X \otimes U$ for X in G-Top or Top/B, U in Top. The definition is straightforward; if X is a G-space define $X \otimes U = X \times U$ with G acting on the factor X, and if (X,f_X) is a space over B define $X \otimes U = (X \times U, \ f_X \circ p_1)$ where p_1 is the projection on the first factor. These operations have various obvious properties which I shall use when they become necessary. They have been formalized in a more general context [5, p.139].

A.4. There is a notion of 'homotopy', an equivalence relation on mappings, defined by:

$f_0, f_1 : X \to Y$ are homotopic if and only if there exists $F: X \otimes I \to Y$ (in the category) such that $F|X \otimes \{i\} = f_i$ $(i = 0,1)$.

Of course $X \otimes \{\text{point}\} = X$ is used here.

In G-Top the relation is homotopy through G-maps, in Top/B fibrewise homotopy.

A.5. The mapping cylinder construction can be defined, and enables us to replace any map by a cofibration relative to the notion of homotopy defined in A.4. To make this explicit it is necessary to have an idea of how to construct an identification space in each category. In G-Top it is easy; an invariant closed subspace R of $X \times X$, for X a G-space, which defines an equivalence relation, gives an identification space in the usual way, on which G still acts. (For details see [18]). In Top/B R must be a subspace of $X \prod_B X$ - a fibrewise relation, that does not identify points in different fibres. For instance, if $f: X \to Y$ is a map and $X \subset U$ an inclusion in Top/B, I can glue U to Y by f in the usual way and map the adjunction space into B unambiguously.

Define, then, the mapping cylinder M_f of $f: X \to Y$ in each case to be the adjunction space

$$Y \cup_f X \otimes I \ .$$

($f(x)$ is identified with $(x,1)$). The result is still in the original category, and in the diagram

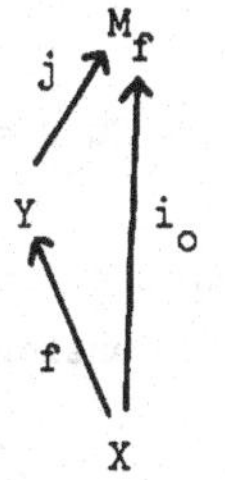

(where $i_o(x) = x \otimes 0$), j is a homotopy equivalence, i_o is a cofibration, and $j \circ f \simeq i_o$ (*) This is what is meant by 'replacing f by a cofibration'.

A.1. - A.5. could, along with a great deal more, be axiomatized and much of the following could go through for an arbitrary category satisfying the required axioms. (See [5]). I prefer to keep up the convention that we are studying G-Top and Top/B, and to appeal to A.1 - A.5 when needed. In what follows, $\mathcal{C}$ will denote a category

(*) The homotopy here is canonical, as are many other features of the construction - homotopy inverse of j, etc - as in the usual mapping cylinder.

of either type, or a subcategory of some suitable kind (say, closed under sums, products and mapping cylinders, and allowing the operation of at least the finite complexes in Top).

A number of obvious functors connect these categories. Given a homomorphism $\rho : H \to G$ of groups define $\rho^* : \underline{G\text{-Top}} \to \underline{H\text{-Top}}$ to be the functor which assigns to each G-space X the space X with the H-action induced by ρ. Its adjoint ρ_* assigns to an H-space Y the G-space $G \times_H Y$, quotient of G by the relation

$$(g \,.\, \rho(h), y) = (g, h.y) \qquad (*)$$

These are analogous to the 'change of rings' functors on modules. If H = 1 (trivial group), ρ^* is the forgetful functor while $\rho_*(Y) = G \times Y$. If on the other hand G = 1, ρ^* is the functor $X \longmapsto \theta \otimes X$ (X with trivial H-action) of A.3. and $\rho_*(Y)$ is the orbit space Y/H.

Mappings of base spaces define analogous functors between different categories Top/B which I leave it to the reader to investigate. Note in particular the fibre functors (induced by inclusions {point} → B) and the functor $X \longmapsto \theta \otimes X$ (induced by B → {point}).

Finally, any G-space X defines a space X_G over the classifying space BG of G, in a well-known way (see [6, p.52]):

$$X_G = EG \times_G X \to EG/G = BG$$

(EG being a fixed universal G-bundle.) This defines for each G a functor $\underline{G\text{-Top}} \to \underline{\text{Top/BG}}$ which is useful in understanding the analogy between the parallel theories being developed here.

In order to do homotopy theory in one of our categories we need to 'introduce basepoints' - to pass to the associated pointed category. If C is as above and θ is its terminal object define C_o to be the category $\theta \backslash C$ of all morphisms $\theta \xrightarrow{k_X} X$ in C . A C_o-morphism $(X, k_X) \to (Y, k_Y)$ is defined by a commutative diagram

---

(*)

More generally, given a right H-space X we define $X \times_H Y = X \times Y/\{(x.h, y) = (x,h.y)\}$. This parallels the tensor product.

(3) 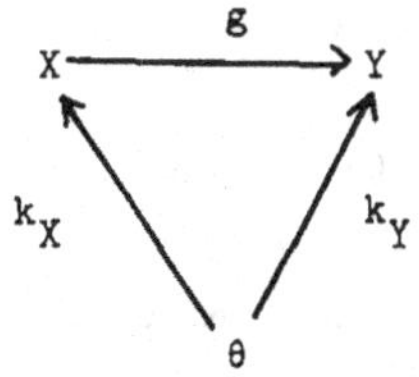

(compare (1)).

If C = G-Top, the interpretation is straightforward; as in Top, since θ is a point, k_X can be identified with its image $x_o = k_X(\theta) \in X$. An object of G-Top_o is a based G-space (X,x_o), a morphism is a basepoint preserving map. Note however that x_o must be a fixed point for k_X to be a G-map; so not all G-spaces admit basepoints.

In the case C = Top/B we have a more complicated idea which has been studied by James and others [14] under the name of 'ex-spaces'. An object of Top/B_o is by definition a triple (X,f_X, k_X) where $f_X: X \to B$, $k_X: B \to X$ are maps and $f_X \circ k_X = 1_B$. In other words k_X is a cross section of the mapping f_X, and a mapping $g: X \to Y$ defines a morphism in Top/B_o only if two triangles are commutative. As in Top we have a functor from C to C_o which assigns to X in C the object $X^+ = (X \cup \theta, i_2)$ where $i_2: \theta \to X \cup \theta$ (disjoint union) is the inclusion.

The categorical sum in C_o generalizes the wedge and will be written $X \vee Y$ as usual (for a finite set of spaces, write $X_1 \vee \ldots \vee X_n = \bigvee_i X_i$). For G-spaces this is the ordinary wedge; for spaces X,Y over B it is the fibrewise wedge obtained from $X \cup Y$ by identifying $k_X(b)$, $k_Y(b)$ for all $b \in B$. The product in C_o is the same as that in C; I shall write it $X \times Y$, or $X \underset{B}{\Pi} Y$ when I wish to emphasize that we are in the Top/B situation.

$X \vee Y \to X \times Y$ is defined in the usual way as a C_o-embedding. As explained in connexion with A.5., we can form the quotient space $X \wedge Y = X \times Y/X \vee Y$, the 'smash product' in C_o. In Top/B this may be written $X \underset{B}{\wedge} Y$, since as a space it is not the ordinary smash product; it is obtained from $X \underset{B}{\Pi} Y$ by identifying $(x,k_Y(b))$ and $(k_X(b),y)$ for all pairs (x,y) with $f_X(x) = f_Y(y) = b$.

The operation (A.3) of Top on C induces an operation of Top_o on C_o corresponding to the smash product. I shall write it as

$$(X,U) \longmapsto X \tilde{\otimes} U \qquad X,U \text{ in } C_o$$

Notice that for Top/B this is not $X \wedge U$ as a topological space (though $X \otimes U$ is the product), but $X \times U$ with the identifications

$$(x,u_o) = (k_X f_X(x),u) \qquad (x \in X, \quad u \in U; \quad u_o \text{ basepoint of } U.)$$

The fibre of this space over any $b \in B$ is $X_b \wedge U$ where X_b is the fibre of X over b. (It is advisable to work out some of these constructions in Top/B to understand the general principle; broadly, to do the standard construction in Top on each fibre separately.)

The above operation defines us the cone and suspension functors:

$$CX = X \tilde{\otimes} I \quad \text{(I has basepoint 1).}$$

$$SX = X \tilde{\otimes} S^1 \quad .$$

More generally, we can define the reduced mapping cylinder and mapping cone of $f\colon X \to Y$ in C_o

$$\tilde{M}_f = Y \cup_f (X \tilde{\otimes} I^+)$$

$$C_f = Y \cup_f (X \tilde{\otimes} I)$$

with appropriate identifications. CX is the mapping cylinder, and SX the mapping cone, of the unique map $X \to \theta$.

All this enables us, and this is the main aim, to reproduce painlessly in C_o the Puppe sequence construction for topological spaces. Let $X \to Y$ be any mapping in C_o; then we have canonically associated a sequence of mappings

(4) $$X \xrightarrow{f} Y \xrightarrow{Pf} C_f \xrightarrow{Qf} SX \xrightarrow{Sf} SY \xrightarrow{SPf} SC_f \to \quad \cdots$$

such that, if we denote the set of C_o-homotopy classes of mappings by $[\ ,\ ]C_o$, the sequence

$$[X,V]_{C_o} \xleftarrow{f^*} [Y,V]_{C_o} \xleftarrow{(Pf)^*} [C_f,V]_{C_o} \leftarrow \cdots$$

is exact, for any V in C_o. In fact each pair of consecutive maps in (4) is equivalent to a cofibration.

We shall need the following properties of the Puppe sequence:

1. If f is an inclusion, C_f is canonically C_o-homotopy equivalent to the quotient space (in C_o) Y/X.
2. The functor $Z \wedge \_$, for Z in C_o, sends the sequence (4) to another Puppe sequence, that of the map $1 \wedge f : Z \wedge X \to Z \wedge Y$.

§2. Negative filtrations

In order to motivate the kind of construction we shall have to introduce here, consider the construction due to Milnor [16] for the universal bundle of a group G, as it has subsequently been explained (cf [11]). The basis of this construction is the following. Let X be a G-space; let the product $G \times X$ be made into a G-space by letting G act on the first factor only: $g' \cdot (g,x) = (g'g,x)$. Then the action map $G \times X \xrightarrow{\mu} X$, $(g,x) \longmapsto g \cdot x$, is a G-map. All this is standard adjoint functor theory related to the adjoint pairs discussed in §1. Now define A(X) to be the G-cofibre of μ, i.e., $A(X) = X \cup_\mu C(G \times X)$; X is included in A(X) as a subspace and the construction can be iterated:

$$X \to A(X) \to A(A(X)) = A^2(X) \to A^3(X) \to \quad \cdots$$

Since $A(X)/X = S(G \times X)$, we arrive at a filtered space $A^\infty(X) = \bigcup_i A^i(X)$ in which the filtration quotients are (mod basepoint) free G-spaces. $A^\infty(X)$ is in fact the space X_G defined in §1, in particular $A^\infty(G) = EG$, $A^\infty(\text{point}) = BG$. The construction of $A^\infty(X)$ is a topological analogue of the bar construction, which finds its reflection in the associated spectral sequence, cf. [19].

Now look at the corresponding attempt at the inverse construction of a loop space for a space B which underlies the theory of the Eilenberg-Moore spectral sequence. Given a space B we have a similar adjoint pair

$$\underline{\text{Top}/B} \xrightarrow{G} \underline{\text{Top}} \xrightarrow{F} \underline{\text{Top}/B}$$

(G is the forgetful functor, F is $- \otimes \theta$). The adjunction $\varepsilon: 1 \to FG$ can be defined by

$$\varepsilon(X) : (X,f_X) \to (X \times B,p_2) \ ; \ \varepsilon(X)(x) = (x,f_X(x)), \quad x \in X .$$

This maps X <u>into</u> a 'free' space over B. The previous construction must therefore be reversed; given X we let D(X) denote the mapping cone of $\varepsilon(X)$ in the sense of <u>Top/B</u>. By a Puppe sequence argument we have a map

$$Q\varepsilon(X) : D(X) \to SX. \quad (*)$$

(*) This is the suspension <u>over B</u> (above, p. 16).

Iterating gives $D^2(X) = D(D(X)) \xrightarrow{Q\varepsilon(D(X))} SD(X) \xrightarrow{SQ\varepsilon(X)} S^2X$ and so on. At the nth stage we have a 'filtration' of S^nX, and we are unable in general to arrive at a space $D^\infty(X)$ analogous to A^∞ above. In the sense of the objects we know how to deal with in homotopy theory (e.g. spectra) the arrows are going the wrong way.

This illustrates a major source of the difficulties we shall find in trying to arrive at Künneth formula spectral sequences in the categories of §1. However, the spaces $D^n(X)$ and maps $D^n(X) \to SD^{n-1}(X)$ give us an important construction, which could be called the geometric cobar construction of X, and which is functorial in X. It is the natural analogue of the Milnor geometric bar construction, and $S^n\Omega^nB$ (the ordinary topological suspension) is filtered by the fibres $\{\text{point}\}\prod_B(S^jD^{n-j}\ (\Lambda B))$, $j = 0,1,\dots,n$. We need to develop a general theory of 'filtrations' like the one defined above.

Definition 2.1. Let C_o be a pointed category as in §1 and X in C_o. A negative filtration of X of degree m is a sequence of spaces X_i and mappings $\phi_i : X_{i+1} \to S^mX_i$ $(i \geq 0)$, with $X_o = X$.

Hence, for any finite n, a negative filtration $X_* = \{X_i,\phi_i\}$ defines a sequence

(1) $$X_n \xrightarrow{\phi_{n-1}} S^mX_{n-1} \xrightarrow{S^m\phi_{n-2}} S^{2m}X_{n-2} \to \dots \xrightarrow{S^{(n-1)m}\phi_o} S^{nm}X_o$$

It would be convenient to turn this into a genuine filtration, i.e., replace the mappings by inclusions: but we can't do it for all mappings simultaneously (modifying ϕ_n involves changing X_n and so spoils our modification of ϕ_{n-1}, etc). But we can do the replacement for any finite n in a canonical way, by the (reduced) mapping cylinder construction, and we can arrange that the replacements fit together well. Suppose in fact that we have a sequence of inclusions homotopy equivalent to (1):

(2) $$X_n^n \overset{\subset}{\to} X_{n-1}^n \overset{\subset}{\to} X_{n-2}^n \overset{\subset}{\to} \dots \overset{\subset}{\to} X_o^n$$

Then $S^mX_{i+1}^n \subset S^mX_i^n$ for $0 \leq i < n$. Via the homotopy equivalence, ϕ_n gives rise to a map $\phi_n' : X_{n+1} \to S^mX_n^n$. Define $X_{n+1}^{n+1} = X_{n+1}$ and for each $i \leq n$ let X_i^{n+1} be the reduced mapping cylinder of the composite $X_{n+1} \xrightarrow{\phi_n'} S^mX_n^n \overset{\subset}{\to} S^mX_i^n$. Then the new sequence

$$X_{n+1}^{n+1} \to X_n^{n+1} \to \dots \to X_o^{n+1}$$

which we obtain is homotopy equivalent to

$$X_{n+1} \xrightarrow{\phi_n'} S^{mn}X_n \overset{\subseteq}{\to} S^{mn}X_{n-1} \overset{\subseteq}{\to} \dots \overset{\subseteq}{\to} S^{mn}X_o .$$

Moreover the maps and homotopies used in showing that this is a homotopy equivalence can, as usual for the mapping cylinder construction, be made canonical.

This is not a full discussion of the extent to which maps in (1) can be taken to be inclusions; however, to go into more detail would, I think, be tedious and would upset the balance of a text which I hope to keep as non-technical as possible. The reader who is doubtful about such details is advised to consult [21] or a text on homotopy theory (for example [25]). I shall not refer to them in future unless special care is required. Where no ambiguity arises I shall, abusively, treat X_{i+1} as a subspace of $S^m X_i$.

Given negative filtrations $X_* = \{X_i, \phi_i\}$ and $Y_* = \{Y_j, \psi_j\}$ of degrees m, m', in C_o, a mapping from the first to the second is (for $m \geq m'$) a sequence of C_o maps $\{f_i : X_i \to S^{i(m-m')} Y_i\}$ such that the diagram

$$\begin{array}{ccc}
X_{i+1} & \xrightarrow{\quad \phi_i \quad} & S^m X_i \\
\big\downarrow {\scriptstyle f_{i+1}} & & \big\downarrow {\scriptstyle S^m f_i} \\
S^{(i+1)(m-m')} Y_{i+1} & \xrightarrow[S^{(i+1)(m-m')}\psi_i]{} & S^{(i+1)m-im'} Y_i
\end{array}$$

is commutative for every i (If $m < m'$ make the obvious adjustments). Negative filtrations and their mappings form a category, $NF(C_o)$; if we restrict attention to those of a given degree m we obtain a subcategory $NF^m(C_o)$.

Many obvious functors can be applied to negative filtrations to obtain others. In particular given Y in C_o and $X_* = \{X_i, \phi_i\}$ in $NF^m(C_o)$, then $X_* \wedge Y = \{X_i \wedge Y, \phi_i \wedge 1_Y\}$ is also in $NF^m(C_o)$. In this way, for example, we can suspend negative filtrations.

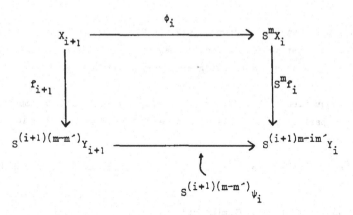

Now given negative filtrations X_* of X and Y_* of Y it is important for the future that I should be able to construct a product filtration of $X \wedge Y$. To do this we go carefully using inclusions (formula (2)). Let the family $\{X_i^n\}$ be defined from X_* as in (2) and similarly $\{Y^n_i\}$ from Y_* $(i \leq n)$. Now filter $X^n_o \wedge Y^n_o = Z^n_o$ in the obvious way:

(3) $$Z^n_k = \bigcup_{i+j=k} X^n_i \wedge Y^n_j \qquad 0 \leq k \leq n.$$

Then $Z^n_n \subset Z^n_{n-1} \subset \ldots \subset Z^n_o$; and the canonical homotopy equivalences of $S^m X^n_i$ with X^{n+1}_i, $S^{m'} Y^n_j$ with Y^{n+1}_j, give equally canonical homotopy equivalences of

$$S^{m+m'} Z^n_k = \bigcup_{i+j=k} S^m X^n_i \wedge S^{m'} Y^n_k$$

with Z^{n+1}_k.

Hence, defining $Z_n = Z^n_n$, we have a mapping χ_n or $Z_{n+1} = Z^{n+1}_{n+1} \to Z^{n+1}_n \to S^{m+m'} Z^n_n = S^{m+m'} Z_n$, which makes $\{Z_n, \chi_n\}$ into a negative filtration of degree $m + m'$, with $Z_o = X_o \wedge Y_o = X \wedge Y$. I shall call this filtration $X_* \otimes Y_*$, and its ith space $(X_* \otimes Y_*)_i$.

Note Because this natural product maps $NF^m \times NF^{m'}$ into $NF^{m+m'}$, it is inconvenient to stick to negative filtrations of degree 1 as I did in [13]. Some unnecessary complication is saved by allowing arbitrary positive degrees.

The filtration quotients $\{S^m X_i / X_{i+1}\}$ $(i = 0,1,\ldots)$ will play an important part in what follows. In particular we shall need a relation between filtration quotients in the product construction.

<u>Lemma 2.1.</u> <u>If X_*, Y_* and $X_* \otimes Y_* = Z_*$ are as above, there is a natural homotopy equivalence</u>:

(4) $$S^{m+m'} Z_k/Z_{k+1} \xrightarrow{\simeq} \bigvee_{i+j=k} \left[(S^{m(k+1-i)} X_i / S^{m(k-i)} X_{i+1}) \wedge S^{m'(k+1-j)} Y_j / S^{m'(k-j)} Y_{j+1})\right]$$

<u>Proof</u> Replace $Z_{k+1} \to S^{m+m'} Z_k$ by the inclusion $Z^{k+1}_{k+1} \subsetneq Z^{k+1}_k$, or

$$\bigcup_{i+j=k+1} X^{k+1}_i \wedge Y^{k+1}_j \subsetneq \bigcup_{i+j=k} X^{k+1}_i \wedge Y^{k+1}_j$$

We now find that for $i + j = k$,

$$(X_i^{k+1} \wedge Y_j^{k+1}) \cap Z_{k+1}^{k+1} = (X_{i+1}^{k+1} \wedge Y_j^{k+1}) \cup (X_i^{k+1} \wedge Y_{j+1}^{k+1})$$

and

$$(X_i^{k+1} \wedge Y_j^{k+1}) \cap (X_{i+1}^{k+1} \wedge Y_{j-1}^{k+1}) = X_{i+1}^{k+1} \wedge Y_j^{k+1}$$

Hence

$$Z_k^{k+1}/Z_{k+1}^{k+1} = \bigvee_{i+j=k} (X_i^{k+1}/X_{i+1}^{k+1} \wedge Y_j^{k+1}/Y_{j+1}^{k+1})$$

which is the right hand side in (4), properly defined.

This is of course a general result about the product of two filtrations in the ordinary sense.

§3. Negative filtration spectral sequences

C continues to denote a category as in §1. We shall define the idea of a 'cohomology theory' on C and show that associated to a cohomology theory h^* and negative filtration X_* there corresponds a spectral sequence $\{E_r(X_*,h^*)\}$. We touch only briefly on convergence, concentrating on the formal properties of the spectral sequence, in particular its multiplication (formula (6), lemma 3.2). There are no surprises.

Either reduced or unreduced type can be made the model for a definition of cohomology theory. Taking the reduced type, say a sequence of contravariant functors $\{\tilde{h}^i : C_o \to Ab\}$ ($i \in Z$ or Z_2) and natural transformations $\{\delta^i\}$ form a cohomology theory on C_o if

C.1. $\delta^i : \tilde{h}^i(A) \to \tilde{h}^{i+1}(X/A)$ is defined for any cofibration $A \xrightarrow{f} X \xrightarrow{g} X/A$ and is natural on the category of such cofibrations.

C.2. In the above situation the sequence

$$\ldots \to \tilde{h}^{i-1}(A) \xrightarrow{\delta^{i-1}} \tilde{h}^i(X/A) \xrightarrow{\tilde{h}^i(g)} \tilde{h}^i(X) \xrightarrow{\tilde{h}^i(f)} \tilde{h}^i(A)$$

$$\xrightarrow{\delta^i} \tilde{h}^{i+1}(X/A) \to \qquad \ldots$$

is exact.

C.3. If $f_o \simeq f_1$, then $\tilde{h}^i(f_o) = \tilde{h}^i(f_1)$ for all i.

Any such $(\tilde{h}^*,\delta)$ defines a corresponding unreduced theory on the category of pairs (X,A) in C by

$$h^i(X,A) = \tilde{h}^i(X \cup CA)$$

with the obvious definition of the connecting morphism.
We shall use reduced theories on the whole when setting up the spectral sequence, and unreduced ones when we have it. The transition from one to the other is always easy.

Example 1. If F is any functor from C_o to Top$_o$ which preserves cofibrations and homotopies and $\tilde{h}^*$ is a reduced cohomology theory on Top$_o$, then $h^* \circ F$ is a cohomology theory on C_o. We could take for F the forgetful functors (G-Top)$_o$ → Top$_o$ and Top/B$_o$ → Top$_o$; the latter, derived from the 'unpointed' forgetful functor, sends (X, f_X, k_X) to $X/k_X(B)$. The cohomology theory on Top/B$_o$ defined in this way by $\tilde{h}^*$ on Top$_o$ is denoted by $\tilde{h}^*_B$, see [21]; we have

$$\tilde{h}^*_B(X) = \tilde{h}^*(X/k_X(B)) = h^*(X, k_X(B))$$

and $\qquad h^*_B(X,A) = h^*(X,A)$ for the unreduced theory.

More generally I can take $F(X) = X \wedge Y$ (the smash product in C_o) followed by the forgetful functor above; on Top/B for example we obtain theories of the type

$$X \longmapsto \tilde{h}^*_B(X \underset{B}{\wedge} Y) = \tilde{h}^*((X \underset{B}{\wedge} Y/k(B))$$

2. As defined by Atiyah and Segal [4; 20], K^*_G; equivariant K-theory, is a cohomology theory on the subcategory of compact spaces in G-Top. Atiyah's theory K^*_R is defined on a similar subcategory of $\mathbb{Z}_2$-Top; and other related theories can be defined.

3. Define a spectrum $\mathbb{E}$ in C_o to be a sequence of objects E_n and morphisms $SE_n \to E_{n+1}$ in C_o. Then we can define a representable cohomology theory in C_o by

$$\tilde{h}^q(X ; \mathbb{E}) = \varinjlim \, [S^n X, E_{n+q}]_{C_o}$$

(Tom Dieck has generalized this idea to provide an equivariant bordism theory in G-Top, see [24].

As a special case of this, one can define a representing space for $\tilde{K}^o_G$ (Tom Dieck, [23]; or, take the space of Fredholm operators in a suitably large G-Hilbert space, i.e., containing every finite-dimensional G-vector space). Hence $\tilde{K}^o_G$ can be made a representable cohomology theory on the larger category G-Top$_o$.

For further study of cohomology theories on these categories, see [7;10;21] .

As indicated in the Introduction, the main application (in part II) will be to theories on $\underline{G\text{-}Top}_o$ associated with equivariant K-theory K_G^*.

Given a negative filtration X_* of degree m, and a cohomology theory theory $\tilde{h}^*$, in C_o, we can define a spectral sequence, which in its finite part can be considered, via the replacement in §2, as the usual spectral sequence of a filtration (See Cartan-Eilenberg [9, XV. 7, example 2] - except that the degrees are wrong for multiplicative structures). The definition is by an H(p,q) system, where the interesting part occurs for p and q negative.

Definition 3.1. $\{H(-p, -q)\} = \{H(-p, -q)\ (X_*, \tilde{h}^*)$ *are defined for* $-\infty \leq q \leq p \leq \infty$ *by*

(i) $$H(-p,-q)^t = \tilde{h}^{(p+1)m+t}(S^{(p-q)m}X_{q+1}/X_{p+1})$$

for $-1 \leq q \leq p < \infty$

(ii) $$H(-p, -q)^t = \tilde{h}^{(p+1)m+t}(S^{(p+1)}X_o/X_{p+1})$$

for $-\infty \leq q < -1 \leq p < \infty$

(iii) $H(-p, -q) = 0$ *for* $-\infty \leq q \leq p < -1$.

(iv) *The restriction and coboundary morphisms arise in the usual way from triple exact sequences of* $(S^{(p-r)m}X_{r+1}, S^{(p-q)m}X_{q+1}, X_{p+1})$, *with suspension isomorphisms used where necessary. Restrictions are of secondary degree* 0, *coboundaries of degree* + 1.

(v) *The groups* $H(-p, -q)$ $(p \geq q)$ *for fixed* q *therefore form a direct system*

$$H(-q, -q) \to H(-q-1, -q) \to H(-q-2, -q) \to \quad \ldots.$$

Define

$$H(-\infty, -q) = \lim_{\substack{\to \\ p}} H(-p, -q)$$

as a ($\mathbb{Z}$ *or* $\mathbb{Z}_2$) - *graded group*.

Note. Part (v) of the definition is unusual; it must be included if Cartan and Eilenberg's condition (S.P.5) is to be satisfied, and then , as we shall see, guarantees automatic strong convergence to the group

(1) $$H(X_*, \tilde{h}^*) = H(-\infty, \infty)\ (X_*, \tilde{h}^*)$$

which may or may not be the group $\tilde{h}^*(X_o)$ to which we want the spectral sequence to converge. This is a further consequence of the fact that the filtration is in the opposite direction to the usual kind. The formal procedure of using (S.P.5) to define $H(-\infty, -q)$ (justified because $\varinjlim$ is exact) will make the convergence problem easier to study.

The spectral sequence $\{E_r(X_*, \tilde{h}^*)\}$ is defined by the above $H(p,q)$ system in the usual way.

<u>Proposition 3.1.</u> <u>The spectral sequence</u> $\{E_r(X_*, \tilde{h}^*)\}$ <u>converges strongly to</u>

$$H = H(-\infty, \infty) = \varinjlim \tilde{h}^*(S^{(p+1)m}X_o/X_{p+1}) .$$

<u>Proof</u> From Cartan and Eilenberg chapter XV we obtain the following necessary and sufficient conditions for strong convergence in an $H(p,q)$ system:

(i) $\operatorname{Im}[H(p,\infty) \to H(p,p+1)] = \bigcap_r \operatorname{Im}[H(p,p+r) \to H(p, p+1)]$

(ii) $H \to \varprojlim (H/\operatorname{Im}(H(p,\infty) \to H) \qquad (p \to \infty)$

is an isomorphism.

Both these conditions are trivially satisfied: the first because for $p + r > 1$, $H(p, p+r) \cong H(p, p+r+1) \cong \dots \cong H(p, \infty)$; the second because for $p > 1$ the image $(H(p,\infty) \to H)$ is zero (since $H(p,\infty) = 0$ by (iii), so

$$H \to H/\operatorname{Im}(H(p, \infty) \to H)$$

is in fact an isomorphism for all $p > 1$.

From this proposition it follows that the main difficulty of negative filtration spectral sequences lies in the study of the objects $H(X_*, \tilde{h}^*)$ to which they converge. We have a natural mapping

(2) $$H(X_*, \tilde{h}^*) = \varinjlim \tilde{h}^*(S^{(p+1)m} X_o/X_{p+1}) \xrightarrow{\Phi} \tilde{h}^*(X_o)$$

and an exact triangle, by exactness of $\varinjlim$

$$\text{(3)} \qquad \begin{array}{ccc} \varinjlim \tilde{h}^*(S^{(p+1)m} X_o/X_{p+1}) & \xrightarrow{\Phi} & \tilde{h}^*(X_o) \\ & \nwarrow \quad \swarrow & \\ & \varinjlim \tilde{h}^*(X_{p+1}) & \end{array}$$

From this follows trivially

Lemma 3.1. Φ *is an isomorphism, i.e.* $\{E_r(X_*,\tilde{h}^*)\}$ *converges strongly to* $\tilde{h}^*(X_o)$, *if and only if the group*

$$\Gamma(X_*,\tilde{h}^*) = \varinjlim \tilde{h}^*(X_p)$$

vanishes.

As a trivial example of bad behaviour we could take $X_p = S^{pm}X_o$ and $X_{p+1} \to S^m X_p$ to be the identity map, for all p. Then we find $H(X_*,\tilde{h}^*) = 0$, $\Gamma(X_*,\tilde{h}^*) = \tilde{h}^*(X_o)$. The question of what constitute good convergence conditions in general is still very badly understood except when h^* is ordinary cohomology; too few examples are known. I leave it as an exercise to the reader to prove that Φ is an isomorphism provided h^* is an ordinary cohomology theory and X_p is $(pm + r)$-connected, where $r \to \infty$ as $p \to \infty$.

We now turn to the multiplicative properties of negative filtration spectral sequences. First, as in the case of Top_o, a cohomology theory $\tilde{h}^*$ on C_o is said to be multiplicative if we have a natural pairing of graded groups

$$\text{(4)} \qquad \bar{\kappa} : \tilde{h}^*(X) \otimes \tilde{h}^*(Y) \to \tilde{h}^*(X \wedge Y) \qquad X,Y \text{ in } C_o$$

which is associative (not necessarily commutative) and admits a unit $1 \in \tilde{h}^o(\theta^+)$ ($\theta^+ = \theta \cup \theta = \theta \otimes S^o$ is the analogue of S^o in C_o; in particular it is the unit for the smash product.) $\tilde{h}^*(\theta^+) = h^*(\theta)$ then has the structure of a ring; and $\tilde{h}^*(X)$ is a two sided module over this ring, for all X. For X in C, $h^*(X) = \tilde{h}^*(X^+)$ is also a ring with unit.

The coefficient ring of the theory $\tilde{h}^*(\theta^+)$ will be written $\bar{h}$.

Now we can use the product (4) to define a pairing of H(p,q) systems, given negative filtrations X_*, Y_*;

$$\{H(p,q)\,(X_*,\tilde{h}^*)\} \otimes \{H(p,q)\,(Y_*,\tilde{h}^*)\} \to$$

(5)
$$\{H(p,q)\,(X_* \otimes Y_*,\, \tilde{h}^*)\}$$

where $X_* \otimes Y_*$ is as defined in §2. (We could introduce a pairing of cohomology theories in (5), but enough is enough.) Pairings of $H(p,q)$ systems are defined by Douady in [8, exp.19] ; for our purposes we require for each p,q and $r > 0$ a pairing of graded groups

(6)
$$H(-p,\,-p+r)\,(X_*,\tilde{h}^*) \otimes H(-q,\,-q+r)\,(Y_*,\tilde{h}^*)$$

$$\xrightarrow{\mu} H(-p-q,\,-p-q+r)\,(X_* \otimes Y_*,\,\tilde{h}^*).$$

We in fact can proceed by replacing the mappings by inclusions as before up to a fixed filtration degree n, and using the standard construction of products in the spectral sequence of a filtration. In our case this means the following. First rewrite the source of μ in (6) as

$$\tilde{h}^*(X^{p+1}_{p-r+1}/X^{p+1}_{p+1}) \otimes \tilde{h}^*(Y^{q+1}_{q-r+1}/Y^{q+1}_{q+1})$$

Map this under $\bar{\kappa}$ to

$$\tilde{h}^*(X^{p+1}_{p-r+1} \wedge Y^{q+1}_{q-r+1}/(X^{p+1}_{p+1} \wedge Y^{q+1}_{q-r+1} \cup X^{p+1}_{p-r+1} \wedge Y^{q+1}_{q+1}))$$

Calling this $\tilde{h}^*(A/B)$ for the moment and setting

$$C = (X^{p+1}_{p+1} \wedge Y^{q+1}_{o}) \cup (X^{p+1}_{o} \wedge Y^{q+1}_{q+1})$$ we see that $A \cap C = B$, hence

$$\tilde{h}^*(A/B) \cong \tilde{h}^*(A \cup C/B \cup C)$$

Now $A \cup C$ contains $(X_* \otimes Y_*)^{p+q+2}_{p+q-r+1}$ and $B \cup C$ contains $(X_* \otimes Y_*)^{p+q+2}_{p+q+1}$. Hence we can map

$$\tilde{h}^*(A \cup C/B \cup C) \text{ into } \tilde{h}^*((X_* \otimes Y_*)^{p+q+2}_{p+q-r+1} \,/\, (X_* \otimes Y_*)^{p+q+2}_{p+q+1})$$

which is identified with $H(-p-q,\,-p-q+r)\,(X_* \otimes Y_*,\tilde{h}^*)$. The composite of these mappings is μ.

It remains to verify Douady's conditions (SPP1) and (SPP 2), i.e., that the pairing commutes with restrictions, and a derivation condition for the coboundary morphism. This is straightforward - it is easier to consider 'ordinary' filtrations by negative degrees and avoid half the indexes above. In Douady's words, 'Le lecteur, s'il existe, le fera lui même, s'il en a envie'. It is a consequence of the way in which the pairings above were constructed that, on the E_1 term, they relate back to the identification of $S^{m+m'}(X_* \otimes Y_*)_k \,/\, (X_* \otimes Y_*)_{k+1}$ in Lemma 2.1. In fact, let us write $A_{p,q}$ for $S^m(X_p/X_{p+1}) \wedge S^{m'}(Y_q/Y_{q+1})$; by Lemma 2.1 the kth filtration quotient of $X_o \otimes Y_*$ is a wedge of suspensions of $A_{p,q}$'s with $p+q=k$; applying $\tilde{h}^*$, and desuspending

$$(7) \qquad \tilde{h}^*(S^{m+m'}(X_* \otimes Y_*)_k/(X_* \otimes Y_*)_{k+1}) \cong \bigoplus_{p+q=k} \tilde{h}^*(A_{p,q})$$

Now the following is easily proved:

<u>Lemma 3.2</u> <u>The diagram</u>

$$\begin{array}{ccc}
E_1^{-p}(X_*,\tilde{h}^*) \otimes E_1^{-q}(Y_*,\tilde{h}^*) & \xrightarrow{\ \mu\ } & E_1^{-k}(X_* \otimes Y_*,\tilde{h}^*) \\
\| & & \| \\
\tilde{h}^*(S^m X_p/X_{p+1}) \otimes \tilde{h}^*(S^{m'} Y_q/Y_{q+1}) & & \bigoplus_{p+q=k} \tilde{h}^*(A_{p,q}) \\
& \searrow^{\bar{\kappa}} \qquad \nearrow^{\subset} & \\
& \tilde{h}^*(A_{p,q}) &
\end{array}$$

is commutative.

Before leaving the general theory, there is a technical result which will be useful when, in §10, we wish to identify elements coming from E_2^{-1} in the spectral sequence, i.e., one stage beyond the edge homomorphism. Suppose we take an element in $E_2^{-1,t}$ represented by a cycle z in $Z_1^{-1,t} \subset E_1^{-1,t} = \tilde{h}^{2m+t-1}(S^m X_1/X_2)$. The cycle condition is that z maps to zero in $E_1^{o,t} = \tilde{h}^{m+t}(S^m X_o/X_1)$; hence for some x_1 in $\tilde{h}^{2m+t-1}(S^{2m}X_o/X_2)$, $z = j(x_1)$ where j is the restriction. Every d_r for $r > 1$ vanishes on E_1^{-1} and so z maps to $E_\infty^{-1,t}$ which can be regarded as a submodule of the quotient of the limit $H(X_*;\tilde{h}^*)^{t-1}$ by its zeroth filtration $E_\infty^{o,t-1}$. Write $\bar{\alpha}$ for the actual edge homomorphism

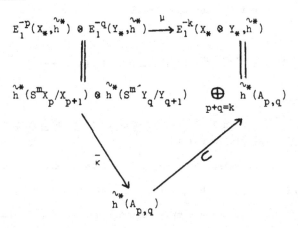

$$\tilde{h}^{m+t-1}(S^m X_o, X_1) = E_1^{o,t-1} \twoheadrightarrow E_\infty^{o,t-1} \subset H(X_*;\tilde{h}^*)^{t-1} \xrightarrow{\Phi} \tilde{h}^{t-1}(X_o)$$

Then we can map z one stage further by Φ and obtain

$$\bar{\beta} : Z_1^{-1,t} \twoheadrightarrow E_\infty^{-1,t} \subset H(X_*;\tilde{h}^*)^{t-1}/E_\infty^{o,t-1} \xrightarrow{\Phi} \tilde{h}^{t-1}(X_o)/\mathrm{Im}\ \bar{\alpha}$$

This is the 'secondary edge homomorphism'; it is immediate from the definition that this simply sends z to the image of x_1 chosen as above under

$$\tilde{h}^{2m+t-1}(S^{2m}X_o/X_2) \to \tilde{h}^{t-1}(X_o) \to \tilde{h}^{t-1}(X_o)/\mathrm{Im}\ \bar{\alpha}$$

and Im $\bar{\alpha}$ corresponds exactly to the indeterminacy in the choice of x_1.

This is not, however, the most convenient definition of $\bar{\beta}$.

To give an alternative, let the maps i_1, i_2 in the diagram

$$E_1^{-1,t} = \tilde{h}^{2m+t-1}(S^m X_1/X_2) \xrightarrow{i_1} \tilde{h}^{m+t-1}(X_1) \xleftarrow{i_2} \tilde{h}^{t-1}(X_o)$$

be the obvious ones.

<u>Lemma 3.3.</u> *$x \in E^{-1,t}$ is in $Z_1^{-1,t}$ if and only if there exists $y \in \tilde{h}^{t-1}(X_o)$ such that $i_1(x) = i_2(y)$; and then $\bar{\beta}(x) = y + \mathrm{Im}\ \bar{\alpha}$.*

<u>Proof</u> $d_1(x) = 0$ is equivalent to $i_1(x) \in \mathrm{Im}(i_2)$ by looking at the composites with $\tilde{h}^*(X_1) \to \tilde{h}^*(S^m X_o/X_1)$. Suppose then, with the previous notation, that $d_1(x) = 0$ and $x = j(x_1)$; $i_1(x) = i_2(y)$. By a simple commutative diagram the image x_2 of x_1 in $\tilde{h}^{t-1}(X_o)$ maps to $i_1(x)$ under i_2; so x_2 and y differ by an element of Ker i_2 = Im $\bar{\alpha}$. This proves the lemma.

§4. Resolutions

Let $\tilde{h}^*$ be a multiplicative cohomology theory on C_o; we wish to investigate the possibility of a generalized 'Künneth formula' holding for $\tilde{h}^*$. For this we first need the product morphism $\bar{\kappa}$ of §3(4) to be in the right form, i.e., a pairing of modules over the coefficient ring $\bar{h} = \tilde{h}^*(\theta^+)$. Given X,Y in C_o, the associativity of the product implies that the diagram

$$\begin{array}{ccc}
\tilde{h}^*(X) \otimes \tilde{h}^*(\theta^+) \otimes \tilde{h}^*(Y) & \xrightarrow{\bar{\kappa} \otimes 1} & \tilde{h}^*(X \wedge \theta^+) \otimes \tilde{h}^*(Y) = \tilde{h}^*(X) \otimes \tilde{h}^*(Y) \\
\Big\downarrow {\scriptstyle 1 \otimes \bar{\kappa}} & & \Big\downarrow {\scriptstyle \bar{\kappa}} \\
\tilde{h}^*(X) \otimes \tilde{h}^*(\theta^+ \wedge Y) = \tilde{h}^*(X) \otimes \tilde{h}^*(Y) & \xrightarrow{\bar{\kappa}} & \tilde{h}^*(X \wedge Y)
\end{array}$$

is commutative. If, then, we give $\tilde{h}^*(X)$ (resp. $\tilde{h}^*(Y)$) its right (resp. left) $\bar{h}$-module structure, the diagram shows that for $a \in h^*(X)$, $s \in \bar{h}$, $b \in \tilde{h}^*(Y)$,

$$\bar{\kappa}(a \,.\, s \otimes b) = \bar{\kappa}(a \otimes s \,.\, b)$$

Hence we have

<u>Lemma 4.1.</u> $\bar{\kappa}$ <u>factors through</u>

(1) $$\kappa : \tilde{h}^*(X) \underset{\bar{h}}{\otimes} h^*(Y) \to \tilde{h}^*(X \wedge Y).$$

κ will be taken as the natural product morphism in $\tilde{h}^*$ from now on. The generalized Künneth formula we shall look for, as in the case of cohomology theories over <u>Top</u>, will involve derived functors Tor over $\bar{h}$, with (1) appearing as the Tor^0 term mapped in by the edge homomorphism.

A difficulty immediately arises in contrast with the case of <u>Top</u>. In that case we have the following, fairly well-known result.

<u>Lemma 4.2.</u> <u>If</u> $\tilde{h}^*$ <u>is a multiplicative cohomology theory on</u> $\underline{\mathrm{Top}}_0$, <u>and</u> $\tilde{h}^*(X)$ <u>is flat as a right</u> $\bar{h}$<u>-module</u>, <u>then</u> (1) <u>is an isomorphism for all finite</u> CW <u>complexes</u> Y.

(The result can be improved, but we shall not need more).

<u>Proof</u> First, without the restriction that we are in <u>Top</u> it is true that the functors $F_1(Y) = \tilde{h}^*(X) \underset{\bar{h}}{\otimes} \tilde{h}^*(Y)$ and $F_2(Y) = \tilde{h}^*(X \wedge Y)$ are both cohomology theories (F_1 is one because $\tilde{h}^*(X)$ is flat, so tensoring with it preserves exactness). And κ gives a natural transformation from F_1 to F_2 which is an isomorphism for $Y = \theta^+$. Now a theorem of Dold (see [12, ch. I] p. 25 , Proposition 1) states that we can deduce κ is an isomorphism for all finite CW complexes Y.

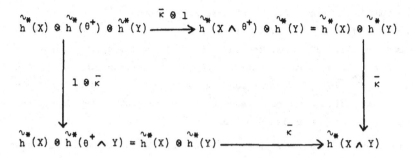

This fact was used by Atiyah in [3], in what has become the standard way of constructing 'geometric' Künneth formulae. In our categories, though, the theorem of Dold is not true in general, so that lemma 4.2 also fails. CW complexes have filtrations in which the filtration quotients are suspensions of S^0's; we could say that they are built up from S^0's, (i.e., from points). In contrast the construction of G-spaces is more complicated. Here the basic unit is the transitive G-space or orbit G/G_x. The way in which they are built up is well illustrated in the spectral sequence of Segal for K_G^* [20]. The corresponding question for spaces over B is less well studied, or even defined. In both cases $\tilde{h}^*(Y)$ can be flat, even free, as an $\bar{h}$ module and κ still fail to be an isomorphism.

<u>Example 4.1.</u> In discussing concrete cases it is almost always easier to omit base-points and deal with unreduced theories. The equivalent pairing to (1) for unbased spaces is

$$h^*(X) \underset{\bar{h}}{\otimes} h^*(Y) \overset{\kappa}{\to} h^*(X \times Y)$$

the product being in the sense of $\mathcal{C}$. In particular we may take in <u>Top/B</u>, $X = b_0$ (a point of B), $Y = \Lambda B$ (paths in B at b_0, say). The product $X \underset{B}{\Pi} Y$ is then ΩB, the fibre of Y over b_0. We have a pairing

$$h^*(b_0) \underset{\bar{h}}{\otimes} h^*(\Lambda B) \to h^*(\Omega B) .$$

If h^* is obtained from a cohomology theory k^* on <u>Top</u> via the forgetful functor, then $h^*(b_0) = h^*(\Lambda B) = \bar{k}$, and $\bar{h} = k^*(B)$. (See example 1, §3.). The pairing is

$$\bar{k} \underset{k^*(B)}{\otimes} \bar{k} \xrightarrow{\kappa} k^*(\Omega B).$$

Now take $k^* = K^*$ (ordinary K-theory), $B = K(\pi,2)$ with π finite. Then $K^*(B) \to K^*(b_0)$ is an isomorphism ([1] theorem I), so $K^*(b_0)$ is free over $K^*(B)$. But $K^*(\Omega B) = K^*(K(\pi,1)) = R(\pi)\hat{}$ (completed representation ring, see [2]), and κ is not an isomorphism unless π is trivial.

I have set out this example in some detail, partly because it is a locus classicus of what may go wrong in this theory, partly to develop some ideas of how abstract theory I have defined feeds back into the real-life world of loop spaces, K-theory and so on.

Having established that flatness is not good enough for an object to behave well under κ , let us define what is.

<u>Definition 4.1.</u> (a) <u>Call Z in $\mathcal{C}_0$ a right Künneth space for $\tilde{h}^*$ if</u>

(i) $\tilde{h}^*(Z)$ <u>is finitely generated and projective as a right $\bar{h}$-module</u>

(ii) $\kappa : \tilde{h}^*(Z) \otimes_h \tilde{h}^*(Y) \to \tilde{h}^*(Z \wedge Y)$ *is an isomorphism for all* Y *in* C_o.

(b) *A Künneth embedding for* $\tilde{h}^*$ *(in* C_o*) is a mapping* $\phi : X \to Z$ *such that* Z *is a right Künneth space and* $\tilde{h}^*(\phi) : \tilde{h}^*(Z) \to \tilde{h}^*(X)$ *is an epimorphism. If for a given* X *such a* ϕ *exists, I say* X *admits a Künneth embedding for* $\tilde{h}^*$.

(c) C_o *has enough Künneth spaces for* $\tilde{h}^*$ *if for every* X *in* C_o *some suspension* $S^m X$ *admits a Künneth embedding for* $\tilde{h}^*$, *and* m *is bounded as* X *runs over* C_o.

Notes 1. Trivially, $\theta \otimes S^m$ is a Künneth space for any $\tilde{h}^*$; and the category of Künneth spaces is closed under suspensions and wedges. (I know of no case where the distinction between 'left' and 'right' Künneth spaces etc. makes any difference, so I shall only observe it when it seems that it might).

2. It may be possible to drop finiteness restrictions and deal with completed tensor products (with respect to some topology). These questions may become particularly important in the case of Top/B.

3. In [13] I called a Künneth space for $\tilde{K}^*_G$ a 'basic G-space' and a Künneth embedding a 'basic G-map'.

Lemma 4.3. (i) *If* $A \subset X$ *and two of* A, X, X/A *are Künneth spaces for* $\tilde{h}^*$, *then so is the third*

(ii) *If* W, Z *are (right) Künneth spaces for* $\tilde{h}^*$ *then so are* $W \wedge Z$, $W \times Z$.

Proof. Part (i) is a simple five-lemma argument. For part (ii), for $W \wedge Z$ use the diagram (which commutes by associativity of κ.)

$$\begin{array}{ccc}
\tilde{h}^*(W \wedge Z) \otimes_h \tilde{h}^*(Y) & \xrightarrow{(1)} & \tilde{h}^*(W \wedge Z \wedge Y) \\
\downarrow{\scriptstyle (2)} & & \downarrow{\scriptstyle (3)} \\
\tilde{h}^*(W) \otimes_h \tilde{h}^*(Z) \otimes_h \tilde{h}^*(Y) & \xrightarrow[(4)]{} & \tilde{h}^*(W) \otimes_h \tilde{h}^*(Z \wedge Y)
\end{array}$$

(2) and (3) are isomorphisms because W is a right Künneth space, (4) because Y is one. Hence (1) is an isomorphism.

Now we can deduce that $W \times Z$ is a right Künneth space by applying part (i) to $(W \vee Z,\ W \times Z,\ W \wedge Z)$.

Suppose we now try, having set up the idea of a Künneth space, to apply it to follow the lines of Atiyah's proof of the Künneth formula for K-theory in [3]. Suppose that C_o has enough Künneth spaces for $\tilde{h}^*$; then for any X we can find a Künneth embedding

$$u_o : S^m X \to Z_o \quad .$$

Let its cofibre (mapping cone) be $v_o : Z_o \to X_1$. If X_1 is itself a Künneth space, we are already in the fortunate situation of Atiyah's proof and can stop. If not, take a Künneth embedding $u_1 : S^m X_1 \to Z_1$, and let its cofibre be $v_1 : Z_1 \to X_2$. At some stage in iterating this procedure we may arrive at a Künneth space, but since there is no guarantee of this it is best to consider an infinite iteration from the start. For this purpose we define a 'complex' or infinite sequence as follows.

Definition 4.2. (i) *For X in C_o, a C_o-complex over X of degree m is a sequence of cofibrations*

$$(X_*,Z_*) : S^m X_o \xrightarrow{u_o} Z_o \xrightarrow{v_o} X_1;\ S^m X_1 \xrightarrow{u_1} Z_1 \xrightarrow{v_1} X_2\ ; \dots$$

with $X_o = X$.

(ii) *Given* $f : X \to Y$ *in* C_o, *a mapping of C_o-complexes over* f, $(f_*,g_*) : (X_*,Z_*) \to (Y_*,W_*)$ *is a sequence* $\{f_i : X_i \to Y_i\}$, $\{g_i : Z_i \to W_i\}$ *defined to make the obvious diagrams commutative.*

(iii) *A C_o-complex (X_*,Z_*) is called acyclic for $\tilde{h}^*$ if, for all* i, $\tilde{h}^*(u_i) : \tilde{h}^*(Z_i) \to \tilde{h}^*(S^m X_i)$ *is an epimorphism. It is called projective for $\tilde{h}^*$ if for all* i, Z_i *is a Künneth space. It is called an $\tilde{h}^*$ resolution of X if it is acyclic and projective, i.e.,* u_i *is a Künneth embedding for all* i.

The imitation of the language of homological algebra here is reasonable in the following terms. We have a diagram of mappings associated with (X_*,Z_*):

(2)

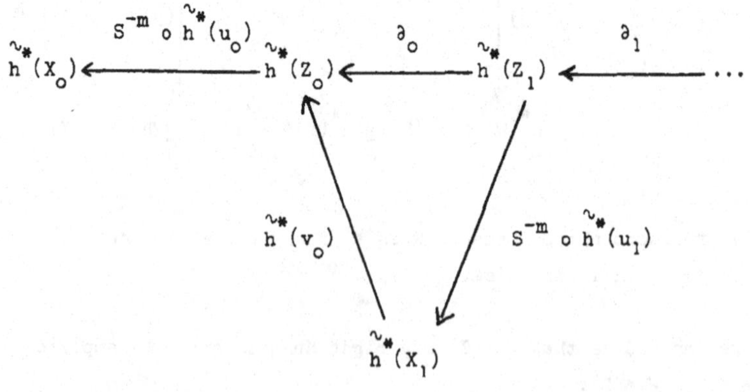

where $\partial_o, \partial_1, \ldots$ are defined to make the triangles commutative. By the cofibration property, $\mathrm{Im}(\tilde{h}^*(v_i)) = \mathrm{Ker}(S^{-m} \circ \tilde{h}^*(u_i))$ for all i; hence, $\partial_i \circ \partial_{i+1} = 0$ automatically. $\{\tilde{h}^*(Z_i), \partial_i\}$ is therefore a chain complex of $\bar{h}$-modules, augmented into $\tilde{h}^*(X_o)$ by $\varepsilon = S^{-m} \circ h^*(u_o)$. If the C_o-complex is acyclic, i.e., $\tilde{h}^*(u_i)$ is an epimorphism, then $\tilde{h}^*(v_i)$ is a monomorphism by the cofibration sequence and we find by a standard argument that $\{\tilde{h}^*(Z_i), \partial_i\}$ is an acyclic chain complex. The conditions that the C_o-complex should be projective and a resolution obviously imply the corresponding conditions for the chain complex; they are stronger because $\tilde{h}^*(Z_i)$ projective does not necessarily mean that Z_i is a Künneth space for $\tilde{h}^*$.

C_o-complexes are closely linked to negative filtrations, a fact which underlies our introduction of the latter idea (compare the example in §2, p. 17). The discussion which follows explains this link.

<u>Lemma 4.4</u>. <u>A C_o-complex (X_*, Z_*) of degree m defines functorially a negative filtration $X_* = \{X_i, \phi_i\}$ of X_o, of degree m + 1; and in the spectral sequence of this negative filtration there is a natural isomorphism</u>

$$E_1(X_*, \tilde{h}^*) \tilde{=} \{\tilde{h}^*(Z_p), \partial_p\}$$

of chain complexes of $\bar{h}$-modules, augmented over $\tilde{h}^*(X_o)$.

Proof If we continue the Puppe sequence of the cofibration $S^m X_p \to Z_p \to X_{p+1}$ two stages we have a sequence

$$S^m X_p \xrightarrow{u_p} Z_p \xrightarrow{v_p} X_{p+1} \xrightarrow{\phi_p} S^{m+1} X_p \xrightarrow{Su_p} SZ_p$$

Defining, then, ϕ_p to be the cofibre of v_p as shown, we obtain a negative filtration of degree m + 1 which is clearly natural. And Su_p is the cofibre of ϕ_p, which gives a natural isomorphism

$$E_1^{-p}(X_*, \tilde{h}^*) = \tilde{h}^*(S^{m+1} X_p/X_{p+1}) \tilde{=} \tilde{h}^*(SZ_p) \tilde{=} \tilde{h}^*(Z_p)$$

Since $d_1 : E_1^{-p}(X_*, \tilde{h}^*) \to E_1^{-p+1}(X_*, \tilde{h}^*)$ is by definition the composite

$$\tilde{h}^*(S^{m+1} X_p/X_{p+1}) \to \tilde{h}^*(S^{m+1} X_p) \to \tilde{h}^*(S^{2m+1} X_{p-1}/S^m X_p) \tilde{=} \tilde{h}^*(S^{m+1} X_{p-1}/X_p)$$

the identification we have defined takes d_1 into the differential ∂_{p-1} of the complex $\{\tilde{h}^*(Z_p), \partial_p\}$. Finally, give the complex $E_1(X_*; \tilde{h}^*)$ an augmentation over $\tilde{h}^*(X_o)$ by

$$\varepsilon^{\sim} : E_1^o(X_*; \tilde{h}^*) = \tilde{h}^*(S^{m+1} X_o/X_1) \to \tilde{h}^*(S^{m+1} X_o) \underset{\cong}{\xrightarrow{S^{-m-1}}} \tilde{h}^*(X_o).$$

The above identification of E_1^o with $\tilde{h}^*(Z_o)$ now clearly sends ε' into $\varepsilon = S^{-m} \circ \tilde{h}^*(u_o) : \tilde{h}^*(Z_o) \to \tilde{h}^*(X_o)$ as defined above.

Conversely, given a negative filtration $\{X_i, \phi_i\}$ of degree m + 1 define $Z_i' = S^{m+1}X_i/X_{i+1}$ to be the filtration quotients. Then clearly

$$S^{m+1}X_i \to Z_i' \to SX_{i+1}$$

is a cofibration and we obtain a C_o-complex of degree m over SX. Moreover the functors from C_o-complexes to negative filtrations and back are, up to suspension, natural equivalences. I shall take advantage of this to relax the notation, writing X_* for a sequence of spaces X_i with either of these two equivalent structures. Here it should be remembered that the constructions of sections 2, 3 are canonical and should be treated as such. See the discussion above, p. 19 for the implications here.

A particular 'trivial' C_o-complex over X, of degree 0, can be obtained by setting

$$X_o = Z_o = X, \qquad u_o = \text{identity}$$

$$X_i = Z_i = \theta \qquad \text{for all } i > 0.$$

This complex will be called $\underline{X}$. The associated negative filtration is the unique one with $X_i = \theta$ for all i > 0. The complex $\underline{X}$ is trivially acyclic, and is an initial object in the category of C_o-complexes over X; the only map from $\underline{X}$ into (X_*, Z_*) being given by

$$\begin{array}{ccccc} X & \xrightarrow{1} & X & \longrightarrow & \theta \\ \Big\| & & \Big\downarrow u_o & & \Big\downarrow \\ X_o & \xrightarrow[u_o]{} & Z_o & \longrightarrow & X_1 \end{array} \tag{3}$$

and trivial mappings $\theta \to X_i$, $\theta \to Z_i$ thereafter. The spectral sequence $\{E_r(X;\tilde{h}^*)\}$ is trivial; $E_1^o(X;\tilde{h}^*) = \tilde{h}^*(X)$ and $E_1^p(X;\tilde{h}^*) = 0$ for all p.

<u>Lemma 4.5</u>. (i) <u>For any C_o-complex (X_*, Z_*) over X the mapping $\underline{X} \to X_*$ induces on the E_1 terms the augmentation of $E_1(X_*\tilde{h}^*)$ defined above</u>.

(ii) <u>If $Y_* = \{Y_i, \psi_i\}$ is any negative filtration, the smash product $\underline{X} \otimes Y_*$ is simply the negative filtration $X \wedge Y_* = \{X \wedge Y_i,\ 1 \wedge \psi_i\}$</u>

Proof The first part is an immediate consequence of the diagram (3). For the second, note that the mappings in $\underline{X}$ are already inclusions

$$\ldots \to \theta \to \theta \to \ldots \to \theta \to X$$

Hence taking a system $\{Y^n_m\}$ as in §2 we find

$$(\underline{X} \otimes Y_*)_n = (\bigcup_{j<n} (\theta \wedge Y^n_j)) \cup (X \wedge Y^n_n) = X \wedge Y^n_n \quad ,$$

from which part (ii) follows.

Finally we have the following simple consequence of Definitions 4.1 and 4.2.

Lemma 4.6. If C_o has enough Künneth spaces for $\tilde{h}^*$, then for any X in C_o there is an $\tilde{h}^*$ resolution of X.

§5. The Künneth formula spectral sequence

Let C_o be a category and $\tilde{h}^*$ a cohomology theory on C_o, as before. In this section $\bar{C}_o \subset C_o$ will be a full subcategory with enough Künneth spaces for $\tilde{h}^*$, and closed under mapping cylinder and cone constructions, but not necessarily under products. This enables us, by lemma 4.6, to obtain $\tilde{h}^*$ resolutions for 'small' spaces X,Y in $\bar{C}_o$ without the product $X \wedge Y$ in which we are interested itself being small. As an obvious example take B to be a finite CW complex $\bar{C}_o \subset \underline{Top/B}_o$ the subcategory of spaces having the homotopy type of a finite CW complex. Then if we choose (see example 4.1.) $X = b^+_o$, $Y = \Lambda B^+$, both are in $\bar{C}_o$, but $X \underset{B}{\wedge} Y = \Omega B^+$ is probably not.

We are now ready to set up the Künneth formula spectral sequence of two spaces in $\bar{C}_o$ and to state its main properties.

Theorem 5.1. Given $\bar{C}_o$, $\tilde{h}^*$ as above, and X,Y in $\bar{C}_o$, let X_*, Y_* be $\tilde{h}^*$ resolutions of X,Y in $\bar{C}_o$. Then the spectral sequences

$$\{E_r(X_* \otimes \underline{Y}\,;\,\tilde{h}^*)\}\,, \quad \{E_r(X_* \otimes Y_*;\tilde{h}^*)\}, \quad \{E_r(\underline{X} \otimes Y_*;\tilde{h}^*)\}$$

are all naturally isomorphic and independent of the resolutions chosen from the E_2 term on, and converge to the same limit. We take advantage of this to write $\{E_r(X,Y;\tilde{h}^*)\}$ for any one of these spectral sequences ($r \geq 2$) and $H(X,Y;\tilde{h}^*)$ for their common limit. They have the following properties:

K.1. $$E_2^{-p,t}(X,Y;\tilde{h}^*) = \mathrm{Tor}_{\underline{h}}^{-p,t}\,(\tilde{h}^*(X), \tilde{h}^*(Y))$$

K.2. $H(X,Y;\tilde{h}^*)$, $\Gamma(X,Y;\tilde{h}^*)$ *are cohomology theories in each variable X,Y* *separately, and*

$$\Phi : H(X,Y;\tilde{h}^*) \to \tilde{h}^*(X \wedge Y)$$

is an isomorphism if for some p X_p *or* Y_p *is a Künneth space.* (Φ *and* Γ *are defined in* §3, pp. 24-5).

K.3. *The edge homomorphism composed with* Φ ,

$$E_2^0 \to H(X,Y;\tilde{h}^*) \xrightarrow{\Phi} \tilde{h}^*(X \wedge Y)$$

goes under the identification K.1. into

$$\mathrm{Tor}^0_{\bar{h}}(\tilde{h}^*(X), \tilde{h}^*(Y)) = \tilde{h}^*(X) \otimes_{\bar{h}} \tilde{h}^*(Y) \xrightarrow{\kappa} \tilde{h}^*(X \wedge Y).$$

K.4. *The spectral sequence, and the identifications made in K.1. and K.3, are natural in X,Y.*

To begin the proof of the theorem I need first a simplified description of the E_1 term in a 'product' spectral sequence, in the case that interests us.

Lemma 5.1. Let X_* *be a projective* C_o *complex,* Y_* *any* C_o *complex. Then the pairing*

$$\mu : E_1(X_*;\tilde{h}^*) \otimes_{\bar{h}} E_1(Y_*;\tilde{h}^*) \to E_1(X_* \otimes Y_*;\tilde{h}^*)$$

is an isomorphism.

Proof If X_* is projective, $S^m X_p/X_{p+1} = SZ_p$ is a Künneth space for all p. Hence (with the notation of lemma 3.2. but replacing $\bar{\kappa}$ by κ),

$$\kappa : \tilde{h}^*(S^m X_p/X_{p+1}) \otimes_{\bar{h}} \tilde{h}^*(S^{m'} Y_q/Y_{q+1}) \to \tilde{h}^*(A_{p,q})$$

is an isomorphism. μ is therefore an isomorphism from each $E_1^{-p} \otimes E_1^{-q}$ to the component $\tilde{h}^*(A_{p,q}) \subset E_1^{-p-q}$, and so an isomorphism overall by (7) of §3.

Now return to Theorem 5.1. Let us write $C_* = E_1(X_*;\tilde{h}^*)$, $D_* = E_1(Y_*;\tilde{h}^*)$, as $\bar{h}$-chain complexes, augmented by ε, ε' over $\tilde{h}^*(X)$, $\tilde{h}^*(Y)$ respectively; by lemma 4.4. and the discussion which precedes it they are actually resolutions. (Here and henceforth we use the standard convention for lowering indices : $C_p = E_1^{-p}$ etc.). Further let $e_X : \underline{X} \to X_*$, $e_Y : \underline{Y} \to Y_*$ be the unique mappings of complexes defined as in §4(3). Then lemma 4.5 (i) tells us that the diagram

$$(1)\qquad \begin{array}{ccccc}
C_* \underset{h}{\otimes} h^*(Y) & \xleftarrow{\;1\,\otimes\,\varepsilon'\;} & C_* \underset{h}{\otimes} D_* & \xrightarrow{\;\varepsilon\,\otimes\,1\;} & \tilde{h}^*(X) \underset{h}{\otimes} D_* \\
\downarrow & & \downarrow & & \downarrow \\
E_1(X_* \otimes \underline{Y};\, \tilde{h}^*) & \xleftarrow{\;E_1(1\,\otimes\,e_Y)\;} & E_1(X_* \otimes Y_*;\tilde{h}^*) & \xrightarrow{\;E_1(e_X\,\otimes\,1)\;} & E_1(\underline{X} \otimes Y_*;\, \tilde{h}^*)
\end{array}$$

is commutative, where the columns are the E_1 pairings μ of spectral sequences. But these are isomorphisms of chain complexes, by lemma 5.1., as X_* and Y_* are both resolutions. Hence they induce isomorphisms of homology. Now $1 \otimes \varepsilon'$, $\varepsilon \otimes 1$ also induce isomorphisms of homology - this is part of the basic definition of Tor, see [9], and the pth homology group of each complex in the top row of (1) is canonically identified with $\mathrm{Tor}_{h}^{-p}(\tilde{h}^*(X), \tilde{h}^*(Y))$. Hence $E_2(1 \otimes e_Y)$, $E_2(e_X \otimes 1)$ are also isomorphisms, and each E_2^{-p} in the bottom row is also canonically identified with $\mathrm{Tor}_{\underline{h}}^{-p}(\tilde{h}^*(X), \tilde{h}^*(Y))$.

Now it follows from standard spectral sequence theory that $1 \otimes e_Y$, $e_X \otimes 1$ induce isomorphisms of all subsequent terms E_r in the three spectral sequences, and (since by Proposition 3.1 they converge strongly) isomorphisms on the limit terms.

Since $\{E_r(X_* \otimes Y;\tilde{h}^*)$ is independent of the resolution of Y chosen, and functorial in Y, we can deduce that the same is true of the other two spectral sequences from the E_2 term on, and of their limit terms. Independence of the resolution of X and functoriality in X follow similarly.

Now let us look at the edge homomorphism composed with Φ for $\{E_r(X_* \otimes \underline{Y};\tilde{h}^*)\}$. On the E_1 term this is $\bar{\varepsilon} = S^{-m} \circ \tilde{h}^*(u_o \wedge 1) : \tilde{h}^*(Z_o \wedge Y) \to \tilde{h}^*(X_o \wedge Y)$. (Here I am using lemma 4.5. (ii) to identify $X_* \otimes \underline{Y}$ and $X_* \wedge Y$). Using (1), the edge homomorphism on $E_2^o = \mathrm{Tor}^o$, composed with Φ, is identified as the unique homomorphism η from $\tilde{h}^*(X) \underset{h}{\otimes} \tilde{h}^*(Y)$ to $h^*(X \wedge Y)$ which makes the diagram

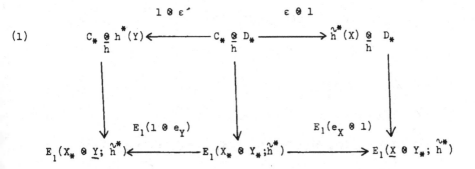

(2)

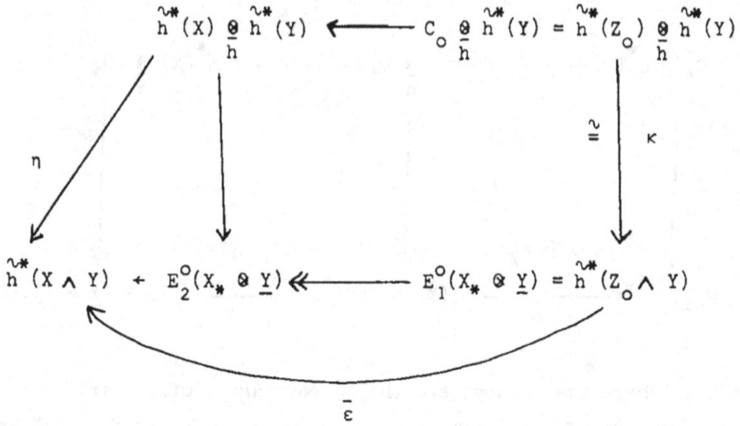

commutative. Since (by naturality of the product) κ does make the diagram commutative, the edge homomorphism composed with Φ must be κ. This proves K.3.

We have now proved all of Theorem 5.1. except for part K.2. which has to do with convergence. We consider $H(X,Y;\tilde{h}^*)$ and the group $\Gamma(X,Y;\tilde{h}^*)$ defined in lemma 3.1. as the obstruction to convergence; a five lemma argument shows that $\Gamma(X,Y;\tilde{h}^*)$ has the same properties of independence of resolution and functoriality as H does. Now for each p, $\tilde{h}^*((S^{mp}X_o/X_p) \wedge Y)$, $\tilde{h}^*(X_p \wedge Y)$ are cohomology theories in the variable Y; and the maps $X_{p+1} \to S^m X_p$ induce morphisms of cohomology theories. Hence, since direct limits are exact,

$$H(X,Y;\tilde{h}^*) = \underset{\rightarrow}{\lim}\ \tilde{h}^*((S^{mp}X_o/X_p) \wedge Y) \qquad \text{and}$$

$$\Gamma(X,Y;\tilde{h}^*) = \underset{\rightarrow}{\lim}\ \tilde{h}^*(X_p \wedge Y)$$

(defined by using the negative filtration $X_* \otimes \underline{Y}$) are cohomology theories in Y. Similarly, using $\underline{X} \otimes Y_*$, we obtain that they are cohomology theories in X.

If X is itself a Künneth space, then Φ is trivially an isomorphism. For X is then already a resolution of X; the spectral sequence is trivial; the edge homomorphism composed with Φ is an isomorphism by K.3. and so Φ is an isomorphism. Equivalently, $\Gamma(X,Y;\tilde{h}^*) = 0$ if X or Y is a Künneth space.

But now we can finish the proof of K.2. with the help of the following result.

<u>Lemma 5.2</u>. $\Gamma(X,Y;\tilde{h}^*) \cong \Gamma(X_p,Y_q;\tilde{h}^*)$ <u>for all</u> p,q.

Proof By induction and symmetry it is enough to prove $\Gamma(X_p,Y;\tilde{h}^*) = \Gamma(X_{p+1},Y;\tilde{h}^*)$. The cofibration $S^m X_p \to Z_p \to X_{p+1}$ gives us an exact triangle, using the cohomology theory properties of Γ,

$$\begin{array}{ccc} \Gamma(X_p,Y) & \rightarrow & \Gamma(X_{p+1},Y) \\ & \nwarrow \quad \swarrow & \\ & \Gamma(Z_p,Y) & \end{array}$$

Z_p is a Künneth space, so $\Gamma(Z_p,Y) = 0$; this proves the lemma. Now, if any X_p is a Künneth space we can deduce $\Gamma(X,Y;\tilde{h}^*) \cong \Gamma(X_p,Y;\tilde{h}^*) = 0$ for all Y and so Φ is an isomorphism. This completes the proof of K.2. and so of Theorem 5.1.

Note I have not taken account in the statement or proof of theorem 5.1. of possible non-commutativity in $\bar{h}$. The reader who wishes to do this, and to require for example that X_* is a right $\tilde{h}^*$ resolution and Y_* a left one, will find that it all works out correctly.

We now need to define the products in the spectral sequences $\{E_r(X,Y;\tilde{h}^*)\}$. These are not nearly so unpleasant as might be supposed, given the good theory of products in negative filtration spectral sequences which already exists from §3. Let X, Y, X´, Y´ be in $\mathcal{C}_o$, and let us take $\tilde{h}^*$ resolutions X_*, X'_* of X,X´ respectively. By §3 we have a pairing

(3)
$$\{E_r(X,Y;\tilde{h}^*)\} \underset{h}{\otimes} \{E_r(X',Y';\tilde{h}^*)\} = \{E_r(X_* \otimes \underline{Y};h^*) \underset{h}{\otimes} \{E_r(X'_* \otimes \underline{Y}';\tilde{h}^*)\}$$

$$\overset{\mu}{\rightarrow} \{E_r(X_* \otimes \underline{Y} \otimes X'_* \otimes \underline{Y}';\tilde{h}^*)\} = \{E_r(X_* \otimes X'_* \otimes \underline{Y} \otimes \underline{Y}';\tilde{h}^*)\}$$

(I have used the obvious commutativity of the $\otimes$ product of negative filtrations). It is an easy consequence of lemma 4.4. that $\underline{Y} \otimes \underline{Y}' = \underline{Y \wedge Y'}$, and we would immediately have a good 'external product' of spectral sequences if $X_* \otimes X'_*$ were an $\tilde{h}^*$ resolution of $X \wedge X'$. This, though, is not generally so (if it were, the Künneth formula spectral sequence of X,X´ would be trivial). The situation is parallel to that found in defining the external product in Tor; $X_* \otimes X'_*$ is an $\tilde{h}^*$ projective C_o-complex over $X \wedge X'$ (by lemmas 2.1 and 4.3. (ii)) and we need to compare it with a resolution.

Finding comparison theorems for $\tilde{h}^*$ resolutions is not as easy as the corresponding results for abelian categories; for a thorough discussion see [21]. I have avoided them up to now (in theorem 5.1) by playing one factor in $X \wedge Y$ off against the other. The one which we require at this point is not very strong. However it is necessary for it to assume that $\bar{C}_o$ is closed under products, which I shall do for the rest of this section. I also assume that the coefficient ring $\bar{h}$ is commutative from now on.

<u>Lemma 5.3</u> <u>Let</u> X <u>be in</u> $\bar{C}_o$, <u>let</u> W <u>be a Künneth space for</u> $\tilde{h}^*$ <u>in</u> $\bar{C}_o$ <u>and</u> $f : X \to W$ <u>a map. Then there exists a Künneth embedding</u> $g : S^m X \to Z$ <u>and a map</u> $h : Z \to S^m W$ <u>such that</u> $h \circ g = S^m f$.

<u>Proof</u> Let $g' : S^m X \to Z'$ be some Künneth embedding for X. Define $Z = S^m W \times Z'$; then Z is a Künneth space by lemma 4.3 (ii). Define h to be left projection $Z \to S^m W$ and g to be $(S^m f, g') : X \to Z$; then $h \circ g = S^m f$. Finally the composite

$$X \xrightarrow{\;g\;} S^m W \times Z' \xrightarrow{\;p_2\;} Z'$$

induces an epimorphism on $\tilde{h}^*$, so g certainly does.

<u>Lemma 5.4.</u> <u>Let</u> X_* <u>be any</u> $\tilde{h}^*$ <u>projective complex over</u> X <u>in</u> $\bar{C}_o$. <u>Then there exists an</u> $\tilde{h}^*$ <u>resolution</u> U_* <u>over</u> X <u>and a map of complexes</u> $U_* \to X_*$ <u>over the identity map of</u> X.

<u>Proof</u> Simply iterate the construction in Lemma 5.3.

Now let X_*, X'_* be as in (3). Then by Lemma 5.4., there exists an $\tilde{h}^*$ resolution U_* of $X \wedge X'$ and a map of complexes, say $f_* : U_* \to X_* \otimes X'_*$ reducing to the identity map on $X \wedge X'$.

This will define us a product as required. To relate it to the algebraic Tor products we must remember that $E_1(U_*;\tilde{h}^*)$ is a resolution <u>not</u> of $\tilde{h}^*(X) \otimes_{\bar{h}} \tilde{h}^*(X')$, but of $\tilde{h}^*(X \wedge X')$; while $E_1(X_*;\tilde{h}^*) \otimes_{\bar{h}} E_1(X'_*;\tilde{h}^*)$ is a projective complex over $\tilde{h}^*(X) \otimes_{\bar{h}} \tilde{h}^*(X')$, if not necessarily a resolution. This gives us the following easy consequence of the Comparison Theorem [15a, III. 6.1].

<u>Lemma 5.5.</u> <u>Given a resolution</u> L_* <u>of</u> $\tilde{h}^*(X) \otimes_{\bar{h}} \tilde{h}^*(X')$, <u>there exist maps of complexes</u> i (<u>lifting the identity</u>) <u>and</u> j (<u>lifting</u> κ) <u>to fill in the following diagram</u>

(4)
$$\begin{array}{ccccc}
E_1(X_*;\tilde{h}^*) \otimes_{\bar{h}} E_1(X'_*;\tilde{h}^*) & \xrightarrow{\;i\;} & L_* & \xrightarrow{\;j\;} & E_1(U_*;\tilde{h}^*) \\
& \searrow^{\varepsilon \otimes \varepsilon} & \downarrow \varepsilon & & \downarrow \varepsilon \\
& & \tilde{h}^*(X) \otimes_{\bar{h}} \tilde{h}^*(X') & \xrightarrow{\;\kappa\;} & \tilde{h}^*(X \wedge X')
\end{array}$$

<u>And</u> $E_1(f_*;\tilde{h}^*)$, <u>which also lifts</u> κ, <u>is homotopic to</u> $j \circ i$.

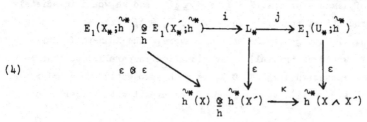

Note that there is no reason for L_* to be described geometrically - $\tilde{h}^*(X) \underset{h}{\otimes} \tilde{h}^*(X')$ is not itself necessarily $\tilde{h}^*$ of any space.

Now define the pairing $\bar{\mu}$ of spectral sequences to be (3) followed by

$$\{E_r(f_* \otimes 1)\} : \{E_r(X_* \otimes X'_* \otimes \underline{Y \wedge Y'};\ \tilde{h}^*)\} \to \{E_r(U_* \otimes \underline{Y \wedge Y'};\ \tilde{h}^*)\}$$

$$= \{E_r(X \wedge X',\ Y \wedge Y';\tilde{h}^*)\}$$

Theorem 5.2. The pairing

$$\bar{\mu} : \{E_r(X,Y;\tilde{h}^*)\} \underset{h}{\otimes} \{E_r(X',Y';\tilde{h}^*)\} \to \{E_r(X \quad X',Y \quad Y';h^*)\}$$

just defined has the following properties

P.1. From the E_2 term on, and on the limit groups H, it is independent of the resolutions chosen, and natural in X,X′,Y,Y′.

P.2. On the E_2 term, $\bar{\mu}$ is identified with the composite

$$\mathrm{Tor}_{\underline{h}}^{-p}(\tilde{h}^*(X),\tilde{h}^*(Y)) \otimes \mathrm{Tor}_{\underline{h}}^{-q}(\tilde{h}^*(X'),\tilde{h}^*(Y'))$$

$$\xrightarrow{\pitchfork} \mathrm{Tor}_{\underline{h}}^{-p-q}\ (\tilde{h}^*(X) \underset{h}{\otimes} \tilde{h}^*(X'),\ \tilde{h}^*(Y) \underset{h}{\otimes} \tilde{h}^*(Y'))$$

$$\xrightarrow{\mathrm{Tor}(\kappa,\kappa)} \mathrm{Tor}_{\underline{h}}^{-p-q}\ (\tilde{h}^*(X \wedge X');\ \tilde{h}^*(Y \wedge Y'))$$

where $\pitchfork$ is the semi-internal Tor product ([9 , XI.4]) and Tor(κ,κ) is induced by the product morphisms κ of h^*.

P.3. The diagram

$$\begin{array}{ccc} H(X,Y;\tilde{h}^*) \underset{h}{\otimes} H(X',Y';\tilde{h}^*) & \xrightarrow{\ \bar{\mu}\ } & H(X \wedge X',\ Y \wedge Y';\tilde{h}^*) \\ \Big\downarrow \Phi \otimes \Phi & & \Big\downarrow \Phi \\ \tilde{h}^*(X \wedge Y) \underset{h}{\otimes} \tilde{h}^*(X' \wedge Y') & \xrightarrow{\ \kappa\ } & \tilde{h}^*(X \wedge X' \wedge Y \wedge Y') \end{array}$$

is commutative.

Proof. We begin with P.2. Write $E_1(X_*,\tilde{h}^*) = C_*$, $E_1(X'_*,\tilde{h}^*) = C'_*$, $E_1(U_*,\tilde{h}^*) = M_*$.

Then $\bar{\mu}$ on the E_1 term is the left vertical column in the diagram

$$
\begin{array}{ccc}
C_* \otimes \tilde{h}^*(Y) \otimes C'_* \otimes \tilde{h}^*(Y') & & \\
\Big\downarrow {\scriptstyle 1 \otimes T \otimes 1} & & \\
C_* \otimes C'_* \otimes \tilde{h}^*(Y) \otimes \tilde{h}^*(Y') & \xrightarrow{\ i \otimes 1\ } & L_* \otimes \tilde{h}^*(Y) \otimes \tilde{h}^*(Y') \\
\Big\downarrow {\scriptstyle 1 \otimes \kappa} & & \Big\downarrow {\scriptstyle 1 \otimes \kappa} \\
C_* \otimes C'_* \otimes \tilde{h}^*(Y \wedge Y') & \xrightarrow{\ i \otimes 1\ } & L_* \otimes \tilde{h}^*(Y \wedge Y') \\
\Big\downarrow {\scriptstyle E_1(f_*)} & \swarrow {\scriptstyle j \otimes 1} & \\
M_* \otimes \tilde{h}^*(Y \wedge Y') & &
\end{array}
$$

In this diagram T is the transposition of second and third factors; i,j are as defined in lemma 5.5; the square is commutative and the triangle (again by lemma 5.5) homotopy commutative. Hence the two maps $E_1(f_*) \circ (1 \otimes \kappa) \circ (1 \otimes T \otimes 1)$, $(j \otimes 1) \circ (1 \otimes \kappa) \circ (i \otimes 1) \circ (1 \otimes T \otimes 1)$ induce the same homology homomorphism, say

$$\alpha: H_*(C_* \otimes \tilde{h}^*(Y) \otimes C'_* \otimes \tilde{h}^*(Y')) \to H_*(M_* \otimes \tilde{h}^*(Y \wedge Y')).$$

Now μ on the E_2 term is therefore α composed with the usual product morphism

$$p: H_*(C_* \otimes \tilde{h}^*(Y)) \otimes H_*(C'_* \otimes \tilde{h}^*(Y')) \to H_*(C_* \otimes \tilde{h}^*(Y) \otimes C'_* \otimes \tilde{h}^*(Y')).$$

It remains to identify $\mathrm{Tor}(\kappa,\kappa) \circ (\pitchfork)$ with $\alpha \circ p$. This follows easily since first $(j \otimes 1) \circ (1 \otimes \kappa)$ has associated homology map $\mathrm{Tor}(\kappa,\kappa)$ by virtue of the fact that j lifts κ; while a standard definition of $\pitchfork$ describes it as p followed by the homology map of $(i \otimes 1) \circ (1 \otimes T \otimes 1)$ (since i lifts the identity).

To get P.1. is similar to the corresponding part of Theorem 5.1. In fact, the product I defined can also be defined using all four resolutions X_*, X'_*, Y_*, Y'_* (if you have the energy) as can the $\pitchfork$ product in Tor. We have, as with Theorem 5.1., three ways of defining the product which agree on the E_2 term; one is independent of

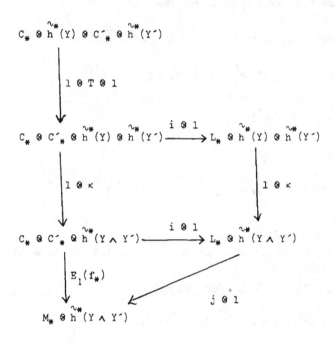

X_*, X_*', the other of Y_*, Y_*'. So, finally, the whole apparatus is independent of the choices made, and functorial.

The proof of P.3. is left as an exercise.

Now let $\Delta : X \to X \wedge X$, $\Delta' : Y \to Y \wedge Y$ be the diagonal maps. By the functorial properties of the spectral sequence we can define an internal product μ_i to make a commutative diagram

$$\begin{array}{ccc} \{E_r(X,Y;\tilde h^*)\} \underset{\bar h}{\otimes} \{E_r(X,Y;\tilde h^*)\} & \xrightarrow{\mu_i} & \{E_r(X,Y;\tilde h^*)\} \\ & \bar\mu \searrow \quad \nearrow \{E_r(\Delta,\Delta')\} & \\ & \{E_r(X \wedge X,\ Y \wedge Y : \tilde h^*)\} & \end{array}$$

The following is a trivial consequence of Theorem 5.2.

<u>Theorem 5.3.</u> <u>The pairing μ_i has the following properties</u>

<u>P.1'</u> <u>It is independent of the resolutions from the E_2 term on, and on the limit groups</u> H, <u>and is natural in</u> X,Y.

<u>P.2'</u> <u>On the E_2 term, μ_i is the internal product in the</u> Tor <u>of algebras</u>

$$\mathrm{Tor}_{\bar h}^{-p}(\tilde h^*(X), \tilde h^*(Y)) \otimes \mathrm{Tor}_{\bar h}^{-q}(\tilde h^*(X), \tilde h^*(Y)) \to \mathrm{Tor}_{\bar h}^{-p-q}(\tilde h^*(X), \tilde h^*(Y))$$

(see [15a, VIII. 2]).

<u>P.3'</u> <u>The diagram</u>

$$\begin{array}{ccc} H(X,Y;\tilde h^*) \underset{\bar h}{\otimes} H(X,Y;\tilde h^*) & \xrightarrow{\mu_i} & H(X,Y;\tilde h^*) \\ \Big\downarrow \Phi \otimes \Phi & & \Big\downarrow \Phi \\ \tilde h^*(X \wedge Y) \otimes \tilde h^*(X \wedge Y) & \xrightarrow{\kappa_i} & \tilde h^*(X \wedge Y) \end{array}$$

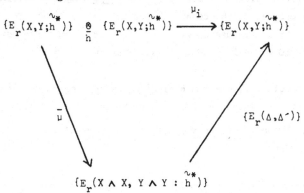

is commutative, where $\kappa_i = \tilde{h}^*(\Delta) \circ \kappa$ *is the internal product in* $\tilde{h}^*$.

§6. A note on the cobar resolution

The geometric cobar construction of §2 (p. 18) associates canonically with any X in *Top/B* a negative filtration of X. In view of the parallelism between this and the geometric bar construction, it is natural to ask when we can use the negative filtration to derive a Künneth formula spectral sequence as defined above. In other words, given the relation between negative filtrations and *Top/B*-complexes (lemma 4.4), when is the complex associated to the cobar construction a resolution for a given cohomology theory? This brief discussion will mark the end of our interest in *Top/B*, apart from the case of BG which is used in an auxiliary way in §8.

Let B be a fixed base space; and to avoid confusion use the *Top/B* notations $X \underset{B}{\Pi} Y$, $X \underset{B}{\wedge} Y$ for the fibred product and smash. Also let us write the cone and suspension of X in $(\underline{Top/B})_o$ as C_BX, S_BX. Then the basic step of the cobar resolution, for X in $(\underline{Top/B})_o$, is a cofibration sequence

$$(1) \qquad X \xrightarrow{\varepsilon_o} X \times B \xrightarrow{r_o} D(X) \xrightarrow{q_o} S_BX \xrightarrow{S_B\varepsilon_o} S_B(X \times B) \rightarrow \cdots$$

Here ε_o is the map called $\varepsilon(X)$ in §2, and q_o is the map called $Q\varepsilon(X)$; r_o is the mapping cone projection. The reader can check that this completes the informal description in §2; in particular note that the distinguished cross-section of X × B is $(k_X, 1) : B \to X \times B$. The cobar construction is obtained by iterating the step (1), obtaining maps

$$q_n : D^{n+1}(X) \longrightarrow S_BD^n(X) \qquad (D^o(X) = X)$$

for $n = 0,1,2,\ldots$; and hence a negative filtration of X in $(\mathrm{Top/B})_o$, by the $\{D^n(X)\}$.

Now introduce a cohomology theory. Following Smith [21], I shall restrict attention to the theories h^*_B, where h^* is a chomology theory on *Top* or some subcategory - see §3, example 1. The $(\underline{Top/B})_o$-complex associated to the negative filtration is

$$(2) \qquad X \xrightarrow{\varepsilon_o} X \times B \xrightarrow{r_o} D(X);\ldots;\ D^n(X) \xrightarrow{\varepsilon_n} D^n(X) \times B \xrightarrow{r_n} D^{n+1}(X);\ldots$$

So the following is an immediate consequence of definition 4.2.

Lemma 6.1. *The complex* (2) *is an* h^*_B*-resolution of* X *if and only if*

$$\varepsilon_n : D^n(X) \to D^n(X) \times B$$

is a Künneth embedding for h_B^*, *for all* $n \geq 0$.

To see when X,B and h_B^* satisfy the condition given by lemma 6.1. we note first that half the condition for a Künneth embedding is always verified. In fact, whatever cohomology theory we choose, if ℓ_n is left projection from $D^n(X) \times B$ to $D^n(X)$, $\ell_n \circ \varepsilon_n$ = identity so ε_n^* is a split epimorphism. It remains, therefore, to find when $(D^n(X) \times B \xrightarrow{p_2} B)$ is a Künneth space for h_B^* in *Top/B*. (Strictly, the question should apply to $\tilde{h}_B^*$ and $(\textit{Top/B})_o$; but the two questions are trivially equivalent.)

I shall give only a necessary condition, which can probably be weakened; and then discuss how inclusive this condition is.

Lemma 6.2. If B and X are both Künneth spaces for h^* *in Top, then* $(D^n(X) \times B \xrightarrow{p_2} B)$ *is a Künneth space for* h_B^*, *in Top/B for all* $n \geq 0$.

Proof. First, let $k: B \to D^n(X)$ be the cross-section; then we can use inductively the isomorphism induced by r_n

$$h^*(D^{n+1}(X), k(B)) \overset{\cong}{\to} h^*(D^n(X) \times B, \ \varepsilon_n(D^n(X)))$$

and standard cohomology theory methods to establish that $D^n(X)$ is a Künneth space for h^* in *Top*, for all n. (The statement is required because $D^n(X)$ is a *Top/B* cofibre, not a *Top* cofibre). Hence it is enough to treat the case n = 0, i.e., to show that if X,B are as stated then $(X \times B \xrightarrow{p_2} B)$ is a Künneth space for h_B^* in *Top/B*.

We have the well known identification

$$(3) \qquad (X \times B) \prod_B Y \xrightarrow{f} X \times Y$$

f is a (*Top/B*)-homeomorphism, where $X \times Y \to B$ is simply right projection followed by p_Y. Writing (as usual) $\bar{h}$ for h^* (point), this gives us a commutative diagram

$$\begin{array}{ccccc}
h^*(X \times B) \otimes_{h^*(B)} h^*(Y) & \xleftarrow[\cong]{\kappa \otimes 1} & h^*(X) \otimes_{\bar{h}} h^*(B) \otimes_{h^*(B)} h^*(Y) & \xrightarrow[\cong]{1 \otimes \mu} & h^*(X) \otimes_{\bar{h}} h^*(Y) \\
\Big\downarrow \kappa_B & & & & \Big\downarrow \kappa \ \cong \\
h^*((X \times B) \prod_B Y) & & \xleftarrow[\cong]{f^*} & & h^*(X \times Y)
\end{array}$$

Here κ_B is the h_B^* cartesian product morphism, and κ is the one for h^*, so by assumption κ, $\kappa \otimes 1$ are isomorphisms. The diagram is commutative from the definitions of the maps involved; in the top row we are using the fact that $h^*(X \times B)$ is not merely isomorphic to $h^*(X) \underset{h}{\otimes} h^*(B)$, but the $h^*(B)$-module structure is identified under $\kappa \otimes 1$ with the action on the right factor in the tensor product; hence the isomorphism μ in the top line. We deduce that κ_B is an isomorphism: in other words, $X \times B$ is a Künneth space for h_B^* in Top/B.

Putting together the two lemmas, and the theory of §5, we can state:

Proposition 6.1. *Suppose* B *is in* Top, X *and* Y *in* Top/B, *and* h^* *is a cohomology theory on* Top. *If* X,B *are Künneth spaces for* h^* *in* Top, *then the above construction gives a 'cobar resolution' Künneth formula spectral sequence*

$$E_2^{p,q} = \mathrm{Tor}^{p,q}_{h^*(B)}(h^*(X),h^*(Y)) \Rightarrow H \xrightarrow{\Phi} h^*(X \underset{B}{\Pi} Y)$$

which is natural in Y *and converges strongly to* H.

Lastly, how restrictive are our conditions on X,B? The first simple point worth noting is that the major application of the standard Eilenberg-Moore sequence is the case $X = \Lambda B$. These are, essentially, the 'base-to-fibre' spectral sequences; if $Y \to B$ is a map then its Serre fibre is $\Lambda B \underset{B}{\Pi} Y = F$. We have therefore (since $\Lambda B \simeq$ point is a Künneth space for any theory):

Cor. 6.1. *Let* $F \to Y \to B$ *be a fibration, with* B *a Künneth space for* h^* *in* Top. *Then there is a strongly convergent spectral sequence, natural in* Y,

$$E_2^{p,q} = \mathrm{Tor}^{p,q}_{h^*(B)}(\bar{h},h^*(Y)) \Rightarrow H \xrightarrow{\Phi} h^*(F).$$

The above is the exact analogue of the case treated in Part II where the cohomology theory is K_G^* and we take $X = G$. (See §9). The analogy is clear when the functor Ψ : G-Top $\to$ Top/BG of p. 14 is applied.

However, even to use Cor. 6.1. we still have to verify that B is a Künneth space for h^*. This is true for example

(i) If B is a finite wedge of spheres

(ii) If h^* is K-theory and $K^*(B)$ is finitely generated and torsion-free (e.g. B a homogeneous space of maximal rank; B a Lie group with Tors $\pi_1(B) = 0$, etc. - see part II).

(iii) If h^* is K-theory mod p and B is any finite complex.

More generally, as we have seen (p. 29) it is always enough for $h^*(B)$ to be finitely generated and flat over $\bar{h}$. Yet further applications can be generated as follows. Suppose $\bar{h}$ is a Prüfer ring (i.e., torsion-free <=> projective for finitely generated $\bar{h}$-modules). Then if $S \subset \bar{h}$ is the subset of elements which annihilate some non-zero $x \in B$, we can form a localized theory $S^{-1} h^*$ and (provided $S^{-1} h^*(B)$ is finitely generated over $S^{-1} \bar{h}$) apply the above spectral sequence. This indicates that there is a fair amount of scope for applications. Note that the question of whether Φ is an isomorphism still remains to be solved even when the spectral sequence has been set up - and my remarks in the Introduction should indicate that this is a non-trivial question!

In cases where $h^*(B)$ is not finitely generated over $\bar{h}$, then the whole theory is wrong (over-simplified) from the outset, in that we should be considering $h^*(B)$ as a topological module, and forming completed tensor products and their derived functors. However, I hope that this outline, in which I have shown the possibility of cobar spectral sequences, will indicate useful ways in which research could proceed.

Part I Bibliography

1. D.W. Anderson and L. Hodgkin, The K-theory of Eilenberg-MacLane complexes, Topology 7 (1968), 317-329.
2. M.F. Atiyah, Characters and cohomology of finite groups, Publ. Math. I H E S 9 (1961), 23-64.
3. ———, Vector bundles and the Künneth formula, Topology 1 (1962), 245-248.
4. ———, K-theory, Benjamin, 1967.
5. J. Beck, On H-spaces and infinite loop spaces. Category Theory, Homology Theory and their Applications III, Lecture Notes in Mathematics no. 99, Springer, 1969.
6. A. Borel et al., Seminar on Transformation Groups, Princeton, Ann. of Math. Studies no. 46, 1960.
7. G. Bredon, Equivariant cohomology theories, Lecture Notes in Mathematics no. 34, Springer 1967.
8. H. Cartan, Séminaire de l'E.N.S. no. 11 (1958-9), Invariant de Hopf.
9. ———, and S. Eilenberg, Homological algebra, Princeton, 1956.
10. A. Dold, Chern classes in general cohomology, Symposia math. vol. v (Geometria), Instituto Naz. di Alta Matematica, Roma (1971).
11. ——— and R. Lashof, Principal quasifibrations and fibre homotopy equivalence of bundles, Ill.J. Math. 3 (1959), 285-305.
12. E. Dyer, Cohomology theories, Benjamin, 1969.
13. L. Hodgkin, An equivariant Künneth formula for K-theory, University of Warwick preprint 1968.
14. I.M. James, Ex-homotopy theory, Illinois J. Math. 15 (1971), 324-337.
15. S. MacLane, Categorical algebra, Bull. Amer. Math. Soc. 71 (1965), 40-106.

15a. ———, Homology, Springer, 1963.

16. J. Milnor, Construction of universal bundles II, Ann. of Math. 63 (1956), 430-6.
17. ———. On spaces having the homotopy type of a CW-complex, Trans. Amer. Math. Soc. 90 (1959), 272-280.
18. R.S. Palais, The classification of G-spaces, Mem. Amer. Math. Soc. 36 (1960).
19. M. Rothenberg and N.E. Steenrod, The cohomology of classifying spaces of H-spaces, Bull. Amer. Math. Soc. 71 (1965), 872-5.
20. G. Segal, Equivariant K-theory, Publ. Math. I H E S 34 (1968), 129-151.
21. L. Smith, Lectures on the Eilenberg-Moore spectral sequence, Lecture Notes in Mathematics No. 134, Springer, 1970.
22. N.E. Steenrod, A convenient category of topological spaces, Mich. Math. J. 14 (1967), 133-152.
23. T. tom Dieck, Faserbündel mit Gruppenoperation, Arch. Math 20 (1969), 136-143.
24. ———, Bordism of G-manifolds and integrality theorems. Topology 9 (1970), 345-358.
25. ———, K.H. Kamps, D. Puppe, Homotopietheorie. Lecture Notes in Mathematics no. 157, Springer 1970.

Part II

General Theory of the K_G^* Künneth Formula

§7. The existence of Künneth spaces

From now on we restrict attention almost entirely to the category G-Top, and within this to the full subcategory $\mathcal{Q}_G$ of compact locally G-contractible G-spaces with finite covering dimension. On this category we shall consider only the cohomology theory K_G^*-equivariant K-theory - regarded as graded by $\mathbb{Z}_2$. For the properties of K_G^* in general the reader is referred to the paper of Segal [20]. I restate the main ones.

B.1. The coefficient ring of K_G^* is K_G^* (point) = R(G), the representation ring (see [1], [3]) of G. More generally, if $H \subset G$ is a closed subgroup and G acts on the coset space G/H by left translations, there is a natural isomorphism $R(H) \to K_G^*(G/H)$ which sends [V] in R(H) to the class in $K_G^0(G/H)$ of the vector bundle $G \times_H V \to G/H$. ([20], p.132).

Here, and from now on, R(H) is given the trivial $\mathbb{Z}_2$-grading; $R(H)^0 = R(H)$, $R(H)^1 = 0$.

B.2. If X is a free G-space, there is a natural isomorphism $K_G^*(X) \to K^*(X/G)$, sending [E] to [E/G] for E a G-vector bundle over X. In particular, $K_G^*(X \times G) \cong K^*(X)$ (diagonal action of G on X × G). ([20], prop. 2.1.). A similar argument shows that if X is in $\mathcal{Q}_{G \times H}$ and the action of H on X is free,

$$K_{G \times H}^*(X) \cong K_G^*(X/H).$$

B.3. For X in $\mathcal{Q}_G$, $K_G^*(X)$ is a finitely generated R(G)-module ([20], Prop. 5.4.).

We shall continue to prefer working in the unreduced cohomology theory; in this the multiplicative structure is a pairing,

$$\kappa: K_G^*(X) \underset{R(G)}{\otimes} K_G^*(Y) \to K_G^*(X \times Y) \quad . \tag{1}$$

The questions which remain before we can use the spectral sequence of §5 are

(i) Are there enough Künneth spaces for K_G^* (Definition 4.1.) in $\mathcal{Q}_G$?

(ii) If so (and so by theorem 5.1. the spectral sequence $\{E_r(X,Y;K_G^*)\}$ can be constructed), does it converge to $K_G^*(X \times Y)$, i.e., is

$$\Phi: H(X,Y;K_G^*) \to K_G^*(X \times Y)$$

an isomorphism?

I shall deal with these questions in this section and the next, respectively. In this one, I shall prove:

Proposition 7.1. Yes, there are enough Künneth spaces for K_G^* in $\mathcal{Q}_G$.

Our basic units in constructing Künneth spaces are just the equivariant analogues of those used by Atiyah in [2], namely the Grassmannians. Let V be a G-vector space; then the space $G_k(V)$ of k-dimensional subspaces of V admits an obvious action of G, and is in $\mathcal{Q}_G$.

Lemma 7.1. For any compact X, and $a \in K_G^0(X)$, there is a map $f: X \to G_k(V)$ for suitable k,V such that $a \in \mathrm{Im}\, f^!$

(I follow the usual practice of writing $f^!$ for $K_G^*(f)$).

Proof. If V is any G-vector space, we shall write $\underline{V}$ for the trivial G-vector bundle $X \times V \to X$ (diagonal action of G) so long as X can be understood. Write $a = [E] - [F]$, where E,F are G-vector bundles over X. Choose $E^{\perp}$, $F^{\perp}$ so that $E \oplus E^{\perp} \cong \underline{V}$, $F \oplus F^{\perp} \cong \underline{W}$ are trivial (see [20], Prop. 2.4.). Then $a = [E \oplus F^{\perp}] - [\underline{W}]$. Since $E \oplus F^{\perp}$ embeds in $\underline{V} \oplus \underline{W}$, there is a map $f: X \to G_k(V \oplus W)$ such that $E \oplus F^{\perp}$ is the pullback under f of the canonical k-plane bundle E' over $G_k(V \oplus W)$, where $k = \dim(E \oplus F^{\perp})$. Hence

$$a = f^!([E'] - [\underline{W}]) .$$

If $X \in (\mathcal{Q}_G)_o$, so $x_o \in X$ is a fixed point, then $f(x_o)$ must be a fixed point which can be taken as a basepoint of $G_k(V)$. Clearly if a is in the subgroup $\tilde{K}_G^0(X) \subset K_G(X)$, the construction of the lemma provides an element of $\tilde{K}_G^0(G_k(V))$ which maps onto a.

In fact, since we shall want to suspend and use the suspension isomorphism $\tilde{K}_G^i(X) \cong \tilde{K}_G^{i+1}(SX)$, $(i \in \mathbb{Z}_2)$, it is more convenient to return to the based category for the moment.

Lemma 7.2. Given any X in $(\mathcal{Q}_G)_o$, there is a based map f from SX to a product of G-Grassmannians and their suspensions which induces an epimorphism on $\tilde{K}_G^*$.

Proof Let $x_1,\dots,x_m$ generate $\tilde{K}_G^0(X)$ and $y_1,\dots,y_n$ generate $\tilde{K}_G^1(X)$. For each x_i, by Lemma 7.1, there is a based map $f_i : X \to G_{k_i}(V_i)$ such that x_i is in $\mathrm{Im}\, f_i^!$. Similarly for each y_j there is a based map $g_j: SX \to G_{\ell_j}(V_j)$ such that Sy_j is in $\mathrm{Im}\, g_j^!$. Now consider

(2) $$f = (\Pi Sf_i) \times (\Pi g_j) : SX \to (\prod_{i=1}^{m} SG_{k_i}(V_i)) \times (\prod_{j=1}^{n} G_{\ell_j}(V_j))$$

By construction $Sx_1,\ldots,Sx_m$, $Sy_1,\ldots,Sy_n$ are in the image of $f^!$; since these generate $\tilde{K}^*_G(SX)$ as an $R(G)$-module, the lemma follows.

Proposition 7.1. will now follow from

<u>Proposition 7.2.</u> *For any G-vector space V and positive integer* $k \leq \dim V$, $G_k(V)$ *is a Künneth space for* K^*_G.

For then, since the property of being a Künneth space is preserved if we add or subtract basepoints, suspend or (lemma 4.1) take products, the target space of the mapping f in (2) is a Künneth space for $\tilde{K}^*_G$ and we have constructed a Künneth embedding for X; Proposition 7.1. follows. The rest of this section is devoted to the proof of Proposition 7.2.

We need to prove that the map

$$\kappa : K^*_G(G_k(V)) \underset{R(G)}{\otimes} K^*_G(X) \to K^*_G(G_k(V) \times X)$$

as an isomorphism for all X in $\mathcal{Q}_G$.

As with many theorems of this kind in K_G, (compare the periodicity theorem in [20], Prop. 3.2.) it is just as simple in proving the theorem to generalize to bundles. $G_k(V) \times X$ is the Grassmannian bundle associated with the trivial G-vector bundle $V \times X \to X$; so the generalization should be a theorem describing the K-theory of the Grassmannian bundle $G_k(E)$ of a G-vector bundle E. (See [3], Prop. 2.7.14.). Let E be an n-dimensional G-vector bundle over B. T, the associated principal bundle, is then a $G \times U(n)$-space on which $U(n)$ acts freely: $T/U(n) = B$ and

(3) $$G_k(E) = (U(n)/U(k) \times U(n-k)) \times_{U(n)} T .$$

Here the identification is G-equivariant, with G acting only on the factor T in the product. We hope, as in the generalized periodicity theorem [3], to show that $K^*_G(G_k(E))$ is a free module over $K^*_G(B)$, with generators corresponding to the integral generators of $K^*(U(n)/U(k) \times U(n-k))$. Now by (B.2), $K^*_G(B) = K^*_{G \times U(n)}(T)$, and this gives $K^*_G(B)$ the structure of an $R(U(n))$-algebra; the homomorphism $R(U(n)) \to K^*_G(B)$ is the one which sends $[\rho]$ to $[\rho(E)]$. We shall prove

<u>Proposition 7.3.</u> *There is a natural isomorphism, for given* n,k,

$$\kappa_1 : R(U(k) \times U(n-k) \otimes_{R(U(n))} K_G^*(B) \to K_G^*(G_k(E)).$$

<u>The naturality is with respect to maps of G-vector bundles.</u>

Note now that, as is well-known (see |16|), $R(U(k) \times U(n-k))$ is a free $R(U(n))$-module on generators $1 = a_0, a_1, \ldots, a_{m-1}$ where $m = \binom{n}{k}$. Hence we have

<u>Corollary 7.1.</u> $K_G^*(G_k(E))$ <u>is a free</u> $K_G^*(B)$<u>-module on generators</u> $\kappa_1(a_i \otimes 1)$ $(i = 0, 1, \ldots, m-1)$.

Before proving Proposition 7.3. let us show how it implies that $G_k(V)$ is a Künneth space. We use the relation $G_k(\underline{\underline{V}}) = G_k(V) \times X$ stated above, where V again means the trivial bundle $V \times X$. Corollary 7.1. now implies that:

$K_G^*(G_k(V))$ is a free $R(G)$-module on generators $\bar{a}_i = \kappa_1(a_i \otimes 1)$

$K_G^*(G_k(\underline{\underline{V}}))$ is a free $K_G^*(X)$-module on generators $\kappa_1(a_i \otimes 1)$

Hence $\kappa : K_G^*(G_k(V)) \otimes_{R(G)} K_G^*(X) \to K_G^*(G_k(V) \times X) = K_G^*(G_k(\underline{\underline{V}}))$ is a map of free $K_G^*(X)$-modules with the same number of generators. It remains to show that $\kappa(\bar{a}_i \otimes 1) = \kappa_1(a_i \otimes 1)$, i.e., the two sets of generators correspond under κ. This follows from the map of G-vector bundles

$$\begin{array}{ccc} \underline{\underline{V}} & \xrightarrow{p_1} & V \\ \downarrow & & \downarrow \\ X & \longrightarrow & \text{point} \end{array}$$

In fact the naturality properties claimed for κ_1 in the proposition imply, writing p for the projection $G_k(V) \times X \to X$:

$$\kappa_1(a_i \otimes 1) = p^!\,(\bar{a}_i) = \kappa(\bar{a}_i \otimes 1) \in K_G^*(G_k(V) \times X).$$

Hence κ is an isomorphism.

We must now return to the general case of a G-vector bundle E, and prove Proposition 7.3. First, to define κ_1, extend the action of $U(n)$ on $U(n)/U(k) \times U(n-k)$ to an action of $G \times U(n)$ by making G act trivially. Then κ_1 is defined by the commutative diagram

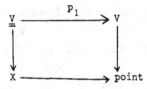

$$\begin{array}{ccc}
K^*_{G\times U(n)}(U(n)/U(k)\times U(n-k)) \underset{R(G\times U(n))}{\otimes} K^*_{G\times U(n)}(T) & \xrightarrow{\kappa} & K^*_{G\times U(n)}((U(n)/U(k)\times U(n-k))\times T) \\
\cong\Big\downarrow ① & & \\
R(G)\otimes R(U(k)\times U(n-k)) \underset{R(G\times U(n))}{\otimes} K^*_G(B) & & ③ \Big\downarrow\cong \\
\cong\Big\downarrow ② & & \\
R(U(k)\times U(n-k)) \underset{R(U(n))}{\otimes} K^*_G(B) & \xrightarrow{\kappa_1} & K^*_G(G_k(E))
\end{array}$$

Here the arrow ① comes from (B.1), (B.2), and the fact that if G acts trivially on X, $K^*_{G\times H}(X) = R(G)\otimes K^*_H(X)$. ② is by identifying $R(G\times U(n)) = R(G)\otimes R(U(n))$ and factoring $R(G)$ through the tensor product; while ③ is from (B.2) again. Essentially, then, κ_1 is simply related to a particular κ. The following description is perhaps even simpler, but will not be needed.

Exercise. Consider T as a $U(k)\times U(n-k)$-bundle (with G-action) over $G_k(E)$; for any $U(k)\times U(n-k)$-vector space V, $\tilde{V} = T\times_{U(k)\times U(n-k)} V$ is a G-vector bundle over $G_k(E)$. Then if $\pi : G_k(E)\to B$ is the projection, for any G-vector bundle F over B we have

$$\kappa_1([V]\otimes[F]) = [\tilde{V}\otimes\pi^! F] \quad .$$

We now need to prove that κ_1 is an isomorphism. As is indicated in [4] and [20], this result, like other similar ones in equivariant K-theory, is a consequence of the equivariant Thom isomorphism (and hence, so far, requires elliptic operator theory for its proof). We have a model in the result on K-theory of projective bundles, proved as Prop. 3.9 of [20] in a form which is quite convenient for our purpose. Observing that $P(E) = G_1(E)$ we shall begin by showing that this description of $K^*_G(P(E))$ is the same as the case $k = 1$ of Proposition 7.3.

<u>Lemma 7.3.</u> κ_1 <u>is an isomorphism when</u> $k = 1$.

Proof Consider

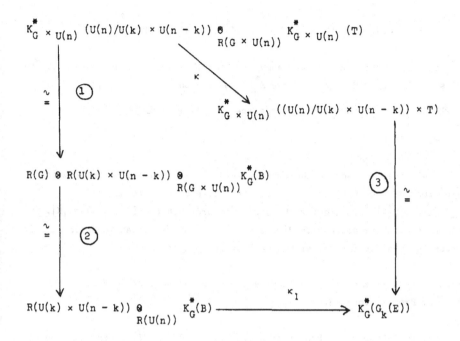

$$\kappa_1 : R(U(1) \times U(n-1)) \otimes_{R(U(n))} K_G^*(B) \to K_G^*(P(E))$$

The result of Atiyah and Segal quoted above implies that there is a line bundle $[H]$ over $P(E)$ such that $K_G^*(P(E))$ is the free $K_G^*(B)$ module on $1, [H], [H]^2, \ldots, [H]^{n-1}$. Write V for the $(U(1) \times U(n-1))$-vector space structure on $\mathbb{C}$ in which

$$(\omega, A) \cdot z = \omega z \qquad (\omega \in U(1),\ A \in U(n-1),\ z \in \quad)$$

(i.e., $U(1)$ alone acts, and in the obvious way). Then standard representation theory [16] implies that $R(U(1) \times U(n-1))$ is a free $R(U(n))$ module on $1, [V], [V]^2, \ldots, [V]^{n-1}$. Hence we have only to prove that $\kappa_1(|V| \otimes 1) = |H|$. Either checking with the definition directly or using the alternative definition in the exercise it is a routine matter to see that this is so.

Hence κ_1 is an isomorphism. (Compare the proof above that $G_k(V)$ is a Künneth space assuming Proposition 7.3.)

To complete the proof of Proposition 7.3. we could use flag manifolds as is done for $G = 1$ (implicitly) in [3], Prop. 2.7.14. It seems a little simpler, though, to stay as close to Grassmannians as we can, and use induction on k - the case $k = 1$ is proved and we can stop when $k = |n/2|$ since $G_k(E) = G_{n-k}(E)$. The induction step is provided by considering the manifold

$$Q_k(E) = (U(n)/U(k-1) \times U(1) \times U(n-k)) \times_{U(n)} T$$

where T is as before the principal bundle associated to E. We have a projective $(k-1)$-space bundle $Q_k(E) \to G_k(E)$ and a projective $(n-k)$-space bundle $Q_k(E) \to G_{k-1}(E)$, depending on whether we take the $U(1)$ with the $U(k-1)$ or the $U(n-k)$. The induction step from G_{k-1} to G_k via Q_k is made possible by

<u>Lemma 7.4.</u> $\kappa_1' : R(U(k-1) \times U(1) \times U(n-k)) \otimes_{R(U(n))} K_G^*(B) \to K_G^*(Q_k(E))$ <u>is an isomorphism if and only if</u>

$$\kappa_1 : R(U(k) \times U(n-k)) \otimes_{R(U(n))} K_G^*(B) \to K_G^*(G_k(E)) \ \ \underline{\text{is one.}}$$

<u>Proof</u> The 'if' part is a direct consequence of the case $(k = 1)$ of projective bundles already proved. For the 'only if' part let $p: U(k)/U(k-1) \times U(1) \times U(n-k) \to U(n)/U(k) \times U(n-k)$ be the projection. By [4], §4 maps $p^!$, $p_!$ are defined to make the diagram

$$\begin{array}{ccc}
R(U(k-1)\times U(1)\times U(n-k)) \otimes_{R(U(n))} K_G^*(B) & \xrightarrow{\kappa_1'} & K_G^*(Q_k(E)) \\
p^! \otimes 1 \Big\downarrow \Big\uparrow p_! \otimes 1 & & p^! \Big\uparrow \Big\downarrow p_! \\
R(U(k)\times U(n-k)) \otimes_{R(U(n))} K_G^*(B) & \xrightarrow{\kappa_1} & K_G^*(G_k(E))
\end{array}$$

commutative; and $p_! \, p^! = 1$. Hence the fact that κ_1' is an isomorphism implies that κ_1 is one.

The proof of Proposition 7.3. is complete; and this implies that we have also completed the proofs of propositions 7.2 and 7.1. As a consequence of Proposition 7.1, the spectral sequence $\{E_r(X,Y;\tilde{K}_G^*)\} \Rightarrow H(X,Y;\tilde{K}_G^*)$ is defined for any X,Y in $(\mathcal{Q}_G)_o$. The unreduced spectral sequence is more interesting to work with, however. Let us therefore, given X,Y in $\mathcal{Q}_G$, define the notation:

$$\{E_r(X,Y)\} = \{E_r(X^+,Y^+;\tilde{K}_G^*)\}$$

$$H(X,Y) = H(X^+,Y^+;\tilde{K}_G^*)$$

The map Φ (§3 (2)) maps H(X,Y) to $\tilde{K}_G^*(X^+ \wedge Y^+) = K_G^*(X \times Y)$. The E_2 term is as stated in the introduction, i.e.,

$$E_2^{p,q} = \mathrm{Tor}_{R(G)}^{p,q} \; (K_G^*(X),\, K_G^*(Y)).$$

All these facts follow from the theory of §5.

§8. Convergence

We now turn to the second question - the convergence of the Künneth formula spectral sequence $\{E_r(X,Y)\}$. The answer to question (ii) of the previous section is not always 'yes'; I give some examples of bad convergence at the end of this section. However, the result obtained is sufficient for the applications I shall make. We have in fact

Theorem 8.1. (i) *Let G be connected and $\pi_1(G)$ torsion-free; then the I(G)-adic completion of Φ (see [5])*

$$\Phi^\wedge : H(X,Y)^\wedge \to K_G^*(X \times Y)^\wedge$$

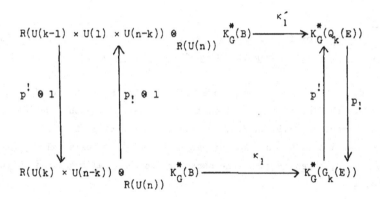

is an isomorphism.

(ii) *If in addition X or Y is a free G-space, then* Φ *is itself an isomorphism*

(iii) *If X or Y is a trivial G-space, then* Φ *is an isomorphism, for any G.*

This result, with those already proved in §7, completes the proof of the theorems claimed in the Introduction. I shall first discuss the obstruction to convergence Γ and prove Proposition 8.1, a generalization of part (iii). Then the main work of the section is a reduction of parts (i), (ii) to the single special case $X = Y = G$; this, together with some necessary homological algebra for $R(G)$, will be left to §11.

Let me recall to begin with that, by theorem 5.1., the 'obstruction to convergence' $\Gamma(X,Y;K_G^*)$, which we shall write $\Gamma(X,Y)$ from now on, is defined; that it fits into an exact triangle of $\mathbb{Z}_2$-graded $R(G)$-modules

(1)
$$\begin{array}{ccc} H(X,Y) & \xrightarrow{\ \Phi\ } & K_G^*(X \times Y) \\ & \nwarrow \quad \swarrow & \\ & \Gamma(X,Y) & \end{array}$$

and is a cohomology theory in X and Y separately. (The $R(G)$-module structure comes from the definition as a direct limit.) However, this cohomology theory has no 'coefficient ring' in the ordinary sense; in fact, we have

Lemma 8.1. $\Gamma(X,Y) = 0$ *if X or Y is a point.*

Proof. If, for example, X is a point, then X is a Künneth space for K_G^*. Hence by K.2 of theorem 5.1., Φ is an isomorphism and $\Gamma(X,Y) = 0$.

If $\Gamma(X,Y)$ were a cohomology theory in the usual sense - i.e., on *Top* - then lemma 8.1 would imply by well known arguments that Γ vanished for all CW complexes. However, this is not the case; we cannot build up G-spaces using spheres and cells, i.e., suspensions and cones of point spaces, as can be done with spaces in *Top*. Instead, the 'building blocks' are homogeneous G-spaces G/H, or orbits, and spaces related to them. This idea, and what can be done with it, finds a particularly clear expression in Segal's spectral sequence for K-theory - Proposition 5.3. of [20]. This admits an immediate formal generalization to any cohomology theory h^* on $\mathcal{O}_G$ satisfying the following continuity condition:

(8.1) *Given a closed G-subset A of X in* $\mathcal{O}_G$, *then the natural map*

$$\lim_{\rightarrow} h^*(V) \to h^*(A)$$

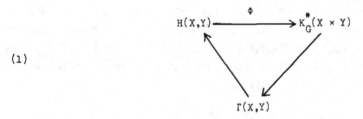

where V *runs over all closed* G-*neighbourhoods of* A, *is an isomorphism*.

I shall state Segal's result, for reference, in the generalized form. Let $p : X \to X/G$ be the projection on the orbit space.

(8.2) *Let* h^* *be a cohomology theory on* $\mathcal{A}_G$ *satisfying the condition* (8.1). *Then there is a strongly convergent spectral sequence, for* X *in* $\mathcal{A}_G$.

$$H^*(X/G;\underline{h}^*) \Rightarrow h^*(X).$$

Here $\underline{h}^*$ *is the sheaf on* X/G *defined by* $U \longmapsto h^*(p^{-1}(\bar{U}))$; *the stalk of* $\underline{h}^*$ *at* $\bar{x} \in X/G$ *is* $h^*(p^{-1}(\bar{x}))$.

It should now be clear that part (iii) of Theorem 8.1. will follow by combining lemma 8.1 and the spectral sequence. In fact, it is worth stating a more general result.

Proposition 8.1. *If* $\Gamma(G.x,Y) = 0$ *for all* $x \in X$, *then* $\Gamma(X,Y) = 0$.

That is, $\Gamma(\_,Y)$ vanishes on X if it vanishes on all orbits in X. Note that if X is a trivial G-space all orbits are points, hence $\Gamma(\_,Y)$ vanishes on orbits by lemma 8.1. The proposition therefore implies Theorem 8.1. (iii).

To prove Proposition 8.1. it is enough to show that the cohomology theory $\Gamma(\_,Y)$ satisfies the condition (8.1). For in that case the spectral sequence of Segal is defined; and the stalks of the sheaf are precisely $\{\Gamma(p^{-1}p(x),Y)\}$, i.e., $\{\Gamma(G.x,Y)\}$. Hence the sheaf $\Gamma(\_,Y)$ is zero, the spectral sequence is zero, and $\Gamma(X,Y) = 0$ by convergence.

This is clearly a useful argument in more general situations. It remains to show that the continuity condition (8.1) is satisfied. Choosing an appropriate K_G^* resolution for Y and negative filtration $\{Y_n\}$, we have

$$\Gamma(X,Y) = \lim_{\substack{\to \\ n}} K_G^*(X \times Y_n, X) \quad .$$

Hence we need only interchange the limits in $\lim\limits_{\substack{\to \\ V}} \Gamma(V,Y) = \lim\limits_{\substack{\to \\ V}} \lim\limits_{\substack{\to \\ n}} K_G^*(V \times Y_n, Y_n)$, and

Use Segal's result ([20], Prop. 2.11) that K_G^* itself satisfies (8.1); we find that the above limit equals $\lim\limits_{\substack{\to \\ n}} \lim\limits_{\substack{\to \\ V}} K_G^*(V \times Y_n, Y_n) = \lim\limits_{\substack{\to \\ n}} K_G^*(A \times Y_n, Y_n) = \Gamma(A,Y)$, as required.

Having dealt with part (iii) of the theorem we now pass on to the main work - parts (i) and (ii).

Completion enters into our results via the functor defined in §1 (2) from G-spaces to spaces over BG. I shall call this functor

$$\Psi : \underline{G - Top} \to \underline{Top/BG}\ ; \qquad \Psi(X) = X_G = X \underset{G}{\times} EG.$$

We have the following relation due to Atiyah and Segal between K_G^* and $K_{BG}^* \circ \Psi$ - see [6].

(8.3). The functors $K_{BG}^* \circ \Psi$ and $K_G^{*\wedge}$ are naturally equivalent on $\mathcal{a}_G$;

$$K^*(X_G) = K^*(\Psi(X)) \cong K_G^*(X)^\wedge$$

as a module over $K_G^*(\text{point})^\wedge = R(G)^\wedge = K^*(BG)$.

We shall also need the facts that R(G) is Noetherian; $R(G)^\wedge$ is flat as an R(G)-module; and for M finitely generated there is a natural isomorphism

(2) $$M \underset{R(G)}{\otimes} R(G)^\wedge \xrightarrow{\cong} M^\wedge$$

See [19] and [8], III §3.

The axiom 'Cyl'.

By using Ψ we transform questions about G-Top into questions about Top/BG. There is one very good reason for doing so; an important result of Dold [12] on cohomology theories on a category Top/B which I shall restate here since we shall use it several times. Consider the following additional axiom on a cohomology theory h^* on Top/B.

Cyl. For any X in Top, and any $\theta : X \times I \to B$,

$$h^*(X \times I,\ X \times \{0\}) = 0$$

Not all cohomology theories satisfy Cyl. However, our theories h_G^* (§3, ex. 1) do; more generally, so do the functors

$$X \mapsto h^*(X \underset{B}{\Pi} Y)$$

provided $Y \xrightarrow{p_Y} B$ is a fibration (See [12], 3.4.) The value of the axiom lies in the following result, for which there seems to be no analogue in G-Top (4.1. of [12]):

(8.4) Let $t : h^* \to k^*$ be a natural transformation of additive cohomology theories which satisfy Cyl. If $t(pt) : h^*(pt) \xrightarrow{\cong} k^*(pt)$ for every point $pt \to B$, then t is an equivalence; $t : h^* \xrightarrow{\cong} k^*$.

If B is arcwise connected, a single point will do; or equivalently a single contractible space in Top/B. And this will always be our method of applying (8.4); to take $B = BG$ and a transformation t such that $t(EG)$ is an isomorphism. Note that (with G acting on G by translations as usual), $EG = \Psi(G)$, so that it is certainly enough to know t is an isomorphism on the image of Ψ. In particular, $h^*(\Psi(G)) = 0$ implies that h^* is identically zero, for h^* as in (8.4).

Our strategy will be, corresponding to the $\mathcal{Q}_G$ spectral sequence of §7, to define a Top/BG spectral sequence using Ψ; to compare the two by (8.3); and to simplify the convergence question (the study of Γ) for the second spectral sequence using (8.4). We shall find that the question of convergence is reduced to the one case $X = Y = G$.

Now, given any space V in Top/BG I can use a K_G^* resolution $\{X_p\}$ for $X_o = X^+$, to define an $H(p,q)$ system and spectral sequence

(3) $$ {}^*H(-p,-q) = K_{BG}^* (S^{2(p-q)} \Psi(X_q) \wedge_{BG} V_o/\Psi(X_p) \wedge_{BG} V_o) $$

with the corresponding H and Γ defined by direct limits as before. I claim that this is a Künneth formula spectra sequence for $\Psi(X) = X_G$ and V in Top/BG. This will clearly be established if I can show that $\{\Psi(X_p)\}$ is a resolution of $\Psi(X)$ for K_{BG}^*. This will follow from

Lemma 8.2. If X is in $\mathcal{Q}_G$ and $X \xrightarrow{j} Z$ is a Künneth embedding in $\mathcal{Q}_G$ for K_G^*, then $\Psi(j)$ is a Künneth embedding in Top/BG for K_{BG}^*.

Proof That $K^*(\Psi(j)) : K^*(\Psi(Z)) \to K^*(\Psi(X))$ is an epimorphism follows from (8.1) and the exactness of completion, since the R(G)-modules in question are finitely generated. It remains to show that $\Psi(Z)$ is a Künneth space in Top/BG for K_{BG}^*. Consider the natural transformation, for V in Top/BG,

$$ \kappa : K_{BG}^*(\Psi(Z)) \otimes_{R(G)^\wedge} K_{BG}^*(V) \to K_{BG}^* (\Psi(Z) \sqcap_{BG} V) $$

Both sides are additive cohomology theories in V, since $K_{BG}^*(\Psi(Z))$ is finitely generated and projective over $R(G)^\wedge$. Furthermore, since Z_G is fibred over BG, both sides satisfy the axiom Cyl (see above). Now a simple application of (8.3) shows that κ is an isomorphism when $V = \Psi(Y)$ for Y in $\mathcal{Q}_G$. In particular this is true if $Y = G$; hence, by the remarks following (8.4), κ is an isomorphism for all V.

The spectral sequence defined by the H(p,q) system (3) is therefore a Künneth formula spectral sequence for K^*_{BG} in Top/BG. We shall accordingly use the notation of §5 from now on, writing it as $\{E_r(\Psi(X),V;K^*_{BG})\}$ etc. The obstruction to its convergence is

$$\Gamma(\Psi(X),V;K^*_{BG}) = \lim_{\rightarrow} \tilde{K}^*_{BG}(\Psi(X_p) \wedge_{BG} V_0)$$

In the particular case when $V = \Psi(Y)$ we have the formula

$$(4) \qquad \Gamma(\Psi(X),\Psi(Y);K^*_{BG}) = \lim_{\rightarrow} \tilde{K}^*_{BG}(\Psi(X_p \wedge Y_0))$$

$$= \lim_{\rightarrow} (\tilde{K}^*_G(X_p \wedge Y_0) \otimes_{R(G)} R(G)\hat{\ })$$

$$= (\lim_{\rightarrow} \tilde{K}^*_G(X_p \wedge Y_0)) \otimes_{R(G)} R(G)\hat{\ }$$

$$= \Gamma(X,Y) \otimes_{R(G)} R(G)\hat{\ }$$

(Note, though, that $\Gamma(X,Y)$ may not itself be a finitely generated R(G)-module, so that we cannot identify this with the I(G)-adic completion of $\Gamma(X,Y)$).

The main aim of this section is to prove the following auxiliary result to Theorem 8.1.

Proposition 8.2. Let gℓ.dim.R(G) be finite. Then

(i) $\Gamma(X,Y)$ is a finitely generated R(G)-module for all X,Y in $\mathcal{Q}_G$.

(ii) The I(G)-adic topology on $\Gamma(X,Y)$ is discrete if X is a free G-space

(iii) If further $\Gamma(G,G) = 0$, then $\Gamma(\Psi(X),\Psi(Y);K^*_{BG}) = 0$ for all X,Y in $\mathcal{Q}_G$.

It should be noted that the condition (finite global dimension of R(G)) is strictly weaker than the condition (G is connected and Tors $\pi_1(G) = 0$) of Theorem 8.1. In §11 we shall see that the second implies the first; on the other hand R(SO(2n + 1)) is polynomial, while $\pi_1(SO(2n + 1))$ is cyclic of order 2.

Part (iii) of proposition 8.2 is the most important; its proof can be reduced by methods we have already used, as follows. $\Gamma(G,G) = 0$ implies $\Gamma(\Psi(G),\Psi(G);K^*_{BG}) = 0$ by (4). But now suppose we have proved the following lemma.

Lemma 8.3. If gℓ.dim. R(G) is finite, then for any X in R(G), $V \mapsto \Gamma(\Psi(X),V;K^*_{BG})$ is an additive cohomology theory satisfying Cyl.

In that case, we can again apply the argument following (8.4); first, to deduce that $\Gamma(\Psi(G),V;K^*_{BG}) = 0$ for all V in Top/BG, and then setting $V = \Psi(Y)$, to prove

$\Gamma(W,\Psi(Y);K^*_{BG}) = 0$ for all W in Top/BG - in particular for $W = \Psi(X)$.

The proof of lemma 8.3 is mostly straightforward. That $\Gamma(\Psi(X),V;K^*_{BG})$ is a cohomology theory in V follows from its definition as a direct limit. Also, since $\Psi(X_p)$ is fibred over BG for each X_p, the functors $V \mapsto K^*_{BG}(\Psi(X_p) \prod_{BG} V, V)$ satisfy Cyl for $p = 0,1,2,\ldots$; hence, again, so does the limit $\Gamma(\Psi(X),V;K^*_{BG})$. The point which remains (and where finite global dimension is used) is the additivity of Γ. For this, as well as for part (i) of Proposition 8.2 I need a more finite description of Γ - one not involving a limit. The arguments apply in the general situation of §5 so I shall state and prove them in that context: suppose therefore that $\tilde{h}^*$ is a reduced cohomology theory on $\mathcal{C}_o$ and X,Y are in $\mathcal{C}_o$, with X admitting a resolution for $\tilde{h}^*$.

<u>Lemma 8.4.</u> *If $\tilde{h}^*(X)$ is $\bar{h}$-flat, there is an exact triangle*

$$\begin{array}{ccc} \tilde{h}^*(X) \otimes_{\bar h} \tilde{h}^*(Y) & \xrightarrow{\ \kappa\ } & \tilde{h}^*(X \wedge Y) \\ {\scriptstyle\delta}\nwarrow & & \swarrow \\ & \Gamma(X,Y;\tilde{h}^*) & \end{array}$$

<u>Proof</u> Flatness implies $\mathrm{Tor}^{-p}_{\bar h}(\tilde{h}^*(X), \tilde{h}^*(Y)) = 0$ for all $p > 0$, so the spectral sequence collapses. The limit is $H(X,Y;\tilde{h}^*) = E^o_2 = \tilde{h}^*(X) \otimes_{\bar h} \tilde{h}^*(Y)$; Φ on this limit is the edge homomorphism κ (by K.2. of Theorem 5.1.). Thus the exact triangle (3) of §3 reduces to the one above.

<u>Lemma 8.5.</u> *If proj. $\dim_{\bar h}(\tilde{h}^*(X))$ is finite, then there is a short exact sequence, for some p,*

$$0 \to \tilde{h}^*(X_p) \otimes_{\bar h} \tilde{h}^*(Y) \xrightarrow{\kappa} \tilde{h}^*(X_p \wedge Y) \to \Gamma(X,Y;\tilde{h}^*) \to 0$$

<u>Proof</u> First, by lemma 5.2., $\Gamma(X,Y;\tilde{h}^*) = \Gamma(X_p,Y;\tilde{h}^*)$ for all p. Since

$$0 \leftarrow \tilde{h}^*(X_o) \leftarrow \tilde{h}^*(Z_o) \leftarrow \ \ldots\ldots\ldots \ \leftarrow \tilde{h}^*(Z_{q-1}) \leftarrow \tilde{h}^*(X_q) \leftarrow 0$$

is exact, and $\tilde{h}^*(Z_i)$ is projective, for $q \geq$ proj.dim.$_{\bar h}\ \tilde{h}^*(X)$ we must have $\tilde{h}^*(X_q)$ projective and therefore flat. For such q the homomorphism δ of lemma 8.4 is defined, and is natural with respect to mappings of spaces on which it is defined. Hence we have a commutative square, for $p - 1 \geq$ proj. $\dim_{\bar h}\ \tilde{h}^*(X)$

$$\begin{array}{ccc}
\Gamma(S^m X_{p-1}, Y; \tilde{h}^*) & \xrightarrow{\delta} & \tilde{h}^*(S^m X_{p-1}) \underset{h}{\otimes} \tilde{h}^*(Y) \\
\Big\downarrow \Gamma(\phi_{p-1},1) & & \Big\downarrow \tilde{h}^*(\phi_{p-1}) \otimes 1 \\
\Gamma(X_p, Y; \tilde{h}^*) & \xrightarrow{\delta} & \tilde{h}^*(X_p) \underset{h}{\otimes} \tilde{h}^*(Y)
\end{array}$$

Since $S^{m-1} X_{p-1} \to Z_{p-1}$ is a Künneth embedding inducing an epimorphism on $\tilde{h}^*$, $\tilde{h}^*(\phi_{p-1}) = 0$. We have by lemma 5.2. that $\Gamma(\phi_{p-1},1)$ is an isomorphism. Hence the lower of the two δ's must be zero, and the exact triangle of lemma 8.4. reduces to a short exact sequence as claimed.

Lemma 8.5 immediately implies Proposition 8.2 (i): for proj. dim. $K_G^*(X)$ is finite and $\tilde{K}_G^*(X_p \times Y, Y)$, which has $\Gamma(X,Y)$ as a quotient, is finitely generated over $R(G)$. Hence the same is true for $\Gamma(X,Y)$. We can also now prove the remaining part of lemma 8.3 - that $\Gamma(\Psi(X),\_; K_{BG}^*)$ is additive. In fact, consider a family of spaces $\{Y_i\}$ in Top/BG; for each i we have an exact sequence

$$0 \to \tilde{K}^*(\Psi(X_p)) \underset{R(G)^\wedge}{\otimes} K^*(Y_i) \to K^*(\Psi(X_p) \times Y_i, Y_i) \to \Gamma(\Psi(X), Y_i; K_{BG}^*) \to 0.$$

Take the direct product, and use the fact [11, II ex. 2] that for A finitely generated $A \otimes \Pi B_i \cong \Pi(A \otimes B_i)$. We deduce that $\Gamma(\Psi(X), \coprod_i Y_i) \cong \prod_i \Gamma(\Psi(X), Y_i)$, i.e., additivity as claimed.

Parts (i) and (iii) of Proposition 8.2 are therefore established. For part (ii) we begin with the following lemma

Lemma 8.6. *If X is any free G-space, there exists an integer q such that $(I(G))^q$ annihilates the R(G)-module $K_G^*(X)$.*

Proof Let us suppose X connected (since X is anyway compact and so $\pi_0(X)$ is finite this is enough).

Choose $x_0 \in X$, let $\bar{x}_0$ be its class in the orbit space X/G, and let $j : G \to X$ be the G-map sending g to $g.x_0$. By B.2 of §1 we have a commutative diagram of ring homomorphisms

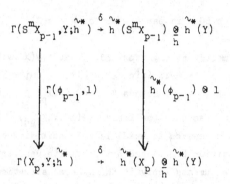

$$\begin{array}{ccc} K_G^*(X) & \xrightarrow{j^!} & K_G^*(G) \\ \Big\downarrow \cong & & \Big\downarrow \cong \\ K^*(X/G) & \xrightarrow{\bar{j}^!} & K^*(\bar{x}_0) \end{array}$$

Consider the composite $R(G) \xrightarrow{i} K_G^*(X) \xrightarrow{\cong} K^*(X/G)$; where i is the coefficient homomorphism. $j^! \circ i$ is easily seen by B.1 to be the augmentation $\varepsilon: R(G) \to \mathbb{Z}$. Hence $j^! i(I(G)) = 0$, so $I(G)$ maps into the kernel of $\bar{j}^!$, i.e., the nilpotent ideal ([3], Cor. 3.1.6) $\tilde{K}^*(X/G) \subset K^*(X/G)$. Hence $i(I(G))^q = 0$ for some q, which proves the lemma.

We see from this that, if X is a free G-space, the I(G)-adic topology on $K_G^*(X)$ is discrete and $K_G^*(X)^\wedge = K_G^*(X)$.

Now return to Proposition 8.2. (ii). Applying lemma 8.5 to Y shows that $\Gamma(X,Y)$ is a quotient of $K_G^*(X_p \times Y, Y)$ for some p. $K_G^*(X_p \times Y, Y)$ has the discrete topology and is a finitely generated R(G)-module; the same is therefore true of $\Gamma(X,Y)$. This proves Prop. 8.2 (ii), hence completes the proof of Proposition 8.2.

In §11 we shall prove

Proposition 8.3. *Let G be a connected group with $\pi_1(G)$ torsion-free. Then*

(i) gℓ.dim. R(G) *is finite* (= rank(G) + 1)

(ii) $\Gamma(G,G) = 0$.

This, with Proposition 8.2, gives us Theorem 8.1. In fact, consider part (ii) of the theorem; under the conditions given, we have by proposition 8.2. (iii) and proposition 8.3 that $\Gamma(\psi(X),\psi(Y);K_{BG}^*) = 0$. From formula (4), then $\Gamma(X,Y) \otimes_{R(G)} R(G)^\wedge = 0$. Now consider the exact triangle (1). $K_G^*(X,Y)$ is finitely generated and by proposition 8.2 (i) so is $\Gamma(X,Y)$. Hence so is $H(X,Y)$ and we can complete the triangle using (2) and deduce that $\Phi^\wedge$ is an isomorphism.

For part (iii) we only use the single further result (Proposition 8.2 (ii)) that $\Gamma(X,Y) \cong \Gamma(X,Y)^\wedge$ if X or Y is free. Hence, again supposing G connected and $\pi_1(G)$ torsion-free, $\Gamma(X,Y)$ is itself zero and the triangle (1) implies Φ is an isomorphism.

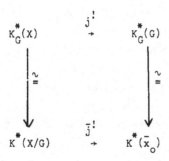

To conclude, here are two cases of non-convergence, of different types, both (since we have Proposition 8.1.) chosen in the key case X = Y = G.

1. $\{E_r^G(G,G)\}$ with G cyclic of order 2. We have $R(G) = \mathbb{Z}[G]$, and this operates trivially on $K_G^*(G) = \mathbb{Z}$. Now, $\mathrm{Tor}_{\mathbb{Z}[G]}(\mathbb{Z},\mathbb{Z})$ is simply $H_*(G;\mathbb{Z})$ which is $\mathbb{Z}_2$ in every odd dimension > 0 and zero in every even one.

Hence the limit of the spectral sequence is $H(G,G)^o = \mathbb{Z}$ and $H(G,G)^1$ a filtered group:

$$A_o \subset A_1 \subset A_2 \subset \dots \subset H(G,G)^1 = \bigcup_i A_i$$

with $A_i/A_{i-1} = \mathbb{Z}_2$. (There are no non-zero differentials as everything is odd-dimensional.)

But $K_G^*(G \times G) = K^*(G)$ which is $\mathbb{Z} \oplus \mathbb{Z}$ in degree 0 and zero in degree 1. Hence there is a serious case of non-convergence - even mod torsion the spectral sequence does not converge.

2. $\{E_r^{SO(3)}(SO(3),SO(3))\}$. Let $\lambda_1 : SO(3) \to U(3)$ be the standard representation. Then $R(SO(3)) = \mathbb{Z}[\lambda_1]$. $E_2 = \mathrm{Tor}_{\mathbb{Z}[\lambda_1]}(\mathbb{Z},\mathbb{Z}) = \Lambda(y_1)$ and differentials again vanish. Hence $H(SO(3),SO(3))^o = H(SO(3),SO(3))^1 = \mathbb{Z}$. But $K_{SO(3)}^*(SO(3) \times SO(3)) = K^*(SO(3))$, which has 2-torsion. Hence the spectral sequence does not converge.

§9. Natural vector bundles

In the spectral sequence $\{E_r(X,Y)\}$ let us now set Y = G. Then $K_G^*(Y) = \mathbb{Z}$, and R(G) acts on $\mathbb{Z}$ via the augmentation

$$\varepsilon : R(G) \to \mathbb{Z} \qquad \varepsilon[V] = \dim V.$$

Moreover, the hoped-for limit of the spectral sequence is $K_G^*(X \times G)$, which by (B.2) of §7 is isomorphic to $K^*(X)$. Hence we have a spectral sequence

$$\{E_r(X,G)\} : \mathrm{Tor}_{R(G)}(K_G^*(X),\mathbb{Z}) \Rightarrow K^*(X).$$

In this section and the next, to simplify exposition, I shall assume that $\{E_r(X,G)\}$ does converge to $K^*(X)$; that is, that we are in the situation of Theorem 8.1 (ii) (G is connected and $\pi_1(G)$ is torsion-free), and that Proposition 8.3. is proved. I promise to be careful about the use of material from this section and the next in the eventual proof (§11) of Proposition 8.3.

I recall from (B.2) that the isomorphism of $K_G^*(X \times G)$ with $K^*(X)$, which will be called ζ, is induced on K_G^o by $[E] \to [E/G]$ (E a G-vector bundle on $X \times G$). Alternatively, take the restriction of E to $X \times \{g\}$ for any $g \in G$ and consider it as a vector bundle over X. ζ^{-1} can be defined as follows. Take $E' \overset{p}{\to} X$, a vector bundle over X, and map $E' \times G \to X \times G$ by $(e,g) \to (g.p(e),g)$. Then this, given a G action by left translation on the second factor in $E' \times G$, is a G-vector bundle over $X \times G$, and represents $\zeta^{-1}[E']$. (All these statements can be easily checked).

Now we specialize further to the case $X = G/H$ where H is a closed subgroup of G. Now $K_G^*(G/H)$, by (B.1.) is identified with $R(H)$, and the identification is induced by

$$R(H) \to K_G^o(G/H), \quad [V] \to [G \times_H V \to G/H]$$

This vector bundle will be very important in what follows; I shall call it $E(V)$ or $E^G(V)$ if I wish to specify G. $R(H)$ inherits its $R(G)$-module structure via the restriction $i^* : R(G) \to R(H)$ which sends $[W]$ to $[W|H]$ (W with the restricted action of H), if W is a G-vector space. The module action seen on $K_G^*(G/H)$ is therefore

(1) $$[E(V)] \,.\, [W] = [E(V \otimes W|H)]$$

Finally, then, with these identifications, we are studying a spectral sequence

(2) $$\{E_r(G/H)\} : \mathrm{Tor}_{R(G)}(R(H),\mathbb{Z}) \Rightarrow K^*(G/H)$$

as stated in the introduction. The notation $\{E_r(G/H)\}$ for $\{E_r(G/H,G)\}$ is a further reasonable abbreviation; the reader will of course not be misled into taking this as a functor of the space (as opposed to the G-space) G/H.

We restrict attention from now on almost exclusively to the sequence $\{E_r(G/H)\}$. We begin by identifying the edge homomorphism $E_2^o(G/H) \overset{\alpha}{\to} K^*(G/H)$. By Theorem 5.1. this can be defined by the diagram

$$\begin{array}{ccc}
\kappa : K_G^*(G/H) \otimes_{R(G)} K_G^*(G) & \to & K_G^*(G/H \times G) . \\
\Vert & & \downarrow \cong \; \zeta \\
R(H) \otimes_{R(G)} \mathbb{Z} & \xrightarrow{\bar{\alpha}} & K^*(G/H)
\end{array}$$

Hence, with the above identifications, $\bar{\alpha}$ is defined by $\bar{\alpha}([V] \otimes 1)$ which comes under ζ from $E(V) \otimes \mathbb{C}$ (external tensor product on $G/H \times G$). This immediately identifies

$\bar{\alpha}([V] \otimes 1)$ (using the description of ζ) with the vector bundle $E(V)$ on G/H. If E is a G-vector bundle write $|E|$ for the underlying vector bundle. Then the above argument shows

Lemma 9.1. *For V an H-vector space $\bar{\alpha}$ is given by*

$$\bar{\alpha}([V] \otimes 1) = [|E(V)|] \in K^0(G/H).$$

Note In the original work of Atiyah and Hirzebruch ([5], 4.5) on K-theory of homogeneous spaces, a homomorphism $\alpha : R(H) \to K^0(G/H)$ was defined by $\alpha([V]) = [|E(V)|]$ (in my notation). (Strictly, this was done for P/H where P is any principal H-bundle; the extension is easily made). $\bar{\alpha}$ is clearly simply α factored through the epimorphism

$$R(H) \to R(H) \otimes_{R(G)} \mathbb{Z}$$

To see why α factors in this way explicitly - it is, of course, the argument leading to (1) in §4 - take V an H-vector space, W a G-vector space. I shall exhibit a vector bundle isomorphism

$$(3) \qquad |E(V \otimes W|H)| \xrightarrow{\eta} |E(V \otimes W_0|H)|$$

where W_0 is the vector space W with the trivial action of G. From this by (1) it follows that $\alpha([V] \cdot [W]) = \alpha([V] \cdot [W_0])$ so that $R(H) \cdot I(G) = 0$ and α factors as required.

The isomorphism η is defined by

$$\eta([g, v \otimes w]) = [g, v \otimes g\cdot w] \qquad g \in G, \quad v \in V, \quad w \in W$$

Then η is well defined since the relation $[g, v \otimes w] = [g\cdot h, h^{-1} v \otimes h^{-1} w]$ $(h \in H)$ in $E(V \otimes W|H)$ is mapped under η into the relation

$$[g, v \otimes g\cdot w] = [g\cdot h, h^{-1} v \otimes g\cdot w]$$

which holds in $E(V \otimes W_0|H)$. On the other hand, η does not commute with the action of G; it is an isomorphism of vector bundles, not of G-vector bundles.

The elements of Im $\bar{\alpha}$ are the natural homogeneous vector bundles over G/H.

In every case that I know of, $\bar{\alpha}$ is a monomorphism, though if some differential d_r $(r \geq 2)$ in $\{E_r(G/H)\}$ mapped non-trivially into E_r^0 it would have a non-trivial kernel. (Conversely in the cases covered by Snaith's vanishing theorem for differentials [21], $\bar{\alpha}$ is a monomorphism.) The simplest case is where R(H) is flat over R(G), when lemma 8.3 and convergence imply the spectral sequence is trivial; we have

Lemma 9.2. *If* R(H) *is* R(G)-*flat then*

$$\bar{\alpha} : R(H) \otimes_{R(G)} \mathbb{Z} \to K^*(G/H)$$

is an isomorphism. In particular, $K^1(G/H) = 0$ *and* $\alpha : R(H) \to K^0(G/H)$ *is an epimorphism ('all bundles over* G/H *are homogeneous').*

This is impossible, consequently, if rank H < rank G; for then the standard theory of rational cohomology for homogeneous spaces [6a] shows that $K^1(G/H; \mathbb{Q}) \cong H^{odd}(G/H; \mathbb{Q})$ is non-trivial. Hence $K^1(G/H) \neq 0$.

On the other hand suppose U is of maximal rank in G, where $\pi_1(G)$ is torsion-free. In this case (at least in the more restricted simply-connected case), Atiyah and Hirzebruch conjectured that α was an epimorphism in [5] , §5, and proved it by classification when G was prime to E_6, E_7, or E_8. Lemma 9.2. reduces the question to the structure of R(H) as an R(G)-module. The solution of this problem is due to Pittie [16], who proves more than mere flatness, namely:

(9.1.) (Pittie) *If* $\pi_1(G)$ *is torsion-free and* H *is of maximal rank in* G, *then* R(H) *is free as an* R(G)-*module.*

A similar result involving $R(G)\hat{} = K^*(BG)$ has been proved by Seymour and Snaith (unpublished). Either of these two implies

Corollary *Under the conditions of* (9.1.),

$$\bar{\alpha} : R(H) \otimes_{R(G)} \mathbb{Z} \to K^*(G/H)$$

is an isomorphism.

Note By concentrating on $\{E_r(G/H)\}$ in this section I have been led to miss out a most elementary point about the edge homomorphism $\bar{\alpha}$ in the general sequence $\{E_r(X,G)\}$. This is a particularly bad omission as the result will be used in §10. The $\bar{\alpha}$ referred to maps $K^i_G(X) \otimes_{R(G)} K^0_G(G) = K^i_G(X) \otimes_{R(G)} \mathbb{Z}$ to $K^i_G(X \times G) = K^i(X)$. We have the following description.

Proposition 9.1. *The edge homomorphism* $\bar{\alpha}$ *is defined by the diagram*

$$\begin{array}{ccc} K_G^i(X) & \xrightarrow{\alpha} & K^i(X) \\ & {\scriptstyle j}\searrow \quad \nearrow{\scriptstyle \bar{\alpha}} & \\ & K_G^i(X) \otimes_{R(G)} \mathbb{Z} & \end{array}$$

where $j(x) = x \otimes 1$, and α is the forgetful map which sends $[E]$ to $[|E|]$.

The proof for $i = 0$ - as usual this is enough - follows by noting that in $K_G(X \times G)$, $\bar{\alpha}([E] \otimes 1)$ is the class of the external tensor product of E and the trivial bundle $\underline{\mathbb{C}}$ over G. From the definition of the definition of the identification ψ above, this corresponds to $[|E|]$ in $K(X)$.

Interpreting proposition 9.1, we see that the image of $\bar{\alpha}$ consists of those bundles over X (or generally, complexes over $X \times \mathbb{R}^n$) which stably admit an action of G. Obviously this makes the spectral sequence more interesting for studying vector bundles over G-spaces.

§10 Generalized difference constructions

Having identified the elements in $K^*(G/H)$ which come from Tor^0 in the spectral sequence, it is natural to go on to Tor^{-1}, which provides the remaining generators in all cases to be dealt with here. In a sense these elements generalize the representation classes $\beta(\rho)$ defined in [13]. The grading in $\{E_r(G/H)\}$ shows that Tor^{-1} leads only to elements in $K^1(G/H)$; I shall first talk about ways of arriving at such elements in a general context.

An element of $K_G^0(X,A)$, when $A \neq \emptyset$, will generally for me be represented by a complex of length 1

$$E_1 \xrightarrow{\phi} E_2$$

where E_1, E_2 are G-vector bundles on X and ϕ is an isomorphism of their restrictions to A. It is unimportant (see [3], 2.6.13) whether we consider ϕ extended to a homomorphism on the whole of X or not.

In particular I choose for $K_G^1(X)$ the model $K_G^0(X \times I, X \times \partial I)$; it should not be hard to see that this agrees with whatever definition you prefer. An element of $K_G^1(X)$

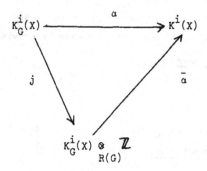

can therefore be represented by a complex over $X \times I$ as above. In particular suppose we have two isomorphisms

$$\phi_0, \phi_1 : E \to E'$$

of G-vector bundles, define $\phi : E \times I \to E' \times I$ so that $\phi(e,t) = \phi_t(e)$ for $t = 0,1$; then this defines an element which I shall write

(1) $$d(E,E'; \phi_0,\phi_1) \in K_G^0(X \times I, X \times \partial I) = K_G^1(X)$$

Exercises 1. $$d(E,E';\phi_0,\phi_1) = d(E,E; \phi_1^{-1} \circ \phi_0, 1)$$

2 If $E \oplus F \xrightarrow{\psi} X \times V$ is an isomorphism with a trivial G-vector bundle then

$$d(E,E;1,\phi) = d(X \times V, X \times V; 1, \psi \circ (\phi \oplus 1_F) \circ \psi^{-1})$$

From these we see that every difference element defined by (1) above can be expressed in a form where the vector bundles E,E' are both the same trivial one.

Let V be a G-vector space and $U(V)$ the unitary group of V w.r.t. some hermitian metric. Give $U(V)$ the action of G defined by

$$(g \cdot A)(v) = g \cdot (A(g^{-1} . v)) \qquad g \in G, \quad A \in U(V), \quad v \in V$$

Then I can define an automorphism of the trivial G-vector bundle $U(V) \times V$ by

(2) $$\phi_V(A,v) = (A, A \cdot v)$$

and so an element

$$x_V = d(U(V) \times V, U(V) \times V; \phi_V, 1)$$

in $K_G^1(U(V))$. This element is universal in the sense that since any difference element of form (1) can be written as $d(X \times V, X \times V; \phi, 1)$, it can be induced from x_V under the map $\tilde{\phi} : X \to U(V)$ defined by

(3) $$\phi(x,v) = (x, \tilde{\phi}(x)(v))$$

We recover in this way the usual correspondence between difference elements and mappings into unitary groups - see [3], 2.4.6. An element so induced by a map $\tilde{\phi} : X \to U(V)$ will be written simply $[\tilde{\phi}]$. From exercise 1, we note the consequence

(4) $$d(X \times V_1, X \times V_2; \phi_0, \phi_1) = [(\phi_1^{-1} \circ \phi_0)^{\sim}] \in K_G^1(X).$$

Now I want to use lemma 3.3. to identify the homomorphism which is there called $\bar{\beta}$ and maps $Z_1^{-1,t}$ into $K^{t-1}(X)/\mathrm{Im}\,\bar{\alpha}$ in our case. The source group $Z_1^{-1,t}$, depends in principle on the particular resolution used. Since the geometric ones are unnecessarily complicated in general, and since the algebraic bar resolution is universal, it will be advantageous to introduce it here.

We need in fact only the very lowest part of the bar resolution for the R(G)-module $\mathbb{Z}$; for a general treatment see [15], IX §6. This is

(5) $$\bar{B} = \bar{B}(R(G), \mathbb{Z}) : \ldots \to R(G) \otimes I(G) \xrightarrow{\partial_1} R(G) \ (\xrightarrow{\varepsilon} \mathbb{Z} \to 0)$$

where ∂_1 is the multiplication of R(G); in terms of representations V, V' and W of G,

$$\partial_1([W] \otimes ([V] - [V'])) = [W \otimes V] - [W \otimes V'] \quad .$$

If M is any R(G)-module, we compute $\mathrm{Tor}_{(R(G)}(M, \mathbb{Z})$ from the complex $M \otimes_{R(G)} \bar{B}$. Using $M \otimes_{R(G)} R(G) = M$, we find for the last terms of this

$$\ldots \to M \otimes I(G) \xrightarrow{\bar{\partial}_1} M \to 0$$

$\bar{\partial}_1$ being induced from the action of R(G) on M. There is an epimorphism $\mathrm{Ker}\,\bar{\partial}_1 = Z_1(M \otimes_{R(G)} \bar{B}) \to \mathrm{Tor}^{-1}_{R(G)}(M, \mathbb{Z})$. I shall take $\mathrm{Ker}\,\bar{\partial}_1$ for the particular version of Z_1^{-1} which will be used. Then we can summarize the state of affairs as follows.

<u>Proposition 10.1.</u> <u>The secondary edge homomorphism in</u> $\{E_r(X,G)\}$ <u>is a map</u>

$$\bar{\beta} : Z_1^{-1,t} \to \mathrm{Tor}^{-1,t}_{R(G)}(K_G^*(X), \ ) \to K^{t-1}(X)/\mathrm{Im}\,\bar{\alpha}$$

<u>where</u>

(i) $Z_1^{-1,t} = \mathrm{Ker}[\bar{\partial}_1 : K_G^t(X) \otimes I(G) \to K_G^t(X)]$, <u>and</u> $\bar{\partial}_1$ <u>is the restricted module action of</u> R(G).

(ii) $\bar{\alpha}$ <u>is defined as in Proposition</u> 9.1.

In order to make my claim about $\bar{\beta}$, let me take an element of $\mathrm{Ker}\,\bar{\partial}_1$ with $t = 0$. Any such can be expressed (as an element of $K_G^0(X) \otimes R(G)$) in the form

(6) $$x = \sum_{i=1}^{n} [E_i] \otimes ([V_i] - [V_i'])$$

where $\{E_i\}$ are G-vector bundles over X and $\{V_i, V_i'\}$ are G-vector spaces with $\dim V_i = \dim V_i'$; and we have an isomorphism (since $\bar{\partial}_1(x) = 0$),

$$\sum E_i \otimes V_i \xrightarrow[\phi]{\cong} \sum E_i \otimes V_i'$$

of G-vector bundles over X. (Only a stable isomorphism in the first place, but we can make it right by adding trivial bundles on both sides).

Now we also have $V_i \cong V_i'$ as vector spaces (forgetting the G-action); let $\psi_i : V_i \to V_i'$ be an isomorphism. I shall also later need the fact that we can use ψ_i to define a G-isomorphism of the vector bundles $V_i \times G$, $V_i' \times G$ over G by

(7) $$\bar{\psi}_i(v,g) = (g \cdot \psi_i(g^{-1}.v), g)$$

<u>Proposition 10.2.</u> <u>For x etc. as above, $\bar{\beta}(x)$ is the class of the difference element</u>

$$d(\sum E_i \otimes V_i, \quad \sum E_i \otimes V_i' \ ; \phi, \quad \sum 1 \otimes \psi_i, \phi) \in K^1(X)/_{\mathrm{Im}\ \bar{\alpha}} \quad .$$

Proof We take $Y_0 = G^+$ and start trying to construct a geometric $\tilde{K}_G^*$ resolution of Y_0, doing as little as we can. $G^+ \to pt^+$ is clearly a Künneth embedding for $\tilde{K}_G^*$; replacing it by an honest embedding we have

$$Z_0 = (CG)^+, \quad Y_1 = CG^+/G^+. \quad \text{(CG means unreduced cone.)}$$

Now let this be continued to any resolution of G^+ for $\tilde{K}_G^*$; to simplify matters I shall suppose I can get a Künneth embedding $Y_1 \to Z_1$ without suspending (it's only a question of indexes).

If I smash with X^+ I get a reasonably unreduced spectral sequence with

$$E_1^0 = \tilde{K}_G^*(X^+ \wedge (CG)^+) = K_G^*(X \times CG).$$

$$E_1^1 = \tilde{K}_G^*(X^+ \wedge Z_1) = K_G^*(X \times Z_1, X).$$

In particular the homomorphism i_2 of lemma 3.3. is the coboundary

$$K_G^1(X \times G) \xrightarrow{\delta} K_G^0(X \times CG, \ X \times G)$$

<u>Lemma 10.1.</u> <u>With the identification of $K_G^1(X \times G)$ used above, let E,E′ be G-vector bundles over X and $\phi_0, \phi_1 : E \times G \to E' \times G$ isomorphisms of G-vector bundles over $X \times G$,</u>

ϕ_1 being of form $\theta \times 1_G$. Then δ sends $d(E \times G, E' \times G; \phi_o, \phi_1)$ to the class represented by

$$E \times CG \overset{\phi}{\rightarrow} E' \times CG$$

where ϕ is any homomorphism restricting to ϕ_o on $E \times G$.

Proof Write $*$ for the basepoint of CG, image of $G \times \{1\}$; $G \subset CG$ is the image of $G \times \{0\}$. Then under our definition of K_G^1, δ is simply the composite

$$K_G^o(X \times G \times I, X \times G \times \partial I) \overset{\sim}{=} K_G^o(X \times CG, X \times G \cup X \times *) \rightarrow K_G^o(X \times CG, X \times G).$$

The isomorphism is a consequence of excision. Since we have required ϕ_1 to come from a bundle isomorphism θ over X, the lemma follows from the definition of relative K-theory in terms of complexes.

The i_2 of lemma 3.3, identified (for the given resolution) with δ, is thus explicitly described - this is the point of lemma 10.1. We now turn our attention to i_1. We have a canonical commutative diagram

(8)

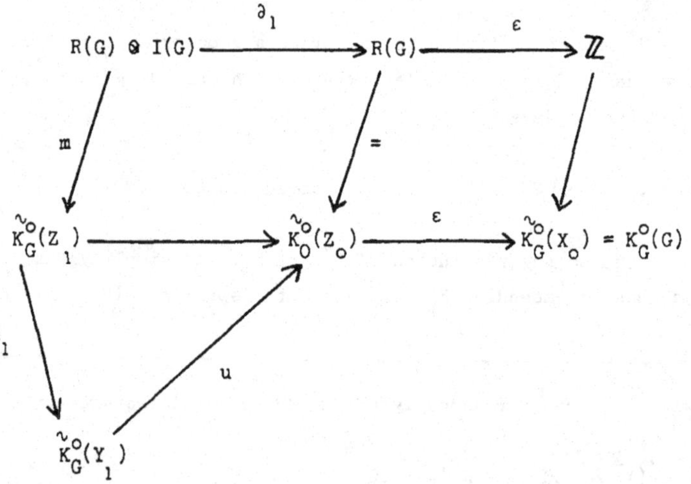

Here m is a map of $R(G)$-modules which makes the diagram commute.

Lemma 10.2. Identifying $\tilde{K}_G^o(Y_1)$ with $K_G^o(CG,G)$, $i_1 \circ m([W] \otimes ([V_i] - [V_i']))$ is represented by $CG \times (W \otimes V_i) \overset{\chi}{\rightarrow} CG \times (W \otimes V_i')$ where χ is any homomorphism extending

$$1_W \otimes \bar{\psi}_i : G \times (W \otimes V_i) \rightarrow G \times (W \otimes V_i')$$

Proof Since ε is an epimorphism, u in (8) is a monomorphism. Hence for any y,

$i_1 \circ m\,(y)$ is determined by $u \circ i_1 \circ m\,(y)$. Now by (8), $u \circ i_1 \circ m$ applied to the given element, is

$$z = [W \otimes V_i] - [W \otimes V_i'] \in R(G) = K_G^0(CG)$$

And the complex above, when u is applied, reduces precisely to the difference of the two vector bundles (the homomorphism ceases to count) i.e., to z. Hence, that complex represents $i_1 \circ m([W] \otimes ([V_i] - [V_i']))$.

But now to know i_1 in general, consider

$$\begin{array}{ccccc} K_G^0(X) \otimes I(G) & \xrightarrow{\bar{m}\ \sim} & \tilde{K}_G^0(X^+ \wedge Z_1) & \xrightarrow{i_1} & \tilde{K}_G^0(X^+ \wedge Y_1) \\ & & & & \| \\ & & & & K_G^0(X \times CG,\ X \times G) \end{array}$$

$\bar{m}$ is the obvious extension of m using tensor and cartesian product. Simply by tensoring on the left by $K_G^0(X)$ we can deduce from lemma 10.2.,

<u>Lemma 10.3.</u> $i_1 \circ \bar{m}\,([E] \otimes ([V_i] - [V_i']))$ <u>is represented by</u>

$$1 \times \chi \ : CG \times (E \otimes V_i) \to CG \times (E \otimes V_i'),$$

χ <u>extending</u> $1_E \otimes \bar{\psi}_i$ <u>as in lemma</u> 10.2.

Now let us return to the situation of Proposition 10.2. and complete the proof. Lemma 3.3 implies that if we have x as given in $K_G^0(X) \otimes I(G)$ and $y \in K_G^1(X \times G)$ such that $\delta(y) = i_1 \circ \bar{m}(x)$, then y represents $\bar{\beta}(x)$. Consider now

$$y = d((\sum E_i \otimes V_i) \times G, (\sum E_i \otimes V_i') \times G; \quad \sum(1 \otimes \bar{\psi}_i),\ \phi \times 1_G)$$

Then by lemma 10.1 and 10.3, $\delta(y) = i_1 \circ \bar{m}\,([E] \otimes ([V_i] - [V_i']))$. This gives us proposition 10.2, given that the functor ζ of §9 from K_G-theory of $X \times G$ to K-theory of X clearly sends y to $d(\sum E_i \otimes V_i,\ \sum E_i \otimes V_i';\ \ \sum(1 \otimes \psi_i),\ \phi)$

Now let us specialize to the case $X = G/H$, $K_G^*(X) = R(H)$. Here everything is simplified since $K_G^1(G/H) = 0$ so that $\text{Im}\ \bar{\alpha} \subset K^1(G/H)$ is zero and $\bar{\beta}$ becomes a map

$$\bar{\beta} : \text{Ker}\ \bar{\partial}_1 \to K^1(G/H)$$

Ker $\bar{\partial}_1$ now consists of elements of $R(H) \otimes I(G)$, of form $\sum_{i=1}^{n} [W_i] \otimes ([V_i] - [V_i'])$, such that

$$\sum [W_i \otimes V_i|H] \cong \sum [W_i \otimes V_i'|H]$$

In this situation, the following is happening. $W_i \otimes V_i|H$ as an element of $K^0(G/H)$ is $|E(W_i \otimes V_i|H)|$ in the notation of §9. And the isomorphism (3) of that section identifies this with $|E(W_i \otimes V_i^0|H)|$ ($V_i^0 = V_i$ with trivial action of G). Now $|E(\sum (W_i \otimes V_i|H))|$ is isomorphic to $|E(\sum (W_i \otimes V_i'|H))|$ for two reasons; first, using the given isomorphism of H-vector spaces between $\sum W_i \otimes V_i|H$ and $\sum W_i \otimes V_i'|H$; and second, using the isomorphism (3) of §9 to consider both $V_i|H$ and $V_i'|H$ as equivalent, in the situation where they appear, to $V_i^0|H$ and so obtaining isomorphisms of each component $|E(W_i \otimes V_i|H)|$ with $|E(W_i \otimes V_i'|H)|$ corresponding to the $1 \otimes \psi_i$'s. Our general philosophy now tells us that when two vector bundles are isomorphic 'for different reasons' we obtain a difference element in $K^1(G/H)$.

Two particular cases deserve attention

1. Suppose V, V' are G-vector spaces which are isomorphic as H-vector spaces. It will simplify matters here to suppose, as we can, that $V = V' = \mathbb{C}^n$ as vector spaces and that the actions of H on V, V' are not just isomorphic but the same. In the language of representations, we have $\rho, \rho' : G \to U(n)$ such that $\rho|H = \rho'|H$.

We are looking for $\bar{\beta}(1 \otimes ([V] - [V'])) \in K^1(G/H)$. $1 \otimes [V] \in R(H) \otimes R(G)$ determines, obviously, the trivial G-vector bundle $G/H \times V$ which as a vector bundle is just $G/H \times \mathbb{C}^n$. For $\psi : G/H \times V \to G/H \times V'$ we can take the identity map. But, as a G-vector bundle, $G/H \times V \cong G \times_H V = E(V|H)$ by the isomorphism η^{-1} of §9; $(gH, v) \to [g, g^{-1}.v]$; and the fact that the actions of H on V, V' are identical means that we can identify $G \times_H V$, $G \times_H V'$.

We arrive at the situation where our two isomorphisms $G/H \times \mathbb{C}^n \to G/H \times \mathbb{C}^n$ are given by ψ = identity

ϕ is the composite;

$$G/H \times \mathbb{C}^n = G/H \times V \xrightarrow{\eta^{-1}} G \times_H V = G \times_H V' \xrightarrow{\eta'} G/H \times V' = G/H \times \mathbb{C}^n$$

where η' comes from the action of G on V' as η did from its action on V. This gives explicitly

$$\phi(gH, v) = (gH, \rho'(g).\rho(g^{-1}).v) \tag{9}$$

Hence Proposition 10.2 tells us that $\bar{\beta}(1 \otimes ([V] - [V']))$ is represented by

(10) $$d(G/H \times \mathbb{C}^n, G/H \times \mathbb{C}^n; 1, \phi)$$

where ϕ is given by (9). Using the description (4) of elements in K^1 given in this form we find

<u>Proposition 10.3.</u> <u>If</u> V, V' <u>are</u> G-<u>vector spaces corresponding to representations</u> ρ, ρ' <u>of</u> G <u>in</u> U(n) <u>which agree on</u> H, <u>then</u>

$$\bar{\beta}(1 \otimes ([V] - [V'])) = [\rho \, . \, (\rho')^{-1}] \in K^1(G/H)$$

<u>where</u> $\rho \, . \, (\rho')^{-1}$ <u>stands for the map</u> $G/H \to U(n)$ <u>sending</u> gH to $\rho(g)\,(\rho'(g))^{-1}$ (<u>well defined because</u> ρ, ρ' <u>agree on</u> H).

<u>Corollary 10.1.</u> <u>If</u> ρ <u>is trivial on</u> H, <u>then</u> $\rho : G \to U(n)$ <u>factors through the projection</u> $G \to G/H$. <u>Call this map</u> $\bar{\rho}$; <u>then</u>

$$\bar{\beta}(1 \otimes ([V] - [\mathbb{C}^n])) = [\bar{\rho}]$$

<u>Corollary 10.2.</u> <u>If</u> H <u>is a normal subgroup of</u> G <u>and</u> $\bar{\rho} : G/H \to U(n)$ <u>is a representation of</u> G/H, <u>let</u> V <u>be the corresponding</u> G/H-<u>vector space regarded as a</u> G-<u>vector space by composition;</u> <u>then</u> $\bar{\beta}(1 \otimes ([V] - [\mathbb{C}^n])) = [\bar{\rho}] = \beta(\bar{\rho})$ <u>as defined in</u> [13], I §4, i.e. <u>the homotopy class of the representation regarded simply as a map.</u>

When we come to do computations it will be convenient to use representations as elements of R(G) rather than G-vector spaces (although the identification of the difference element above shows that it is useful to keep them apart at present.) In that notation, Proposition 10.2 assumes the more obvious form

(11) $$\bar{\beta}(1 \otimes (\rho - \rho')) = [\rho \, . \, (\rho')^{-1}]$$

We shall allow ourselves the notation

$$\beta(\rho - \rho')$$

for this element, which clearly does not depend on the precise expression in the form $\rho - \rho'$.

2. The second example which I want to look at applies only in the case we are dealing with a discrete (so finite) subgroup of G. Let Γ be such a subgroup; then the regular representation ρ_Γ annihilates the augmentation ideal $I(\Gamma)$, i.e., $\rho_\Gamma \, . \, \rho = \rho_\Gamma \, . \, \varepsilon(\rho)$ for all $\rho \in I(\Gamma)$. (Because the character of ρ_Γ is zero except at

the identity element). Hence for any representation σ of G, $\bar{\partial}(\rho_\Gamma \otimes \tilde{\sigma}) = \rho_\Gamma \cdot \tilde{\sigma} = 0$

<u>Proposition 10.4.</u> $\bar{\beta}(\rho_\Gamma \otimes \tilde{\sigma}) = \pi_!(\beta(\sigma))$ <u>where</u> β <u>is as above and</u> $\pi_! : K^*(G) \to K^*(G/\Gamma)$ <u>is defined in</u> [1], §2.

<u>Proof</u> Regarding $\pi_!(\beta(\sigma))$ as a complex E over $(G/\Gamma \times I,\ G/\Gamma \times \partial I)$ as before, it is enough to give for each $g\Gamma \in G/\Gamma$ an isomorphism of $\bar{\beta}(\rho_\Gamma \otimes \tilde{\sigma})_{g\Gamma}$ with $\bigoplus_{g' \in g\Gamma} \beta(\sigma)_{g'}$ which depends continuously on $g \in G$.

The representation space of ρ_Γ is the group ring $\mathbb{C}\Gamma$ on which Γ acts by left translations. Let σ be a representation of degree n; then by the above

$$\beta(\sigma) = d(G \times \mathbb{C}^n,\ G \times \mathbb{C}^n;\ \phi, 1)$$

where
$$\phi(g,v) = (g, \sigma(g)(v))$$

On the other hand we can describe $\bar{\beta}(\rho_\Gamma \otimes \tilde{\sigma})$ by a method similar to that used in example (1) above; we find that

$$\bar{\beta}(\rho_\Gamma \otimes \tilde{\sigma}) = d(G \times_\Gamma (\mathbb{C}\Gamma \otimes \mathbb{C}^n),\ G \times_\Gamma (\mathbb{C}\Gamma \otimes \mathbb{C}^n);\ \psi, 1)$$

where ψ is described as follows. Let V denote $\mathbb{C}^n$ regarded as a G-vector space with the action induced by σ; then $\mathbb{C}\Gamma \otimes \mathbb{C}^n$, $\mathbb{C}\Gamma \otimes V$ are isomorphic Γ-vector spaces by the remark preceding proposition 10.4. Let ψ_1 be an isomorphism between them. We have the standard identification

$$G \times_\Gamma (\mathbb{C}\Gamma \otimes V) \xrightarrow{\psi_2} G \times_\Gamma (\mathbb{C}\Gamma \otimes \mathbb{C}^n)$$

$$\psi_2\ [g,\ u \otimes v] = [g,\ u \otimes \sigma(g).v]$$

already used several times. Then

$$\psi[g,\ u \otimes v] = \psi_2\ \psi_1\ [g,\ u \otimes v] = \psi_2\ [g,\ \psi_1(u \otimes v)]$$

Now an explicit isomorphism ψ_1 is given by

$$\psi_1(y \otimes v) = y \otimes \sigma(y)v$$

where $y \in \Gamma$, $v \in \mathbb{C}^n$. (A general element of $\mathbb{C}\Gamma \otimes \mathbb{C}^n$ has form $\sum_i y_i \otimes v_i$ where $v_i \in \mathbb{C}^n$; it is enough to describe ψ_1 on $y \otimes v$ and extend linearly). Hence finally

$$\psi[g, y \otimes v] = [g,\ y \otimes \sigma(g)\ \sigma(y)v] = [g,\ y \otimes \sigma(gy)v] \quad .$$

Now the map

$$\bar{\beta}(\rho_\Gamma \otimes \tilde{\sigma})_{g\Gamma} \to \bigoplus_{y\in\Gamma} \beta(\sigma)_{gy}$$

defined by $[g,y \otimes v] \to (gy,v)$ is clearly an isomorphism of complexes and therefore the elements $\bar{\beta}(\rho_\Gamma \otimes \sigma)$ and $\pi_!\beta(\sigma)$ coincide as claimed.

§11 The case where π_1 is torsion-free

We now have a general description of how to get at elements in $K^*(G/H)$ corresponding to terms in the spectral sequence. In this section we investigate the homological algebra of the ring R(G) when G is connected and $\pi_1(G)$ is torsion-free and, using the good result on the K-theory of these groups which is available ([13], Theorem A), prove the outstanding result from §8, Proposition 8.3. This will complete the proof of theorem 8.1 and allow us to use $\{E_r(G/H)\}$ for such groups G with no further scruples.

Proposition 11.1. *Let G be a compact connected Lie group with $\pi_1(G)$ torsion-free and G_o a maximal connected semisimple subgroup.*[*] *Then there is an isomorphism*

(1) $$R(G) \cong R(G_o) \otimes R(G/G_o).$$

Before starting to prove this, note that G_o is simply-connected; for the exact homotopy sequence shows that $\pi_1(G_o) \to \pi_1(G)$ is a monomorphism, $\pi_1(G)$ is torsion-free by hypothesis, and $\pi_1(G_o)$ is finite. Hence by I. lemma 3.3. of [13], $R(G_o)$ is a polynomial ring on k = rank G_o generators; G/G_o is compact, connected, with no semi-simple component, and so a torus. Hence $R(G/G_o)$ is the group ring of the free abelian group of rank ℓ, $\mathrm{Hom}(G/G_o,S^1)$. Explicitly, therefore, (1) implies that we can write, for G as above,

Corollary 11.1. $R(G) = \mathbb{Z}[\rho_1,\ldots,\rho_k] \otimes \mathbb{Z}[\theta_1,\ldots,\theta_\ell;\theta_1^{-1},\ldots,\theta_\ell^{-1}]$ *for suitable elements ρ_i, θ_j in R(G).*

Corollary 11.2. $g\ell.\dim.R(G) = \text{rank } G + 1 < \infty$

This follows from standard results (e.g. [11], VIII. 4.2, [15], VII. 4.2.) on the global dimension of polynomial algebras and free abelian group rings. I shall consider an explicit finite resolution later on.

Now let us begin the proof of Proposition 11.1. There is a covering (see [7], 2.9) $G_o \times T \xrightarrow{f} G$ such that the kernel Γ of f is a discrete subgroup intersecting T and G_o at the identity element only. Let $p_1 : G_o \times T \to G_o$, $p_2 : G_o \times T \to T$ be the projections; then $p_1|\Gamma$, $p_2|\Gamma$ are monomorphisms. Since $G_o \to G \to G/f(T)$ has kernel exactly $p_1(\Gamma)$, $p_1(\Gamma)$ must be in the centre of G_o.

[*] So that G_o is normal and G/G_o is a group.

We need the following lemma

Lemma 11.1. *If $\rho : G_o \to U(n)$ is an irreducible unitary representation of G_o, there exists a 1-dimensional representation σ of T such that $\rho \circ p_1|\Gamma$ is equal to the sum of n copies of $\sigma \circ p_2|\Gamma$.*

Proof By Schur's lemma, since $p_1(\Gamma)$ is in the centre of G_o, $\rho(p_1(y))$ is a multiple of the identity matrix for any $y \in \Gamma$. Hence $\rho \circ p_1|\Gamma$ is the sum of n copies of some 1-dimensional representation σ' of Γ. Since $\Gamma \xrightarrow{p_2|\Gamma} T$ is a monomorphism and $\mathrm{Hom}(\_,S^1)$ is an exact functor, there is a 1-dimensional representation σ of T such that $\sigma \circ p_2|\Gamma$ is equal to σ'. This proves the lemma.

Now let us pass to the representation rings. We need the following facts (see [13] I, §3)

(i) Every irreducible representation of $G_o \times T$ is of form $\rho \otimes \theta$ where ρ, θ are irreducible representations of G_o,T respectively; hence

$$R(G_o \times T) \cong R(G_o) \otimes R(T)$$

(ii) If $\rho_1', \ldots, \rho_k'$ are the basic representations of G_o, $R(G_o) = \mathbb{Z}[\rho_1', \ldots, \rho_k']$.

(iii) $R(G) \xrightarrow{f^*} R(G_o \times T)$ is a monomorphism whose image is generated by those irreducible representations of $G_o \times T$ which are trivial on Γ.

It will be more convenient to use the basis of $R(G_o)$ constituted by the monomials $m(\rho') = \rho_1'^{\,i_1} \ldots \rho_k'^{\,i_k}$; there is a natural 1 - 1 correspondence (determined by the highest weight) between these representations and the irreducible ones. An inductive argument similar to that for [13] , I. Lemma 3.3. shows that we can change basis in (iii) above, and conclude that the image of f^* is generated by those elements of form $m(\rho') \otimes \theta$ ($\theta \in \mathrm{Hom}(T,S^1)$) which are trivial on Γ .

For each $j = 1,\ldots,k$, lemma 11.1 proves us with an element $\sigma_j \in \mathrm{Hom}(T,S^1)$ such that the representation $\rho_j' \otimes \sigma_j^{-1}$ of $G_o \times T$ is trivial on Γ. This therefore defines a unique representation ρ_j of G (actually an irreducible representation) such that $\rho_j \circ f = \rho_j' \otimes \sigma_j^{-1}$. The elements ρ_j generate a polynomial subalgebra of R(G), since the ρ_j''s do in $R(G_o \times T)$.

Write $\bar{\Gamma}$ for $p_2(\Gamma) \subset T$. Then the homomorphisms $G_o \times T \to G \to G/G_o = T/\bar{\Gamma}$, together with the embedding of $\mathbb{Z}[\rho_1,\ldots,\rho_k]$ in R(G), define homomorphisms

$$\mathbb{Z}[\rho_1,\ldots,\rho_k] \otimes R(T/\bar{\Gamma}) \xrightarrow{u} R(G) \xrightarrow{v} R(G_o) \otimes R(T).$$

Proposition 11.1. will be proved if we can show that u is an isomorphism.

First, since $T \to T/\bar{\Gamma}$ is an epimorphism, $\mathrm{Hom}(T/\bar{\Gamma},S^1) \to \mathrm{Hom}(T,S^1)$ is a monomorphism; so v o u is injective on the generators $m(\rho) \otimes \theta$, ($m(\rho)$ is a monomial in the ρ_i's and $\theta \in \mathrm{Hom}(T/\bar{\Gamma},S^1)$), and hence a monomorphism. u is therefore also a monomorphism. To show that u is an epimorphism take a generator of Im(v), i.e., an element of form $m(\rho') \otimes \theta$, (where $\theta \in \mathrm{Hom}(T,S^1)$), which is trivial on Γ. We can rewrite this as $m(\rho_i' \otimes \sigma_i^{-1})$. $(1 \otimes \theta_1)$ with $\sigma_1,\ldots,\sigma_k$ as above, for some $\theta_1 \in \mathrm{Hom}(T,S^1)$. Since $\rho_i' \otimes \sigma_i^{-1}$ is trivial on Γ, $1 \otimes \theta_1$ must also be. That is, θ_1 is trivial on $p_2(\Gamma) = \bar{\Gamma}$, and so comes from an element $\tilde{\theta}_1$ in $\mathrm{Hom}(T/\bar{\Gamma},S^1) \subset R(T/\bar{\Gamma})$. Hence

$$m(\rho') \otimes \theta = v \circ u(m(\rho) \otimes \tilde{\theta}_1)$$

and u is an epimorphism, so an isomorphism. This proves Proposition 11.1., hence (via the second corollary), Proposition 8.3.(i).

Now let us look at the homological algebra of R(G). Since it is not important here to keep the distinction between the ρ_i's and the θ_i's, I shall let $\rho_1,\ldots,\rho_\ell$ denote a set of generators, some of one kind and some of the other. Let $\varepsilon : R(G) \to \mathbb{Z}$ be the augmentation and write $\tilde{\rho}$ for $\rho - \varepsilon(\rho)$, for ρ a representation of G. Then I(G) is a free R(G)-module on $\tilde{\rho}_1,\ldots,\tilde{\rho}_\ell$ and $(\tilde{\rho}_1,\ldots,\tilde{\rho}_\ell)$ satisfy the condition of [11], VIII, 4, so that the associated 'Koszul complex' is an R(G)-resolution of $\mathbb{Z}$. This complex is defined by

(2) $$L_*(R(G)) = (\Lambda_{R(G)}(y_1,\ldots,y_\ell), d)$$

where Λ denotes exterior algebra, $\dim(y_i) = 1$, and $d: L_i(R(G)) \to L_{i-1}(R(G))$ is the derivation of the exterior algebra determined by

(3) $$d(y_i) = \tilde{\rho}_i \qquad i = 1,\ldots,\ell .$$

The augmentation of the complex is the usual one: $L_o(R(G)) = R(G) \xrightarrow{\varepsilon} \mathbb{Z}$; it induces a homology isomorphism. If M is an R(G)-module write $L_*(R(G);M)$ for the complex $M \otimes_{R(G)} L_*(R(G))$. Then (from the definition of Tor), $H_i(L_*(R(G);M)) = \mathrm{Tor}^{-i}_{R(G)}(M,\mathbb{Z})$.

If M is an R(G)-algebra it follows from (2) that $L_*(R(G);M) = \Lambda_M(y_1,\ldots,y_\ell)$ is a differential graded algebra and the product structure induced on Tor is the usual one, see §5, [15], VII.2.

The formula for the differential d in $L_*(R(G); R(H))$ (our means of computing the E_2 term of $\{E_r(G/H)\}$) will be important in what follows. Using (3), and the fact that the R(G)-module structure of R(H) is induced by $i^* : R(G) \to R(H)$, we have

$$(4) \qquad d(1 \otimes y_i) = i^*(\tilde{\rho}_i) \otimes 1 \in R(H) \underset{R(G)}{\otimes} \Lambda_{R(G)}(y_1,\dots,y_\ell)$$

and d is extended to be a derivation of R(H)-algebras. On a generator of $L_k(R(G); R(H))$ we have

$$d(\sigma \otimes y_{i_1} \dots y_{i_k}) = \sum_{j=1}^{k} (-1)^j \sigma . i^*(\tilde{\rho}_j) \otimes y_{i_1} \dots \hat{y}_{i_j} \dots y_{i_k} .$$

(As usual, $\hat{}$ means that y_{i_j} is left out of the product.)

Since $L_*(R(G))$ is a resolution, the bar resolution $\bar{B}(R(G),Z)$ maps into $L_*(R(G))$. (Compare §10). If we consider the diagram

$$\begin{array}{ccccccc}
R(G) \otimes I(G) & \xrightarrow{\partial_1} & R(G) & \xrightarrow{\varepsilon} & \mathbb{Z} & \longrightarrow & 0 \\
\Big\downarrow m_1 & & \Big\downarrow = & & \Big\downarrow = & & \\
R(G) . \{y_1,\dots,y_\ell\} & \xrightarrow{d} & R(G) & \xrightarrow{\varepsilon} & \mathbb{Z} & \longrightarrow & 0
\end{array}$$

we can define $m_1 : \bar{B}_1 \to L_1(R(G))$ as follows. Given $\rho \otimes \tilde{\rho}'$ in $R(G) \otimes I(G)$, $\partial_1(\rho \otimes \tilde{\rho}') = \rho . \tilde{\rho}' \in I(G)$. Since $\tilde{\rho}_1,\dots,\tilde{\rho}_\ell$ form a free basis for I(G) we can write

$$\rho . \tilde{\rho}' = \sum \xi_i \tilde{\rho}_i = d(\sum \xi_i \otimes y_i)$$

for unique $\xi_1,\dots,\xi_\ell \in R(G)$. Define $m_1(\rho \otimes \tilde{\rho}') = \sum \xi_i \otimes y_i$, and then m_1 is the required mapping. If $\tilde{\rho}' = \sum \xi_i^0 \tilde{\rho}_i$ then $\xi_i = \rho\, \xi_i^0$.

We can now more generally compare the bar resolution (used in §10 for computing $\{E_r(G/H)\}$) and the Koszul complex, described by (4) and much easier for calculations. The mapping which compares the two complexes $R(H) \underset{R(G)}{\otimes} \bar{B}(R(G),\ )$ and $R(H) \underset{R(G)}{\otimes} L_*(R(G))$ is, in degree 1, a map

$$\bar{m}_1 : R(H) \underset{R(G)}{\otimes} (R(G) \otimes I(G)) = R(H) \otimes I(G) \longrightarrow R(H) . \{y_1,\dots,y_\ell\} .$$

(Compare §10). And it is easily verified that for $\sigma \in R(H)$, $\tilde{\rho}'$ as above in I(G),

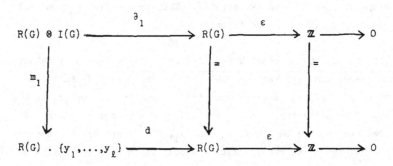

$\bar{m}_1$ satisfies the formula

$$(5) \qquad \bar{m}_1(\sigma \otimes \tilde{\rho}') = \sum \sigma \cdot (\xi_i^0 | H) \otimes y_i \quad .$$

We can now prove Proposition 8.3. (ii), i.e., that Φ is an isomorphism for the given class of groups G. The proof depends on the result of [13] (rephrasing theorem A in our terminology);

(11.1). Let G be connected with $\pi_1(G)$ torsion-free; let $\rho_1,\ldots,\rho_\ell$ be generators of R(G) as above. Then as a ring,

$$K^*(G) = \Lambda_{\mathbb{Z}}(\beta(\rho_1),\ldots,\beta(\rho_\ell))$$

where β is defined as in Corollary 10.2.

To use this, let us consider the spectral sequence which in the notation of §7 is $\{E_r(G,G)\}$, or in that of §9 (2) is $\{E_r(G/1)\}$. Its E_2-term is $\mathrm{Tor}_{R(G)}(K_G^*(G), K_G^*(G)) = \mathrm{Tor}_{R(G)}(\mathbb{Z},\mathbb{Z})$, its limit is called H(G,G) and maps under Φ into $K^*(G/1) = K^*(G)$. In the Koszul complex for Tor, $L_*(R(G);\mathbb{Z}) = \Lambda_{\mathbb{Z}}(y_1,\ldots,y_\ell)$, we have from (4) that $d(1 \otimes y_i) = 0$ for all i; so, by the derivation condition d vanishes identically and we have the result (familiar for rings of this type.

$$(6) \qquad E_2(G/1) = \mathrm{Tor}_{R(G)}(\mathbb{Z},\mathbb{Z}) = \Lambda_{\mathbb{Z}}(y_1,\ldots,y_\ell)$$

as an algebra. (Compare [15], VII, Theorem 2.2). By (5), $y_i = \bar{m}_1(1 \otimes \tilde{\rho}_i)$; by Corollary 10.2., the image of y_i in $K^1(G)$ under the secondary edge homomorphism $E_2^{-1} \to H(G,G)$ followed by Φ is just $\beta(\rho_i)$. (This is where it is necessary to note that the identification of difference elements in §10 did not require $\{E_r(G/H)\}$ to be convergent; the same results hold for any group G if we compose the edge homomorphism with Φ.)

The differentials d_r in $\{E_r(G)\}$ are derivations (a consequence of Theorem 5.3) which vanish on $y_1,\ldots,y_\ell \in E_2^{-1}$ for dimensional reasons and hence vanish identically. Hence $E_2(G/1) = E_\infty(G/1) = \mathrm{Gr}\, H(G,G)$. Filter $K^*(G)$ by the usual exterior algebra filtration in which the monomials of weight $\leq i$ in the $\beta(\rho_j)$'s generate the ith filtration subgroup. Then Φ maps the zeroth filtration $H(G,G)_0 \to K^*(G)_0$ isomorphically, trivially; and the first filtration isomorphically since, by what has been said $E_2^{-1} = H(G,G)_1/H(G,G)_0$ maps isomorphically to $K^*(G)_1/K^*(G)_0$. It is now an easy induction on the filtration degree to deduce from (11.1) and (6) that Φ is an isomorphism. This completes the proof of Proposition 8.3. (iii), and so, finally, of Theorem 8.1. We can now therefore use

$$\{E_r(G/H)\} : \mathrm{Tor}_{R(G)}(R(H),\mathbb{Z}) \Rightarrow K^*(G/H)$$

with a clear conscience whenever G is connected and $\pi_1(G)$ is torsion-free.

§12. Non-simply-connected groups

We now begin the computation of particular spectral sequences $\{E_r(G/H)\}$ using the methods derived above. In this section we shall consider the use of the spectral sequence to find the K-theory of a group which is simple but not simply-connected. Such a group is a quotient G/Γ where G is simple and simply-connected and Γ is a discrete subgroup of its centre; hence we are investigating $\{E_r(G/\Gamma)\}$. Γ is abelian so by [13], I. 3.1. $R(\Gamma)$ is the group ring of its character group Γ^*. If ℓ = rank G and $\rho_1,\ldots,\rho_\ell$ are the basic representations, E_2 can be computed (see §11) by using the Koszul complex:

$$E_2^{-p}(G/\Gamma) = H_p(\Lambda_{R(\Gamma)}(y_1,\ldots,y_\ell))$$

where $d(y_i) = \tilde{\rho}_i|\Gamma$. The cases which occur are the following:

1. (Lie algebra type) A_ℓ. $G = SU(\ell+1)$, Γ is a subgroup of $\mathbb{Z}_{\ell+1}$ (the centre of G) and G/Γ is a covering of the projective unitary group $PU(\ell+1)$.

   I shall not deal with this case, which is the most difficult - although particular cases of it are tractable, see below.
2. Type B_ℓ ; $G = \mathrm{Spin}(2\ell+1)$. The centre of G is $\mathbb{Z}_2 = \{\pm 1\}$, and if $\Gamma = \mathbb{Z}_2$, $G/\Gamma = SO(2\ell+1)$ (compare case 4.)
3. Type C_ℓ : $G = Sp(\ell)$. The centre of G is $\mathbb{Z}_2$ and if $\Gamma = \mathbb{Z}_2$, $G/\Gamma = PSp(\ell)$ is the projective symplectic group.
4. Type D_ℓ : $G = \mathrm{Spin}(2\ell)$ ($\ell > 1$). The centre of G is generated by - 1 and $e_1 \ldots e_{2\ell}$ in the Clifford algebra - see [10] ; it is $\mathbb{Z}_4$ for ℓ odd, $\mathbb{Z}_2 + \mathbb{Z}_2$ for ℓ even. If Γ is the whole centre, G/Γ is the <u>projective</u> orthogonal group $PO(2n)$; if $\Gamma = \{\pm 1\}$, G/Γ is $SO(2n)$. If ℓ is even there is a third possibility, $\Gamma = \{1, e_1 \ldots e_{2\ell}\}$ (and the isomorphic case $\Gamma = \{1, -e_1 \ldots e_{2\ell}\}$).
5. Type E_ℓ ($\ell = 6,7$). In each case the centre is cyclic of order 3,2 respectively; and the only quotient which is not simply-connected is $G/\Gamma = \mathrm{Ad}\, E_\ell$, the quotient by the full centre.

(The exceptional groups E_8, F_4, G_2 have trivial centre.)

In all of these cases except the first two the group Γ is necessarily of prime order. This suggests considering separately this case, which is less complicated than the general one. Consider therefore the case where Γ is cylic of order p, p prime. If $\theta \in \Gamma^*$ is a non-trivial character we have $R(\Gamma) = \mathbb{Z}[\theta]/(\theta^p - 1)$ (Γ^* is also cyclic of order p.)

Lemma 12.1. *In the Koszul complex $\Lambda_{R(\Gamma)}(y_1,\ldots,y_\ell)$ we can choose a new basis $y_1',\ldots,y_\ell'$ for the $R(\Gamma)$-module $R(\Gamma)$. $\{y_1,\ldots,y_\ell\}$ such that*

$$d(y_i') = 0 \quad (i = 1,\ldots,\ell-1), \quad d(y_\ell) = m(\theta - 1)$$

for some integer m.

Proof. Let $\rho_1,\ldots,\rho_\ell$ be the basic representations of G. By Schur's Lemma, $\rho_i|\Gamma$ is of the form $m_i\,\theta^{k_i}$ where m_i is the degree of ρ_i; hence $d(y_i) = m_i\,(\theta^{k_i} - 1)$. Let us number the ρ_i's so that $\rho_1,\ldots,\rho_r$ are trivial on Γ and the rest are not. (Necessarily $r < \ell$ since G admits a faithful representation). Then for $i \leq r$, $d(y_i) = 0$ and we can choose $y_i' = y_i$.

Next I need the following easily proved fact:

Lemma 12.2. *In $\mathbb{Z}[\theta]/(\theta^m - 1)$, $\theta^i - 1$ divides $\theta^j - 1$ if and only if i divides j mod m.*

In particular if p is prime and $k_i \not\equiv 0 \bmod p$,

$$\theta^{k_i} - 1 = a_i(\theta - 1) \in \mathbb{Z}[\theta]/(\theta^p - 1)$$

where a_i is a unit of the ring. Hence, in the proof of lemma 12.1, we can find a_i for $i > r$ such that $d(a_i^{-1}\,y_i) = m_i(\theta - 1)$.

It is now a standard process to find a new basis $y_{r+1}',\ldots,y_\ell'$ for the $\mathbb{Z}$-module generated by $a_{r+1}^{-1}\,y_{r+1},\ldots,a_\ell^{-1}y_\ell$ such that

$$d(y_i') = 0 \qquad (i = r+1,\ldots,\ell-1)$$

$$d(y_\ell') = m.(\theta - 1)$$

where m is the g. c. d. of $m_{r+1},\ldots,m_\ell$. The set $y_1',\ldots,y_\ell'$ now satisfy the conditions of lemma 12.1.

In the normal form given by lemma 12.1, the homology of the Koszul complex is now easy to find. We can split off the trivial complex $\Lambda_{R(\Gamma)}(y_1',\ldots,y_{\ell-1}')$ as a factor:

$$H(\Lambda_{R(\Gamma)}(y_1',\ldots,y_\ell')) = \Lambda_{R(\Gamma)}(y_1',\ldots,y_{\ell-1}') \otimes_{R(\Gamma)} H(\Lambda_{R(\Gamma)}(y_\ell'))$$

And $\Lambda_{R(\Gamma)}(y_\ell')$, with $d(y_\ell') = m.(\theta - 1)$, is the complex

$$0 \to R(\Gamma) \xrightarrow{m(\theta-1)} R(\Gamma) \to 0$$

The annihilator of $\theta - 1$ in $R(\Gamma)$ is the ideal generated by the regular representation ρ_Γ or $1 + \theta + \ldots + \theta^{p-1}$. Hence if $y = [\rho_\Gamma . y_\ell'] \in H(\Lambda_{R(\Gamma)}(y_\ell'))$

$$H_o(\Lambda_{R(\Gamma)}(y_\ell')) = R(\Gamma)/_{(m(\theta - 1))}$$

and

$$H_1(\Lambda_{R(\Gamma)}(y_\ell')) = \mathbb{Z} . y$$

with the trivial action of $R(\Gamma)$ on $\mathbb{Z}$, i.e., $(\theta - 1).y = 0$.

Set $R = R(\Gamma)/_{(m(\theta - 1))}$; we have shown that

$$H(\Lambda_{R(\Gamma)}(y_\ell')) = \Lambda_R(y)/_{(y(\theta - 1))}.$$

From this it is easy to conclude

<u>Proposition 12.1.</u> <u>With the above notation,</u>

$$\mathrm{Tor}_{R(G)}(R(\Gamma), \mathbb{Z}) = \Lambda_R(y_1', \ldots, y_{\ell-1}', y)/_{(y(\theta-1))}$$

Our main structural result follows:

<u>Proposition 12.2.</u> <u>For</u> Γ <u>as above</u> $\{E_r(G/\Gamma)\}$ <u>collapses, and there are classes</u> $z_1, \ldots, z_\ell$, <u>which can be expressed as difference elements in</u> $K^1(G/\Gamma)$, <u>such that</u>

$$K^*(G/\Gamma) = \Lambda_R(z_1, \ldots, z_\ell)/_{(z_\ell(\theta - 1))}$$

<u>where</u>

$$R = R(\Gamma)/_{(p^k(\theta - 1))}, \quad k \geq 1.$$

<u>Proof.</u> The E_2 term in $\{E_r(G/\Gamma)\}$, given by Proposition 12.1., is generated by classes in degrees 0 and 1. Since d_r for $r > 1$ is a derivation and must vanish on these classes, it vanishes on the whole of E_r; inductively, we see $d_r = 0$ for all r. We have a map from $R = R(\Gamma) \otimes_{R(G)} \mathbb{Z}$ into $K^o(G/\Gamma)$, the edge homomorphism (§9) and we have generalized difference elements (§10) $z_1, \ldots, z_\ell$, corresponding to $y_1', \ldots, y_{\ell-1}', y$.

The relations

$$z^2 = 0, \quad zz' + z'z = 0$$

hold for all $z,z' \in K^1(X)$; this follows immediately by considering the universal elements in $K^1(U) = \varprojlim K^1(U(n))$ and in $K^1(U \times U)$ (a well-known folk-theorem). Hence there is a map of rings

$$\Lambda_R(z_1,\ldots,z_\ell) \xrightarrow{\phi} K^*(G/\Gamma)$$

with $\phi(z_i) = z_i$. Next, since $E_\infty^{0,1}(G/\Gamma) = 0$ for reasons of grading, $E_\infty^{1,0}(G/\Gamma)$ is contained in $K^1(G/\Gamma)$ as a sub-R-module, and the relation $y.(\theta - 1) = 0$ in E_2 (and so in E_∞) implies $z_\ell.(\theta - 1) = 0$ in K^1. Hence ϕ factors through the quotient ring $\Lambda_R(z_1,\ldots,z_\ell)/(z_\ell.(\theta - 1))$. Now filter $\Lambda_R(z_1,\ldots,z_\ell)/(z_\ell.(\theta - 1))$ by the filtration defined by the ideal $(z_1,\ldots,z_\ell)$; we find that the map from this ring to $K^*(G/\Gamma)$ with its spectral-sequence filtration is a map of filtered rings which is an isomorphism on the associated graded rings, so an isomorphism.

It remains to clear up the statement in Proposition 12.2. that the integer m in the definition of R is in fact a power of p. This is an immediate consequence of the presence in $K^0(G/\Gamma)$ of a summand $\mathbb{Z}_m.(\theta - 1) \subset R$; since from the 'Serre spectral sequence' [5]

$$H^*(B\Gamma; K^*(G)) \Rightarrow K^*(G/\Gamma)$$

all torsion in $K^*(G/\Gamma)$ must be p-primary. We can now illustrate the content of Proposition 12.2. by going through the particular cases in ascending order of difficulty.

<u>Case 2, type B_ℓ</u> : $G = \mathrm{Spin}(2\ell + 1)$, $\Gamma = Z_2$, $G/\Gamma = SO(2\ell + 1)$.

We have (cf. [14], 13.10.3.):

$R(G) = \mathbb{Z}[\lambda_1,\ldots,\lambda_{\ell-1},\Delta]$, where λ_i is the ith exterior power of

$\mathrm{Spin}(2\ell + 1) \xrightarrow{\pi} SO(2\ell + 1) \to U(2\ell + 1)$

and Δ is the 'spin representation'. λ_i is trivial on Γ by definition ($1 \leq i \leq \ell - 1$), while Δ is non-trivial of degree 2^ℓ. Hence the generators y_i for the Koszul complex corresponding to $\lambda_1,\ldots,\lambda_{\ell-1},\Delta$ are already in the form given by lemma 12.1. and we can apply Proposition 12.2 immediately. Moreover the difference elements corresponding to the y_i's are $\beta(\lambda_i)$ for $i \leq \ell - 1$ (Corollary 10.2) and $\pi_!(\beta(\Delta))$ for $i = \ell$ (Proposition 10.4). The conclusion is

<u>Proposition 12.3</u>. <u>Let</u> $R = R(\mathbb{Z}_2) \otimes_{R(\mathrm{Spin}(2\ell+1)} \mathbb{Z} = \mathbb{Z}[\theta]/_{(\theta^2-1,2^\ell(\theta-1))}$.

<u>Then</u>

$$K^*(SO(2\ell+1)) = \Lambda_R(\beta(\lambda_1),\ldots,\beta(\lambda_{\ell-1}),\pi_!\beta(\Delta))/_{((\theta-1).\pi_!\beta(\Delta))}.$$

<u>Note</u>. Alternatively we could describe the additive structure rather than the multiplicative, noting that $R = \mathbb{Z}.1 \oplus \mathbb{Z}_{2^\ell}.(\theta-1)$:

$$K^*(SO(2\ell+1)) = \Lambda_{\mathbb{Z}}(\beta(\lambda_1),\ldots,\beta(\lambda_{\ell-1}),\ \pi_!\beta(\Delta))$$

$$\oplus\ (\Lambda_{\mathbb{Z}}(\beta(\lambda_1),\ldots,\beta(\lambda_{\ell-1})) \otimes \mathbb{Z}_{2^\ell}.(\theta-1)$$

$$(= 2^\ell.\mathbb{Z} \oplus 2^{\ell-1}.\mathbb{Z}_{2^\ell})$$

with the multiplicative relations

$$(\theta-1)^2 = -2(\theta-1),\ \pi_!\beta(\Delta).(\theta-1) = 0$$

In future I shall give the multiplicative structure, as it is found in Proposition 12.2, and leave it to the reader to deduce the additive structure if she needs it. The point to note is that

(1) $$R = \mathbb{Z}[\theta]/_{(\theta^p-1,p^k(\theta-1))}$$

$$= \mathbb{Z}.1 \oplus \sum_{i=0}^{p-1} \mathbb{Z}_{p^k}.(\theta^{i+1}-\theta^i) \cong \mathbb{Z} \oplus (p-1)\mathbb{Z}_{p^k}.$$

<u>Case 4, type D_ℓ</u> : $G = \mathrm{Spin}(2\ell)$, $\Gamma = \{\pm 1\} \cong \mathbb{Z}_2$,

$$G/\Gamma = SO(2\ell).$$

(I shall leave the other quotients (see remarks on case 2 above) of $\mathrm{Spin}(4\ell)$ by a $\mathbb{Z}_2$ subgroup until we consider the quotient $PO(4\ell)$ by the full centre).

$$R(G) = \mathbb{Z}[\lambda_1,\ldots,\lambda_{\ell-2},\ \Delta_+,\Delta_-]$$

- see [14], 13.10.3 - where λ_i is as in the preceding case, and Δ_+, Δ_- are the 'half-spin' representations. λ_i is trivial on Γ as before, while

$$\Delta_+|\Gamma = \Delta_-|\Gamma = 2^{\ell-1}\ \theta$$

(θ is as usual the faithful 1-dimensional representation of Γ). In this case we take the Koszul complex corresponding to the set of generators $\lambda_1,\ldots,\lambda_{\ell-2}$, $\Delta_+ - \Delta_-$, Δ_+ ; we are again in the form given by lemma 12.1. and $d(y_\ell) = 2^{\ell-1}\ (\theta - 1)$. We can deduce as before

<u>Proposition 12.4.</u> <u>If</u> $R = \mathbb{Z}[\theta]/(\theta^2 - 1,\ 2^{\ell-1}(\theta - 1))$,

$K^*(SO(2\ell)) = \Lambda_R(\beta(\lambda_1),\ldots,\beta(\lambda_{\ell-2}),\beta(\Delta_+ - \Delta_-),\ \pi_!\beta(\Delta_+))/((\theta - 1)\ \pi_!\beta(\Delta_+))$.

<u>Case 3, type C_ℓ</u> : $G = Sp(\ell)$, $\Gamma = \mathbb{Z}_2$, $G/\Gamma = PSp(\ell)$.

Let λ_i denote the ith exterior power of the standard representation $Sp(\ell) \xrightarrow{\lambda_1} U(2\ell)$. Then $\lambda_1,\ldots,\lambda_\ell$ are not the basic representations of $Sp(\ell)$, but we still have (cf [14], 13.5.4.),

$$R(Sp(\ell)) = \mathbb{Z}[\lambda_1,\ldots,\lambda_\ell]\ .$$

Now let us look at $\lambda_i|\Gamma$. We know λ_1 is faithful, so $\lambda_1|\Gamma = 2\ell.\theta$. Hence for $i = 1,\ldots,\ell$

(2) $$\lambda_i|\Gamma = \binom{2\ell}{i}.\ \theta^i$$

and, using $\theta^2 = 1$, we find that for the corresponding generators of the Koszul complex

$$d(y_i) = 0 \qquad \text{(i even)}$$

$$= \binom{2\ell}{i}\ (\theta - 1) \qquad \text{(i odd)}$$

We use now the procedure of lemma 12.1 to find new generators $y_1',\ldots,y_\ell'$. For i even we can take $y_i' = y_i$; for i odd, y_i' is an integral linear combination of $y_1, y_3, y_5,\ldots$. It is convenient in this case to choose the indexing so that

$$d(y_1') = m.(\theta - 1)$$

$$d(y_i') = 0 \qquad \text{for} \quad i > 1$$

where m is the g. c. d. of $\binom{2\ell}{1}$, $\binom{2\ell}{3}$, ..., $\binom{2\ell}{\ell'}$, ℓ' being the greatest odd number $\leq \ell$.

Lemma 12.4. m as defined above is equal to the greatest power of 2 which divides 2ℓ. i.e.,

$$m = 2^s \text{ where } 2\ell = 2^s a, \text{ and } a \text{ is odd.}$$

Proof By standard arithmetic if a,b are prime to p, $p^{k-\ell}$ divides $\binom{p^k a}{p^\ell b}$, and $p^{k-\ell+1}$ does not divide $\binom{p^k a}{p^\ell}$. Hence if a,b are odd, 2^s divides $\binom{2^s a}{b}$ while 2^{s+1} does not divide $\binom{2^s a}{1} = 2^s a$. Also, p does not divide $\binom{2^s a}{p^k}$ where p^k is the greatest power of p dividing a, for p odd. The lemma follows from these facts.

Now we can calculate the Tor algebra and spectral sequence as in Propositions 12.1., 12.2. The generators y_i for i even correspond to the elements $\beta(\lambda_i)$ of $K^*(PSp(\ell))$. For i odd we have integral linear combinations $y_i' = \sum a_{ij} y_j$. Set $\zeta_i = \sum a_{ij} \lambda_j$; then $\zeta_3, \zeta_5, \ldots$ are elements of $R(Sp(\ell))$ trivial on Γ, and so define elements $\beta(\zeta_i)$ in $K^1(PSp(\ell))$. $\zeta_1|\Gamma$ on the other hand is $2^s\,\theta$. We deduce as before

Proposition 12.5. $K^*(P\,Sp(\ell)) = \Lambda_R(\beta(\lambda_2), \beta(\lambda_4), \ldots, \beta(\lambda_{2[\ell/2]}),$

$$\beta(\zeta_3), \beta(\zeta_5), \ldots, \beta(\zeta_{2\left[\frac{\ell+1}{2}\right]-1}), \pi_!\, \beta(\zeta_1))/((\theta-1).\ \ \pi_!\beta(\zeta_1))$$

where $R = \mathbb{Z}[\theta]/_{(\theta^2-1, 2^s(\theta-1))}$, 2^s being the highest power of 2 dividing 2ℓ.

Note The ζ_i's as given above have not been precisely defined, their choice depending on the particular method chosen for putting the generators $y_1, y_3, \ldots$ in normal form. For example in $Sp(2^s)$ we can simplify matters by setting $\zeta_1 = \lambda_1$, $\zeta_i = \lambda_i - \frac{1}{2^s}\binom{2^s}{i}\lambda_1$. On the other hand in $Sp(6)$ a typical choice would be (given that $\dim \lambda_1 = 6$, $\dim \lambda_3 = 20$)

$$\zeta_1 = \lambda_3 - 3\lambda_1,\quad \zeta_3 = 3\lambda_3 - 10\lambda_1$$

Case 4 $G = E_\ell$ $(\ell = 6,7)$; Γ = centre of E_ℓ, $G/\Gamma = \mathrm{Ad}\, E_\ell$.

I shall give these results in outline only, since I expect that they won't be very often used.

E_6 has centre $\mathbb{Z}_3$. We use the notation of [9], Appendix, and let ρ_i (i = 1,...,6) be the basic representation corresponding to the weight $\bar{\omega}_i$. We find that

ρ_2, ρ_4 are trivial on the centre; ρ_1, ρ_6 are conjugate nontrivial of dimension 27; ρ_3, ρ_5 are conjugate nontrivial of dimension 351 = 27.13. Writing θ for a nontrivial 1-dimensional representation of $\mathbb{Z}_3$, we get

$$dy_1 = 27(\theta - 1), \quad dy_6 = 27(\theta^2 - 1) \text{ and}$$

$$d(y_5 - 13y_1) = d(y_3 - 13y_6) = 0 .$$

Hence, by the usual methods, $R = \mathbb{Z}[\theta]/(\theta^3-1, \; 27(\theta-1))$ and

(3) $$K^*(\mathrm{Ad}\, E_6) = \Lambda_R(\beta(\rho_2), \beta(\rho_3 - 13\rho_6), \beta(\rho_4), \beta(\rho_5 - 13\rho_1),$$

$$\bar{\beta}(\theta^2\tilde{\rho}_1 + \tilde{\rho}_6), \; \pi_!\beta(\rho_1)))/((\theta - 1)\; \pi_!\beta(\rho_1))$$

where, as in §10 we use $\bar{\beta}$ for the mapping from $\mathrm{Ker}\, d_1 \subset R(\mathbb{Z}_3) \otimes I(E_6)$ to $K^1(\mathrm{Ad}\, E_6)$.

E_7 has centre $\mathbb{Z}_2$. With the notation of the previous case, $\rho_1, \rho_3, \rho_4, \rho_6$ are trivial on the centre, while ρ_2, ρ_5, ρ_7 are non-trivial of dimensions 8.95. 16.5187 and 8.7 respectively. Write $\rho = 2\rho_2 - 27\rho_7$, then $\rho | \mathbb{Z}_2$ is equal to 8θ, and we find by the previous methods

(4) $$K^*(\mathrm{Ad}\, E_7) = \Lambda_R(\beta(\rho_1), \beta(\rho_3), \beta(\rho_4), \beta(\rho_5 - 2.5187\rho),$$

$$\beta(7\rho_2 - 95\rho_7), \quad \pi_!\beta(\rho))/((\theta - 1)\; \pi_!\beta(\rho))$$

Proposition 12.2 will also apply to give the K-theory of the projective unitary groups PU(p) for p prime - and, more generally, the quotients $SU(n)/\mathbb{Z}_p$ whenever p is a prime dividing n. I shall leave out the details of this - it seems to me that it would be better to try to fit these cases into an overall understanding of the K-theory of $SU(n)/\Gamma$, Γ finite; which, as I have already said, is a good deal more difficult.

We conclude by dealing with two cases where the group Γ is not of prime order - the projective orthogonal groups, which represent the next stage of difficulty. Let us start with the case of $PO(4k + 2)$, quotient of Spin $(4k + 2)$ by its full centre $\Gamma = \mathbb{Z}_4$. Γ is generated by $e_1 \ldots e_{4k+2} \in$ Spin $(4k + 2)$ and $R(\Gamma) = \mathbb{Z}[\theta]/(\theta^4 - 1)$ where $\theta(e_1 \ldots e_{4k+2}) = \sqrt{-1}$. We have (see case 2 above),

$$R(\mathrm{Spin}\,(4k + 2)) = \mathbb{Z}[\lambda_1, \ldots, \lambda_{2k-1}, \Delta^+, \Delta^-]$$

The projection $\pi_1 : Spin(4k+2) \to SO(4k+2)$ sends $e_1 \dots e_{4k+2}$ to -1; hence $\lambda_1|\Gamma = (4k+2)\theta^2$, and for $i \geq 1$, $\lambda_i|\Gamma = \binom{4k+2}{i}\theta^{2i}$ (compare the case of $Sp(\ell)$). Also, $\Delta^+|\Gamma = 2^{2k}\theta$, $\Delta^-|\Gamma = 2^{2k}\theta^3$.

We can now follow a similar procedure to that for $Sp(\ell)$. Let $y_1, \dots, y_{2k+1}$ be Koszul complex generators corresponding to the basic representations so that

$$d(y_i) = 0 \qquad i \text{ even}, \qquad i < 2k-1$$

$$= \binom{4k+2}{i}(\theta^2 - 1) \qquad i \text{ odd}, \quad i \leq 2k-1$$

$$d(y_{2k}) = 2^{2k}(\theta - 1), \quad d(y_{2k+1}) = 2^{2k}(\theta^3 - 1).$$

<u>Lemma 12.5.</u> *We can choose a basis $y_1', \dots, y_{2k+1}'$ for the $R(\Gamma)$-module generated by $y_1, \dots, y_{2k+1}$ so that $d(y_1') = 2(\theta^2 - 1)$, $d(y_{2k+1}') = 2^{2k}(\theta - 1)$, $d(y_i') = 0$ otherwise.*

<u>Proof</u> First, the following definitions are easy:

$$y_i' = y_i \ (i \text{ even and } i < 2k-1); \quad y_{2k+1}' = y_{2k}; \ y_{2k}' = \theta^3 y_{2k} + y_{2k+1}.$$

Next, I claim that I can find integers a,b such that

$$(5) \qquad d(a.y_3 + b.(\theta+1).y_{2k}) = 4k(\theta^2 - 1)$$

First, if $k = 1$, $d(\theta+1)y_2 = 4(\theta^2 - 1)$ so $a = 0$, $b = 1$ will do. If $k > 1$, $d(y_3) = \binom{4k+2}{3}(\theta^2 - 1)$; $d((\theta+1)y_{2k}) = 2^{2k}(\theta^2 - 1)$; and the g.c.d. of $\binom{4k+2}{3}$ and 2^{2k} is the highest power of 2 dividing $4k$. Hence the g.c.d divides $4k$, and (5) holds. Now set

$$y_1' = y_1 - ay_3 - b(\theta+1)y_{2k}, \quad \text{and}$$

$$y_i' = y_i - \tfrac{1}{2}\binom{4k+2}{i} y_1', \quad \text{for } i \text{ odd}, \ i \leq 2k-1$$

This is the required basis.

Next, with the basis we have defined, it is easy to split off a trivial tensor factor from the Koszul complex:

$$(6) \qquad H(\Lambda_{R(\Gamma)}(y_1', \dots, y_{2k+1}')) = \Lambda_{R(\Gamma)}(y_2', \dots, y_{2k}') \otimes_{R(\Gamma)} H(A)$$

where $A = \Lambda_{R(\Gamma)}(y_1', y_{2k+1}')$ with the differential given by lemma 12.5. In A define

$$z_1 = (1 + \theta^2)y_1' \; ; \; z_2 = 2^{2k-1}y_1' - (1 + \theta)y_{2k+1}'$$

Then z_1, z_2 are clearly cycles.

<u>Lemma 12.6</u>. z_1, z_2 <u>generate the</u> 1-<u>cycles of</u> A, <u>and</u> $z_1z_2 = -(1 + \theta + \theta^2 + \theta^3)y_1'y_{2k+1}'$ <u>generates the</u> 2-<u>cycles</u>. <u>And if</u> $R = R(\Gamma)/(2(\theta^2 - 1), 2^{2k}(\theta - 1))$,

$$H(A) = \Lambda_R(z_1, z_2)/(2(\theta - 1)z_2, (\theta^2 - 1)z_1, (\theta - 1)z_1z_2).$$

Proof Suppose $z = \alpha y_1' + \beta y_{2k+1}'$ is a cycle in A, where $\alpha, \beta \in R(\Gamma)$. Then

$$2(\theta - 1) \;\; [(\theta + 1)\alpha + 2^{2k-1}\beta] = 0.$$

Since $(\theta - 1)$ annihilates only multiples of $(1 + \theta + \theta^2 + \theta^3)$ this tells us that for some integer m,

$$(\theta + 1)\alpha + 2^{2k-1}\beta = (\theta + 1)\,(\theta^2 + 1)m$$

But $(\theta + 1)$ is a prime ideal in $R(\Gamma)$; so $\beta = (\theta + 1)\beta'$ for some β'. From this we find immediately that $z + \beta' z_2 = (\alpha + 2^{2k-1}\beta')y_1'$. But $\alpha + 2^{2k-1}\beta' - m(\theta^2 + 1)$ is annihilated by $1 + \theta$, and so must be a multiple of $1 + \theta^2$. Hence $z + \beta' z_2$ is a multiple of z_1.

More simply, $d(\gamma . y_1' \, y_{2k+1}') = 2\gamma((\theta^2 - 1)y_{2k+1}' - 2^{2k-1}(\theta - 1)y_1') = -2\gamma(\theta-1)z_2$.

From this we deduce (i) that $\gamma . y_1' \, y_{2k+1}'$ is a cycle if and only if $\gamma(\theta - 1) = 0$, i.e., γ is a multiple of $1 + \theta + \theta^2 + \theta^3$; (ii) that $2(\theta - 1)z_2$ generates the boundaries in degree 1. The relations $(\theta^2 - 1)z_1 = 0 = (\theta - 1)z_1z_2$ are immediate from the structure of $R(\Gamma)$, and all relations in H(A) must be consequences of these, and of the relations in degree 0

$$[2(\theta^2 - 1)] = [d(y_1')] = 0$$

$$[2^{2k}(\theta - 1)] = [d(y_{2k+1}')] = 0$$

defining the ring R. This completes the proof of the lemma.

It remains to carry these results over to get the structure of $K^*(PO(4k + 2))$.

Note that E_2 as we have defined it is generated by elements in degree -1; hence (cf. Proposition 12.2) the spectral sequence collapses and Lemma 12.6 gives us the structure of the graded algebra associated to the filtration of $K^*(PO(4k+2))$, using (6).

Next, associated to the generators $y_2,\ldots,y_{2k-2}$ (even) we have elements $\beta(\lambda_2),\ldots,\beta(\lambda_{2k-2})$ of K^1, while associated to $y_3',y_5',\ldots,y_{2k-1}'$ we have more complicated difference elements, say $\bar\beta(\eta_3),\ldots,\bar\beta(\eta_{2k-1})$. And the cycles z_1,z_2 in $E_\infty^{-1,0}$ in turn give rise to difference elements in $K^1(PO(4k+2))$ which we shall write $\bar\beta(\zeta_1),\bar\beta(\zeta_2)$. As in Proposition 12.2 the relations defining R and the relations

$$(\theta^2-1)\bar\beta(\zeta_1) = 2(\theta-1)\,\bar\beta(\zeta_2) = 0$$

hold in $K^1(PO(4k+2))$. On the other hand the relation $(\theta-1)z_1z_2 = 0$ only tells us that $(\theta-1)\bar\beta(\zeta_1)\,\bar\beta(\zeta_2)$ is some element of filtration zero in $K^0(PO(4k+2))$, i.e., an element of R.

<u>Lemma 12.7</u> $(\theta-1)\,\bar\beta(\zeta_1)\,\bar\beta(\zeta_2) = 0$.

<u>Proof</u> Consider the homomorphisms

$$\mathrm{Spin}(4k+2) \xrightarrow{\pi_1} SO(4k+2) \xrightarrow{\pi_2} PO(4k+2).$$

These give rise to obvious homomorphisms of spectral sequences induced on the E_2 term by

$$\mathbb{Z} = R(1) \longleftarrow R(\mathbb{Z}_2) \longleftarrow R(\mathbb{Z}_4)$$

Let $u \in K^1(SO(4k+2))$ be the element which corresponds to

$$y_1' = y_1 - a.y_3 - b.(1+\theta)y_{2k}$$

in the E_2-term of the $SO(4k+2)$-spectral sequence (a, b as in (5)). $d(y_1') = 2(\theta^2-1) = 0$ in this spectral sequence, so u is defined. And as in Proposition 10.3 we have

(7) $$\bar\beta(\zeta_1) = (\pi_2)_!\,(u)$$

But the usual 'Frobenius reciprocity' for $\pi_!$, $\pi^!$, gives

$$(\pi_2)_!(u).\,(\theta-1)\,\bar\beta(\zeta_2) = (\pi_2)_!\,(u\,.\,(\pi_2)^!\,(\theta-1)\,\bar\beta(\zeta_2))$$

Next applying $(\pi_2)^!$ to the spectral sequences we find that $(\pi_2)^!\,(\bar\beta(\zeta_2)) = 2^{2k-1}\,u - v$

where v corresponds to $(1 + \theta)y_{2k}$ in the spectral sequence of $SO(4k + 2)$. We have seen above that this v is $(\pi_1)_!\,(\beta(\Delta^+))$. On the coefficient rings $(\pi_2)^!$ maps θ to θ, so

$$(\pi_2)^!\,((\theta - 1)\,\bar{\beta}(\zeta_2)) = (\theta - 1)\,(2^{2k-1}\,u - (\pi_1)_!\,\beta(\Delta^+)) = (\theta - 1)2^{2k-1}\,u$$

since $(\theta - 1)$ annihilates $\pi_!(\beta(\Delta^+))$, see Proposition 12.4. Hence

$$(\theta - 1)\,\bar{\beta}(\zeta_1)\,\bar{\beta}(\zeta_2) = (\pi_2)_!\,(u).\,(\theta - 1)\,\bar{\beta}(\zeta_2)$$

$$= (\pi_2)_!\,(u.\,(\theta - 1).\,2^{2k-1}\,u)$$

$$= 0 \text{ since } u^2 = 0.$$

This proves lemma 12.7. All other relations in $K^*(PO(4k + 2))$ which come from E_∞ hold for obvious reasons, as we have seen; and we find that $K^*(PO(4k + 2))$ is isomorphic to E_∞ as an $R(\Gamma)$-algebra. Our structural result is

Proposition 12.6. With the above notation

$$K^*(PO(4k + 2)) = \Lambda_R(\bar{\beta}(\eta_3),\ldots,\bar{\beta}(\eta_{2k-1}),\beta(\lambda_2),\ldots,\beta(\lambda_{2k-2}),$$

$$\bar{\beta}(\theta^3\Delta^+ + \Delta^-),\ \bar{\beta}(\zeta_1),\ \bar{\beta}(\zeta_2))/_I$$

where

$$R = \mathbb{Z}\,[\theta]/(\theta^4 - 1,\ 2(\theta^2 - 1),\ 2^{2k}(\theta - 1))$$

and I is the ideal generated by $(\theta^2 - 1)\,\bar{\beta}(\zeta_1)$, $2(\theta - 1)\,\bar{\beta}(\zeta_2)$ and $(\theta - 1)\,\bar{\beta}(\zeta_1)\,\bar{\beta}(\zeta_2)$.

Lastly we consider $PO(4k)$, quotient of Spin $(4k)$ by its full centre $\Gamma = \mathbb{Z}_2 \times \mathbb{Z}_2$. Γ is generated by -1 and by $e_1 \ldots e_{4k}$. Define representations θ_1, θ_2 of Γ by

$$\theta_1(-1) = \theta_2(-1) = -1$$

$$\theta_1(e_1 \ldots e_{4k}) = -1 = -\theta_2(e_1 \ldots e_{4k})$$

Then $R(\Gamma) = \mathbb{Z}\,[\theta_1,\theta_2]/(\theta_1^2 - 1,\ \theta_2^2 - 1)$.

The representations of Spin $(4k)$ mentioned earlier restrict as follows:

$\lambda_1|\Gamma = 4k.\ \theta_1\theta_2$ and so

$$\lambda_i|\Gamma = \binom{4k}{i}\ \theta_1\ \theta_2 \qquad \text{(i odd)}$$

$$= \binom{4k}{i}.1 \qquad \text{(i even)}$$

$$\Delta^+|\Gamma = 2^{2k-1}\ \theta_1,\ \Delta^-|\Gamma = 2^{2k-1}\ \theta_2 \quad .$$

Hence, if we choose generators $y_1,\ldots,y_{2k}$ for the Koszul complex corresponding to the basic representations, we shall have

$$d(y_i) = 0 \qquad (\text{i even},\quad i < 2k)$$

$$= \binom{4k}{i}\ (\theta_1\ \theta_2 - 1) \qquad (\text{i odd},\ i < 2k - 1)$$

$$d(y_{2k-1}) = 2^{2k-1}\ (\theta_1 - 1),\ \ d(y_{2k}) = 2^{2k-1}\ (\theta_2 - 1)$$

Lemma 12.8 Let 2^s be the highest power of 2 dividing $4k$, $k > 1$. Then we can choose a new basis, $y_1',\ldots,y_{2k}'$, for the $R(\Gamma)$-module generated by $y_1,\ldots,y_{2k}$, such that

$$d(y_1') = 2^s(\theta_1\ \theta_2 - 1);\ \ d(y_{2k}') = 2^{2k-1}\ (\theta_1 - 1);$$

$$d(y_i') = 0 \qquad \text{otherwise}.$$

The proof of this is approximately that of lemma 12.5: We find a linear combination of y_3 and $\theta_2(y_{2k-1} - y_{2k})$ whose image under d is precisely $2^s(\theta_1\ \theta_2 - 1)$, and subtract a multiple from y_1 to obtain y_1'. Then we subtract suitable multiples of y_1' from $y_3,y_5,\ldots,y_{2k-3}$ as in lemma 12.6. The last two y_i's are given by

$$y_{2k-1}' = y_{2k} - y_{2k-1} - 2^{2k-s-1}\ \theta_2\ y_1'$$

$$y_{2k}' = y_{2k-1}$$

(Note that $s \leq 2k - 1$ for $k > 1$).

Now again we have

$$E_2(\mathrm{Spin}\ (4k)/\Gamma) = \Lambda_{R(\Gamma)}(y_2',\ldots,y_{2k-1}') \otimes_{R(\Gamma)} H(A)$$

where $H(A) = \Lambda_{R(\Gamma)}(y_1', y_{2k}')$ with differentials given by lemma 12.8. Define

$$z_1 = (\theta_1 \theta_2 + 1)y_1', \quad z_2 = (\theta_1 + 1)y_{2k}' \ .$$

<u>Lemma 12.9.</u> $\quad H(A) = \Lambda_R(z_1, z_2)/((\theta_1 \theta_2 - 1)z_1, \ (\theta_1 - 1)z_2)$

<u>where</u> $\quad R = R(\Gamma)/(2^s(\theta_1 \theta_2 - 1), \ 2^{2k-1}(\theta_1 - 1)).$

<u>Proof</u> This is similar to lemma 12.6 but simpler. First if $z = \alpha y_1' + \beta y_{2k}'$ is a cycle we find

$$\alpha(\theta_1 \theta_2 - 1) + 2^{2k-s-1} \beta(\theta_1 - 1) = 0$$

Hence $\alpha(\theta_1 \theta_2 - 1)$ is a multiple of $\theta_1 \theta_2 - 1$ <u>and</u> of $\theta_1 - 1$. It must therefore be of form

$$\alpha(\theta_1 \theta_2 - 1) = m(1 - \theta_1 + \theta_2 - \theta_1 \theta_2) \qquad m \in \mathbb{Z}$$

$$= m(\theta_1 - 1)(\theta_1 \theta_2 - 1).$$

Hence $\alpha - m(\theta_1 - 1)$ is a multiple of $\theta_1 \theta_2 + 1$. Similarly $\beta + 2^{2k-s-1} m(\theta_1 \theta_2 - 1)$ is a multiple of $\theta_1 + 1$. This shows that z is equal to

(8) $$m((\theta_1 - 1)y_1' - 2^{2k-s-1}(\theta_1 \theta_2 - 1) y_{2k}')$$

modulo linear combinations of z_1, z_2. But (8) is precisely $m \cdot d(y_1' y_{2k}')$; hence z_1, z_2 generate 1-cycles modulo boundaries. Again, $z_1 z_2 = (1 + \theta_1 + \theta_2 + \theta_1 \theta_2) y_1' y_{2k}'$ generates 2-cycles. It is easy to see that no linear combination of z_1, z_2 can be a boundary and that the relations given are the only ones; this time, all relations for $z_1 z_2$ follow from those for z_1, z_2.

The last statement implies that the one real difficulty that we had with $K^*(PO(4k + 2))$ - for which Lemma 12.7 was needed - is absent here. The generators of E_2 (Spin $(4k)/\Gamma$) are in degree 1 and so are the relations. Hence

$$E_2 \cong E_\infty \cong K^*(PO(4k)).$$

To make the generators explicit we again take $\eta_1, \eta_3, \ldots, \eta_{2k-3}, \eta_{2k-1}$ in $R(\mathrm{Spin}\,(4k)) \otimes I(\Gamma)$ corresponding to $y_1', \ldots, y_{2k-1}'$; and ζ_1, ζ_2 corresponding to z_1, z_2 respectively.

It is at this point that we can also deal with the quotients of Spin $(4k)$ by the cyclic subgroups $\Gamma_{\pm}$ of Γ generated by $\pm e_1 \dots e_{4k}$. Taking for example $\mathrm{Spin}\ (4k)/(- e_1 \dots e_{4k}) = \Gamma_-$; the ring $R(\Gamma_-)$ is the quotient of $R(\Gamma)$ by $(\theta_1 - 1)$. If we call the (usual) generator of $R(\Gamma_-)$ 'θ', then θ is the image of θ_2; we find that in the resulting spectral sequence

$$d(y_1') = 2^s(\theta - 1), \quad d(y_i') = 0 \qquad \text{for } i \neq 1$$

so that we can apply Proposition 12.2 and the previous methods. Γ_+ requires a change in basis, but is essentially the same. And if $\pi : \mathrm{Spin}\ (4k)/\Gamma_- \to PO(4k)$ is the covering, the element $\bar\beta(\zeta_2)$ of $K^1(PO(4k))$ is equal to $\pi_!\ \beta(\eta_{2k}') = \pi_!\beta(\Delta_+)$ (since Δ_+ factors through $\mathrm{Spin}\ (4k)/\Gamma_-$.

Now for the final structural result.

<u>Proposition 12.7.</u> <u>With the above notation</u>

$$K^*(\mathrm{Spin}\ (4k)/\Gamma_-) = \Lambda_{R_-}\ (\bar\beta(\eta_3),\dots,\bar\beta(\eta_{2k-1}),\ \beta(\lambda_2),\dots,\beta(\lambda_{2k-2}),$$

$$\beta(\Delta_+),\ \bar\beta(\zeta_1))/((\theta - 1)\ \bar\beta(\zeta_1))$$

<u>where</u> $R_- = \mathbb{Z}[\theta]/(\theta^2 - 1,\ 2^s(\theta - 1))$.

$$K^*(PO(4k)) = \Lambda_R(\bar\beta(\eta_3),\dots,\bar\beta(\eta_{2k-1}),\ \beta(\lambda_2),\dots,\beta(\lambda_{2k-2}),$$

$$\bar\beta(\zeta_1),\ \bar\beta(\zeta_2))/((\theta_1\ \theta_2 - 1)\ \bar\beta(\zeta_1),\ (\theta_1 - 1)\ \bar\beta(\zeta_2))$$

<u>where</u> $$R = \mathbb{Z}[\theta_1,\theta_2]/(\theta_1^2 - 1,\ \theta_2^2 - 1,\ 2^s(\theta_1\ \theta_2 - 1),$$

$$2^k(\theta_1 - 1)).$$

Note that additively, $R = \mathbb{Z}.\ 1 \oplus \mathbb{Z}_{2^s}\ .\ (\theta_1\ \theta_2 - 1)$

$$\oplus\ \mathbb{Z}_{2^s}\ .\ (\theta_1 - \theta_2) \oplus \mathbb{Z}_{2^k}\ .\ (\theta_1 - 1).$$

Part II Bibliography

1. M.F. Atiyah, Characters and cohomology of finite groups, Publ. Math. IHES 9 (1961), 23 - 64.
2. ———, Vector bundles and the Künneth formula, Topology 1 (1962), 245-8.
3. ——— , K-theory, Benjamin, New York, 1967.
4. ——— , Bott periodicity and the index of elliptic operators, Quarterly J. Math. 19 (1968), 113-140.
5. ——— , and F. Hirzebruch, Vector bundles and homogeneous spaces, Proc. Sympos. Pure Math. AMS 3 (1961), 7-38.
6. ——— , and G. Segal, Equivariant K-theory and completion, J. Diff. Geom. 3 (1969), 1 - 19.

6a. A. Borel, Sur la cohomologie des espaces fibrés principaux et des espaces homogenes de groupes de Lie compacts, Ann. Math. 57 (1953), 115-207.

7. ——— , and F. Hirzebruch, Characteristic classes and homogeneous spaces, Amer. J. Math. 80 (1958), 458-538.
8. N. Bourbaki, Algebre commutative (Eléments de mathématique, fasc. 28), Hermann, Paris, 1961.
9. ———, Groupes et algebres de Lie (Eléments de mathématique, fasc. 34), Hermann, Paris, 1968.
10. E. Cartan, La géométrie des groupes simples, Annali di Mat. 4 (1927), 209-256.
11. H. Cartan and S. Eilenberg , Homological algebra, Princeton, 1956.
12. A. Dold, Chern classes in general cohomology, Symposia math. 5 (INDAM, Rome, 1970), 385-410.
13. L.H. Hodgkin, The K-theory of Lie groups, Topology 6 (1967) 1 - 36.
14 D. Husemoller, Fibre bundles, McGraw-Hill, New York, 1966.
15. S. MacLane, Homology, Springer-Verlag, Berlin, 1963.
16. H. Pittie, Homogeneous vector bundles on homogeneous spaces, Topology 11 (1972), 199-204.
18. A. Roux, Application de la suite spectrale d'Hodgkin au calcul de la K-théorie des variétés de Stiefel, Bull. Soc. Math. France 99 (1971), 345-368.
19. G. Segal, The representation ring of a compact Lie group, Publ. Math. IHES 34 (1968), 113-128.
20. ——— , Equivariant K-theory, Publ. Math. IHES 34 (1968), 129-151.
21. V.P. Snaith, Massey products in K-theory II, Proc. Camb. Phil. Soc. 71 (1969), 259-289.

DYER-LASHOF OPERATIONS IN K-THEORY

V.P. Snaith

Table of Contents

Introduction

The purpose of this paper is to study $K_*(X;Z/_p)$ when X is an infinite loopspace. Let $\mathrm{Ind}_p(X) \subset K_*(X;Z/_p)$ be the submodule generated by elements of the form

$$\{x^p \mid x \in K_\alpha(X;Z/_p)\} \quad \text{if} \quad p \neq 2$$

and $\{x^2 \mid x \in \ker \beta_2 \subset K_\alpha(X;Z/_2)\}$ if $p = 2$,

(where β_2 is the Bockstein and p is a prime).
The main theorem is as follows:

Theorem 5.1:

Let X be an H^∞-space and p a prime. There exist operations

$$Q: K_\alpha(X;Z/_p) \to K_\alpha(X;Z/_p)/\mathrm{Ind}_p(X), \quad \text{if } p \neq 2,$$

$$Q: \ker \beta_2 \to K_\alpha(X;Z/_2)/\mathrm{Ind}_2(X), \quad \text{if } p = 2$$

$$\ker \beta_2 \subset K_\alpha(X,Z/_2)$$

satisfying the following conditions.

(i) Q is natural for H^∞-maps.

(ii) Let $x,y \in K_\alpha(X;Z/_p)$.

$$Q(x+y) = Q(x) + Q(y) + \sum_{i=1}^{p-1} \left\{-\binom{p}{i}/_p\right\} x^i . y^{p-i}$$

if $\alpha \equiv 0 \pmod 2$

$$Q(x+y) = Q(x) + Q(y) \quad \text{if} \quad \alpha \equiv 1 \pmod 2.$$

(iii) Cartan formula

Let $x \in K_\alpha(X;Z/_p)$ and $y \in K_\beta(X;Z/_p)$.

(a) If $\alpha + \beta \equiv 1 \pmod 2$

$$Q(x.y) = Q(x)y^p + x^p.Q(y) \in K_{\alpha+\beta}(X;Z/_p) .$$

(b) If $\alpha \equiv \beta \pmod 2$

$$\Omega(x.y) = \begin{cases} \Omega(x)y^p + x^p\Omega(y) & \text{if } \alpha \equiv 0 \pmod 2 \\ \lambda\Omega(x).\Omega(y) & \text{if } \alpha \equiv 1 \pmod 2). \end{cases}$$

in $K_o(X;Z/_p)/_{[\mathrm{Ind}_p(X)]^2}$, $(0 \neq \lambda \in Z/_p)$.

(c) If $z \in \ker \beta_2 \subset K_1(X,Z/_2)$ then $z^2 = 0$

and if $z \in K_1(X,Z/_p)$ then $z^p = 0$.

(iv) If $(\psi^k)_*: K_*(X;Z/_p) \to K_*(X;Z/_p)$

is the dual of the Adams operation, ψ^k, for k prime to

p then

$$\Omega\,\psi^k_*(x) = \psi^k_*\Omega(x) \in K_*(X;Z/_p)/_{\mathrm{Ind}_p(X)}.$$

(v) (a) If $p \neq 2$ there exists $\gamma_i \in Z/_p$, independent of

X, such that

$$\beta_p(\Omega(x)) = \begin{cases} \gamma_1\Omega(\beta_p(x)) - x^{p-1}\beta_p x, & \text{if deg } x \equiv 0 \pmod 2) \\ \gamma_o(\beta_p(x))^p & \text{, if deg } x \equiv 1 \pmod 2). \end{cases}$$

(b) Let B_2 be the second mod 2 Bockstein.

For $x \in \ker \beta_2 \subset K_\alpha(X;Z/_2)$

$\beta_2\Omega(x) = \Omega(B_2x) \in K_1(X;Z/_2)$ if $\alpha \equiv 0 \pmod 2)$

and

$\beta_2\Omega(x) = B_2(x)^2$ if $\alpha \equiv 1 \pmod 2)$.

[Here $B_2(x)^2$ means z^2 for any $z \in B_2(x)$.]

(vi) Let $\sigma: K_\alpha(\Omega X;Z/_p) \to K_{\alpha-1}(X;Z/_p)$ be the suspension

homomorphism. Let $x \in K_\alpha(\Omega X;Z/_p)$.

(a) If $p = 2$, $\alpha \equiv 0 \pmod 2)$

$$\sigma\Omega(x) = \Omega\sigma(x) \in K_1(X;Z/_2).$$

(b) If $p = 2$, $\alpha \equiv 1 (\mathrm{mod}\ 2)$

$$\sigma Q(x) = \sigma(x)^2 \in K_0(X; Z/_2).$$

(c) If $p \neq 2$ there exists $0 \neq \lambda$, $n_0 \in Z/_p$ independent of X (c.f. §A3.5) such that

$$\sigma Q(x) = \begin{cases} (-\lambda/n_0) Q(\sigma(x)), & \text{if } \alpha \equiv 0 (\mathrm{mod}\ 2) \\ n_0 \sigma(x)^p, & \text{if } \alpha \equiv 1 (\mathrm{mod}\ 2). \end{cases}$$

In [H1] a preliminary study was made of Dyer-Lashof operations in $K_*(X; Z/_p)$ ($p \neq 2$). It will be noticed that the operations of Theorem 5.1 have smaller indeterminacy than those of [H1]. It is clear (§4) that any worthwhile operation must have some indeterminacy. The smaller indeterminacy arises naturally if one first studies the case $p = 2$, which was how I started. I am very much indebted to Luke Hodgkin for enlightening me upon the obligatory nature of the indeterminacy. At the time of the Symposium in which [H1] was given I had the mod 2 Q-operation and since in odd degree this has no indeterminacy I would doubtless have spent a lot of time in attempts to remove the indeterminacy in even degree had it not been for Luke Hodgkin's help. Similarly his understanding of $K_*(QS^0 : Z/_p)$ provided the enlightenment (§4) that **Q**-operations with good indeterminacy could not be additive. I am also grateful to Stewart Priddy for pointing out to me the existence of the homotopy commutative diagrams of Boardman (§6) which reduce the homotopy theory of H^∞-action maps on $Z \times BU$ to group theory. Using this I was able considerably to shorten my computations of Q in $K_*(Z \times BU; Z/_p)$ which yield the results of [H2] on $K_*(QS^0; Z/_p)$. Incidentally, from this result it emerges that there do not exist Adem relations for Q, that is no polynomial is satisfied by Q mod $\mathrm{Ind}_p(-)$. The paper is arranged in the following manner.

There are nine sections and three Appendices, which contain long technical proofs and pictures of K-theory elements. The theory of the Q-operation is applied to certain calculations in §§6-9. The main results in these sections are concerning the splitting phenomenon of the image of the J-homomorphism. We show that in a splitting $SG \simeq \mathrm{Im}J \times \mathrm{Coker}\, J$ the K-theory of Coker J is zero and also that for reasonable choices of ImJ no such splitting is a splitting of threefold loopspaces.

§1: Review of K-theory mod p.

Description of G-resolutions, Milnor's G-resolution and wreath product G-resolutions.

The Rothenberg-Steenrod spectral sequence,

$$\mathrm{Tor}^{Z/_p[G]}(K_*(X,Y;Z/_p),Z/_p) = E^2(X,Y;G;Z/_p) \Longrightarrow K^G_*(X,Y;Z/_p).$$

Computation of $\{E^2((U,V)^p;\pi_p;Z/_p)\}$.

For $i:H \subset G$, description of forgetful and transfer homomorphis

$$\mathrm{Tor}^{Z/_p[H]}(-,Z/_p) \underset{\mathrm{Tor}(\bar{\mu})}{\overset{\mathrm{Tor}(i)}{\rightleftarrows}} \mathrm{Tor}^{Z/_p[G]}(-,Z/_p).$$

Computation of $\{E^2((U,V)^p;\ \Sigma_p;Z/_p)\}$.

§2: Description of the transfer in K-theory for finite coverings and inclusions $i:H \subset G$ of finite index.

Identifications of $\mathrm{Tor}(\bar{\mu})$ with the E^2-map induced by the transfer.

Periodicity of $E^2_{t,*}(X,Y;\Sigma_2;Z/_2)$ in t.

Identification of maps in the mod 2 exact couple as transfers and forgetful maps.

Determination of $\{E^r(pt,\phi;G;Z/_p)\}$, $G = \pi_p$ or Σ_p.

Description of $\mathrm{Tor}(\bar{\mu})$ on the chain level.

Computation of $\mathrm{Tor}(\bar{\mu})\circ\mathrm{Tor}(i)$ for $\Sigma_2 \int \Sigma_2 \subset \Sigma_4$ and examination of this on $\mathrm{Tor}_j^{Z/_2[\Sigma_2\int\Sigma_2]}$ for $0 \leq j \leq 2$.

§3: Determination of $E^r((U,V)^p;\Sigma_p;Z/_p)$ and corollaries.

§4: Definition of $q(x) \subset K_*^{\Sigma_p}((U,V)^p;Z/_p)$.

Indeterminacy and non-additivity of q.

Formulae for $q(x+y), q(i_*(x\otimes y)), \psi_*^k\circ q, \beta_p\circ q$ $(p \neq 2)$.

Computation of β_2 on $K^*_{\Sigma_2}(-;Z/_2)$.

Formula for $\beta_2\circ q$ and effect of B_i (i-th order Bockstein) on q.

Description of generators on $K_*^{\Sigma_2\int\Sigma_2}((U,V)^4;Z/_2)$.

Decomposition of $(j_2)_*q(i_*(y\otimes\beta_2 y))$ and $(j_2)_*q(i_*(y\otimes y))$.

§5: Definition of $Q(x) \subset K_*(X,Z/_p)$.

Proof of Theorem 5.1 and decomposition of $Q(x.x)$ and $Q(x.\beta_2 x)$ mod 2.

§6: Computation of the Q-algebras $K_*(Z \times BU;Z/_p)$ and $K_*(QS^0;Z/_p)$.

§7: K-theory spherical characteristic classes mod p.

$K_*(SG;Z/_p)$ is computed

§8: The Q-algebras $K_*(BSO;Z/_2)$, $K_*(BO \times Z;Z/_2)$, $K_*(BSpin;Z/_2)$, $K_*(SO;Z/_2)$ and $K_*(Spin;Z/_2)$ are computed.

§9: The J-homomorphisms.

$K_*(ImJ;Z/_p)$ is computed and it is shown that $\tilde{K}^*(\mathrm{Coker}\ J) = 0$

for suitable definitions of ImJ.

Maps $\gamma:\mathrm{ImJ} \to SG$ are studied and it is shown that no splitting of threefold loopspaces

$$SG \simeq \mathrm{ImJ} \times \mathrm{Coker}\, J \quad \text{exists at}$$

the prime, p = 2.

§§AI,II and III contain technical results on $\{E^r((U,V)^n;\Sigma_p;Z/_p)\}$ and analysis of a coboundary homomorphism used in obtaining the formula for $\sigma Q(x)$.

Finally I would like to express my gratitude to Mrs. Joan Scutt for converting a poor manuscript into such a good typescript.

Emmanuel College,
Cambridge.

March 1974.

§1:

Throughout this paper all spaces will be CW complexes. When we work in the based category the point $x_o \in X$ will denote the basepoint of X. The functor

$( )^+$: {CW complexes} → {CW complexes with basepoint}

is given by $X^+ = X \cup x_o$, (disjoint union). By a group, G, we will mean an inclusion, $G \subset \Sigma_n$, where Σ_n is the symmetric group on n letters.

If X is a compact G-space let $K_G^*(X)$ denote the equivariant, $Z/_2$-graded, complex K-theory ring of X, [Se]. If X has a G-invariant base-point, $x_o \in X$, let $\tilde{K}_G^*(X)$ denote the reduced theory. When G = {1} extend the K-theory to the CW category, using the unitary spectrum, by defining

$$\tilde{K}^o(X) = [X, Z \times BU],\ \tilde{K}^1(X) = [X, U]$$

([-,-] denotes based homotopy classes of maps), and $K^*(X) = \tilde{K}^*(X^+)$. Now let $p \in Z$, the integers. Define $\tilde{K}_G^*(X;Z/_p) = \tilde{K}_G^*(X \wedge M_p)$ where $M_p = S^1 \cup_p CS^1$, with trivial G-action. Hence $\tilde{K}^*(-;Z/_p)$ is represented by the $Z/_2$-graded Ω-spectrum of based maps from M_p into the spaces of the unitary spectrum. Let $\tilde{K}_*(-;Z/_p)$ denote the $Z/_2$-graded homology theory associated with this spectrum, in the sense of [W]. Thus $\tilde{K}_o(X;Z/_p) = \varinjlim_n [S^n, X \wedge U_n^{M_p}]$,

$$\tilde{K}_1(X;Z/_p) = \varinjlim_n [S^{n+1}, X \wedge U_n^{M_p}]$$

where
$$U_n = \begin{cases} Z \times BU & (n \text{ even}) \\ U & (n \text{ odd}) \end{cases}$$

and $\{SU_n \to U_{n+1}\}$ are the Bott maps.

Now let p be prime. A multiplication map $\tilde{K}^q(X;Z/_p) \otimes \tilde{K}^r(Y;Z/_p) \to \tilde{K}^{r+q}(X \wedge Y;Z/_p)$ may be defined in the following manner [A-T, I §5]. For $x \in \tilde{K}^q(X \wedge M_p)$ and $y \in \tilde{K}^r(Y \wedge M_p)$ we may form the external product

$x.y \in \tilde{K}^{q+r}(X \wedge Y \wedge M_p \wedge M_p)$. In [A-T, I §4.3] a complex, $N_p = S^2 \cup_g C(SM_p)$, and a map $\alpha: N_p \to M_p \wedge M_p$, are defined with the property that the cofibration sequence $0 \to \tilde{K}^*(X \wedge S^2M_p) \to \tilde{K}^*(X \wedge N_p) \to \tilde{K}^*(X \wedge S^2) \to 0$ is naturally split exact for all X. Define the product $x.y \in \tilde{K}^{q+r}(X \wedge Y; Z/_p) \cong \tilde{K}^{q+r}(X \wedge Y \wedge S^2M_p)$ as the component of $\alpha^*(x.y)$ in this group. If p is odd the map $g: SM_p \to S^2$ is nullhomotopic and one has the following simpler description of the multiplication [An, §1; A-T, I §4]. By [B] $p \wedge 1: S^1 \wedge M_p \to S^1 \wedge M_p$ is nullhomotopic and hence extends to $D^2 \wedge M_p$ giving the following diagram of Puppe sequences

$$\begin{array}{ccccc} S^1 \wedge M_p & \longrightarrow & D^2 \wedge M_p & \longrightarrow & S^2 \wedge M_p \\ \downarrow 1 & & \downarrow & & \downarrow \bar{\alpha} \\ S^1 \wedge M_p & \xrightarrow[p \wedge 1]{} & S^1 \wedge M_p & \longrightarrow & M_p \wedge M_p . \end{array}$$

Define $x.y \in \tilde{K}^{q+r}(X \wedge Y; Z/_p) \cong \tilde{K}^{q+r}(X \wedge Y \wedge S^2 \wedge M_p)$ by $x.y = \bar{\alpha}^*(x.y)$. When p is odd the multiplication obtained in this manner is associative and commutative. When $p = 2$ the multiplication obtained is associative but not commutative. In fact, if $\beta_p: \tilde{K}^q(-;Z/_p) \to \tilde{K}^{q+1}(-;Z/_p)$ is the Bockstein and $T: X \wedge Y \to Y \wedge X$ is the switching map then [A-T; II §7.9] $T^*(x.y) = y.x + \beta_2(y).\beta_2(x) \in \tilde{K}^*(Y \wedge X; Z/_2)$. It is these multiplications and their dual comultiplications which will be used throughout this paper.

Let G be a subgroup of Σ_n and let X be a compact G-space with a closed, G-invariant subspace, Y. Let $G \to EG \to EG/_G = BG$ be the universal principal G-bundle. By

[At-Se,§2.1]

$$K_G^*(X,Y;Z/_p)^\wedge \cong K^*((X,Y)_G;Z/_p)$$

where $(X,Y)_G = (X \times_G EG, Y \times_G EG)$ and $(\ )^\wedge$ denotes completion with respect to the $(I(G) \otimes Z/_p)$-adic topology. However the $(I(G) \otimes Z/_p)$-adic topology on $R(G) \otimes Z/_p \cong K^*(pt;Z/_p)$ is discrete so we have $K_G^*(X,Y;Z/_p) \cong K_G^*(X,Y;Z/_p)^\wedge$. These isomorphisms justify the definition

$$K_*^G(X,Y;Z/_p) = K_*((X,Y)_G;Z/_p).$$

A <u>G-resolution</u> is a realisation of EG as a filtered space satisfying the following conditions.

(a) EG is a free, right G-space filtered by closed subspaces,

$(pt) = D_0 \subset E_0 \subset D_1 \subset E_1 \subset \ldots\ldots$

(b) $E_i\ (i \geq 0)$ is G-invariant and $EG = \bigcup_{i\geq 0} E_i$ with the topology of the union.

(c) For each n, E_n is contractible in E_{n+1}.

(d) For each n the action map $\psi_n\colon E_n \times G \to E_n$ restricts to a relative homeomorphism

$$\phi_n\colon (D_n,E_{n-1}) \times G \to (E_n,E_{n-1}),$$

$(n = 0,\ \phi_0\colon G \to E_0$ is a homeomorphism).

(e) For each n there exists $u_n\colon D_n \to I$ (the unit interval) and $h_n\colon I \times D_n \to D_n$ representing (D_n,E_{n-1}) as an NDR pair [Ma, Appendix; R-S] and $u_n'\colon E_n \to I$, $h_n'\colon I \times E_n \to E_n$ such that

$$u_n \circ (\mathrm{proj})_1 = u_n' \circ \phi_n\colon D_n \times G \to I$$

and $\phi_n \circ (h_n \times 1) = h_n' \circ (1 \times \phi_n)\colon I \times D_n \times G \to E_n$.

The <u>Milnor G-resolution</u> is defined in the following manner.

Put $D_o = \{1\} \subset E_o = G$. Inductively define $D_n = CE_{n-1}$, the cone on E_{n-1},

$$E_n = D_n \times G \cup_{\psi_{n-1}} E_{n-1},$$

and $\psi_n : E_n \times G \to E_n$ by $\psi_n | E_{n-1} \times G = \psi_{n-1}$,

$$\psi_{n-1} = \psi_n | \{t\} \times E_{n-1} \times G \to \{t\} \times E_{n-1} \quad \text{for} \quad t \in I.$$

For example, when $G = \Sigma_2$ the Milnor resolution is just the filtration given by $\ldots . \subset S^n = E_n \subset S^{n+1} = E_{n+1} \subset \ldots \subset S^\infty \doteq E\Sigma_2$ where Σ_2 acts by the antipodal action on each sphere.

Recall that if $G \subset \Sigma_m$ and $H \subset \Sigma_e$ then the wreath product, $G \int H \subset \Sigma_{me}$ is defined in the following manner. $G \int H$ is generated by $G^e \subset (\Sigma_m)^e = \Sigma_m \times \ldots \times \Sigma_m \subset \Sigma_{me}$ and H, where an element of H acts by permuting the e blocks of m ordered integers of the form

$$(j.m + 1,\ j.m + 2, \ldots, (j+1).m) \quad , \quad (0 \le j \le e-1).$$

With the convention that if $\alpha, \beta \in \Sigma_t$ then $(x)(\alpha.\beta) = ((x)\alpha)\beta$ then the multiplication in the group $G \int H$ is given by

$$h_1.(s_1, \ldots, s_e).h.(t_1, \ldots, t_e) = h_1.h.(s_{(1)h^{-1}}, \ldots, s_{(e)h^{-1}}).(t_1, \ldots,$$

$$(h, h_1 \in H;\ s_i, t_i \in G).$$

Suppose that $\{EH = \bigcup_n EH_n; \ldots \subset DH_n \subset EH_n \subset \ldots .\}$ and $\{EG = \bigcup_n EG_n; \ldots \subset DG_n \subset EG_n \subset \ldots .\}$ are respectively an H-resolution and a G-resolution. On the space $EH \times (EG)^e$ put the $G \int H$-action

$$(x, y_1, \ldots, y_e).(h_1(s_1, \ldots, s_e)) = (x.h, (y_{(1)h^{-1}})s_1, \ldots, (y_{(e)h^{-1}})s_e).$$

Filter this space in the following manner.

If $\{\ldots \subset DA_n \subset EA_n \subset DA_{n+1} \subset \ldots\}$ and $\{\ldots \subset DB_n \subset EB_n \subset DB_{n+1} \subset \ldots\}$

are filtered spaces filter the product $(\bigcup_{o \leq n} EA_n) \times (\bigcup_{o \leq m} EB_m) = E(A \times B)$

by $E(A\times B)_n = \bigcup_{i=o}^{n} EA_i \times EB_{n-i}$

and $D(A\times B)_n = E(A\times B)_{n-1} \cup (\bigcup_{i=o}^{n} DA_i \times DB_{n-i})$. Apply this process inductively to filter $EH \times (EG)^e = E(G/H)$ as $\{\ldots \subset D(G/H)_n \subset E(G/H)_n \subset \ldots\}$.

Proposition 1.1

$$\{E(G/H) = \bigcup_n E(G/H)_n; \ldots \subset D(G/H)_n \subset E(G/H)_n \subset \ldots\}$$

is a (G/H)-resolution.

Proof: G/H clearly acts freely on $E(G/H)$ and the topological conditions are the same as those proved for a product of groups in [R-S, §6.1]. Condition (d) also follows from the fact that it is true for the action of $H \times G^e$ on $EH \times (EG)^e$ with this filtration.

Now let X be a G-space and let $Y \subset X$ be a closed G-subspace. Applying $\tilde{K}^*(-;Z/_p)$ or $\tilde{K}_*(-;Z/_p)$ to the filtered space

$$\{\ldots \subset (X/_Y \wedge EG_n^+)/_G \subset (X/_Y \wedge EG_{n+1}^+)/_G \subset \ldots \subset (X/_Y \wedge EG^+)/_G\}$$

we obtain spectral sequences convergent to

$$\tilde{K}^*((X/_Y \wedge EG^+)/_G; Z/_p) \cong K^*((X,Y)_G; Z/_p) \cong K_G^*(X,Y;Z/_p)$$

and $\tilde{K}_*((X/_Y \wedge EG^+)/_G; Z/_p) \cong K_*^G(X,Y;Z/_p)$.

These spectral sequences could equally well have been obtained from constructions in the unbased category and the one convergent to $K_G^*(X,Y;Z/_p)$ could have been obtained by applying $K_G^*(-,-;Z/_p)$ to the G-pairs, $\{(X,Y) \times EG_n\}$, since

$$K^*_G((X,Y) \times EG_n; Z/_p) \cong K^*((X,Y) \times_G EG_n; Z/_p)$$

and $\lim\limits_{\overleftarrow{n}} K^*_G\,((X,Y) \times EG_n; Z/_p) \cong K^*_G(X,Y;Z/_p)$.

To compute the E_1-term observe that conditions (c) and (d) imply that

$$(1.2)\quad Z/_p \xrightarrow{\varepsilon} K^*(E_o;Z/_p) \xrightarrow{d_I} K^*(E_1,E_o;Z/_p) \xrightarrow{d_I} K^*(E_2,E_1;Z/_p) \xrightarrow{d_I} \cdots$$

and

$$(1.3)\quad Z/_p \xleftarrow[\eta]{} K_*(E_o;Z/_p) \xleftarrow[d_{II}]{} K_*(E_1,E_o;Z/_p) \xleftarrow[d_{II}]{} K_*(E_2,E_1;Z/_p) \xleftarrow[d_{II}]{} \cdots$$

are respectively free $K^*(G;Z/_p)$-comodule and $K_*(G;Z/_p) = Z/_p[G]$-module resolutions of $Z/_p$. From [A_n], there is a natural isomorphism

$$K^*(-;Z/_p) \cong \mathrm{Hom}(K_*(-;Z/_p),Z/_p)$$

on the category of pairs of finite complexes. Thus (1.2) and (1.3) are dual chain complexes. Condition (d) implies isomorphisms

$$\begin{aligned} & \tilde{K}^*((^X/_Y \wedge EG_n^+)/_G,\ (^X/_Y \wedge EG_{n-1}^+)/_G; Z/_p) \\ \cong\ & K^*([(X,Y) \times (EG_n,EG_{n-1})]/_G; Z/_p) \\ \cong\ & K^*(X,Y;Z/_p)\ \square_{K^*(G;Z/_p)}\ K^*(EG_n,EG_{n-1};Z/_p) \end{aligned}$$

and dually

$$\begin{aligned} & \tilde{K}_*((^X/_Y \wedge EG_n^+)/_G,\ (^X/_Y \wedge EG_{n-1}^+)/_G; Z/_p) \\ \cong\ & K_*(X,Y;Z/_p) \otimes_{Z/_p[G]} K_*(EG_n,EG_{n-1};Z/_p). \end{aligned}$$

Under these identifications the E_1-differentials are $1 \,\square\, d_I$ and $1 \otimes d_{II}$. The spectral sequences satisfy the following properties, for details see [An-H;E-Mo; R-S;Sn 1].

<u>Theorem 1.4</u>

Let p be a prime.

(a) There is a spectral sequence

$$\{E^r_{*,*}(X,Y;G;Z/_p),\ d^r\} \qquad (r \geq 2),$$

natural in the G-pair (X,Y) and satisfying the following properties.

(i) $E^r_{*,*}(X,Y;G;Z/_p)$ is a $Z \times Z/_2$-bigraded $Z/_p$-coalgebra and $E^r_{*,*}(pt,\phi;G;Z/_p)$-comodule.

(ii) $d^r\colon E^r_{q,\alpha} \to E^r_{q-r,\alpha+r-1}$ is a derivation with respect to the co-actions in (i).

(iii) There is a natural isomorphism

$$E^2_{q,\alpha}(X,Y;G;Z/_p) \cong \mathrm{Tor}^{Z/_p[G]}_{q,\alpha}(K_*(X,Y;Z/_p),Z/_p)$$

(iv) The spectral sequence converges strongly to $K^G_*(X,Y;Z/_p)$.

(b) There is a spectral sequence

$$\{E_s^{*,*}(X,Y;G;Z/_p),\ d_s\} \qquad (s \geq 2),$$

natural in the G-pair (X,Y) and satisfying the following properties.

(i) $E_s^{*,*}(X,Y;G;Z/_p)$ is a $Z \times Z/_2$-bigraded $Z/_p$-algebra and $E_s^{*,*}(pt,\phi;G;Z/_p)$-module.

(ii) $d_s\colon E_s^{q,\alpha} \to E_s^{q+r,\alpha-s+1}$ is a derivation with respect to the actions in (i).

(iii) There is a natural isomorphism

$$E_2^{q,\alpha}(X,Y;G;Z/_p) \cong \mathrm{Cotor}^{q,\alpha}_{K^*(G;Z/_p)}(K^*(X,Y;Z/_p),Z/_p).$$

(iv) If X is a finite complex the spectral sequence converges strongly to $K^*_G(X,Y;Z/_p)$.

We now describe generators for the E_2-terms of these spectral sequences in the case when $G = \pi_p$, the cyclic group of order p (p prime) and (X,Y) is the p-fold product, $(U,V)^p$, acted upon in the standard manner by $\pi_p \subset \Sigma_p$. A free $Z/_p[\pi_p]$-resolution of $Z/_p$ is given by the following complex. Let D_q be the free, left $Z/_p[\pi_p]$-module on generator, $e_q (q \geq o)$, and define the free resolution

(1.5) $$Z/_p \xleftarrow{\varepsilon} D_o \xleftarrow{d_o} D_1 \xleftarrow{d_1} D_2 \longleftarrow \cdots$$

by
$$\varepsilon(e_o) = 1,$$
$$d_{2k}(e_{2k+1}) = (1-\tau).e_{2k},$$
$$d_{2k+1}(e_{2k+2}) = \sum_{i=o}^{p-1} \tau^i.e_{2k+1}$$

where $\tau \in \pi_p$ is the canonical p-cycle and $Z/_p$ has the trivial π_p-module structure. Using the isomorphisms

$$K_*((U,V)^p;Z/_p) \cong K_*(U,V;Z/_p)^{\otimes p}$$

and

$$K_*(U,V;Z/_p)^{\otimes p} \otimes_{Z/_p[\pi_p]} D_i \cong K_*(U,V;Z/_p)^{\otimes p}$$

we have that $\mathrm{Tor}_{*,*}^{Z/_p[\pi p]}(K_*((U,V)^p;Z/_p),Z/_p)$ is computable as the homology of the complex

(1.6) $$0 \leftarrow K_*(U,V;Z/_p)^{\otimes p} \xleftarrow[(1-\tau_*)]{} K_*(U,V;Z/_p)^{\otimes p} \xleftarrow[(\Sigma\tau_*^i)]{} K_*(U,V;Z/_p)^{\otimes p} \longleftarrow \cdots$$

Let β_p denote the $Z/_p \to Z/_{p^2} \to Z/_p$ Bockstein in $K_*(-;Z/_p)$ or $K^*(-;Z/_p)$-theory and let $\{u_\alpha\}$ be a $Z/_p$-basis for $K_*(U,V;Z/_p)$.

Let p be an odd prime. The π_p-module, $K_*(U,V;Z/_p)^{\otimes p}$, is the direct sum of π_p-submodules generated over $Z/_p$ by the orbit of elements of the form $(u_{\alpha_1} \otimes u_{\alpha_2} \otimes \cdots \otimes u_{\alpha_p}) \in K_*(U,V;Z/_p)^{\otimes p}$.

Let $F(\alpha_1,\alpha_2,\ldots,\alpha_p)$ be the free π_p-module generated by the orbit of $(u_{\alpha_1}\otimes\ldots\otimes u_{\alpha_p})$ in the case when not all the $\{u_{\alpha_i}\}$ are equal and let $M(u_\alpha)$ be the trivial π_p-module generated by $(u_\alpha)^{\otimes p}$. We have $\mathrm{Tor}_{q,*}^{Z/_p[\pi p]}(F(\alpha_1,\ldots,\alpha_p),Z/_p) = 0$ for $q > o$ and $\mathrm{Tor}_{o,*}^{Z/_p[\pi p]}(F(\alpha_1,\ldots,\alpha_p),Z/_p) \cong F(\alpha_1,\ldots,\alpha_p)/(\mathrm{im}(1-\tau_*))$ the module of coinvariants of $F(\alpha_1,\ldots,\alpha_p)$. Also $\mathrm{Tor}_{q,\deg(u_\alpha)}^{Z/_p[\pi p]}(M(u_\alpha),Z/_p) \cong Z/_p$ generated by the class of $(u_\alpha)^{\otimes p}\otimes e_q$ and $0 = \mathrm{Tor}_{q,\deg(u_\alpha)+1}^{Z/_p[\pi p]}(M(u_\alpha),Z/_p)$.

For $p = 2$ the π_2-module, $K_*(U,V;Z/_2)^{\otimes 2}$, has the π_2-action $\tau_*(x\otimes y) = y\otimes x + \beta_2 y\otimes\beta_2 x$. Choosing a basis for $K_*(U,V;Z/_2)$ as in [A-Y, §6.14] it is clear that the π_2-module $K_*(U,V;Z/_2)^{\otimes 2}$, is a direct sum of π_2-submodules of the following types.

<u>Type (i)</u> Two dimensional modules of the form

$$\{u_{\alpha_1}\otimes u_{\alpha_2},\ u_{\alpha_2}\otimes u_{\alpha_1} + \beta_2 u_{\alpha_2}\otimes\beta_2 u_{\alpha_1}\}.$$

<u>Type (ii)</u> One dimensional modules of the form

$$M(u_\alpha) \text{ where } \beta_2 u_\alpha = 0 \text{ and } u_\alpha \notin \mathrm{im}\,\beta_2.$$

$\mathrm{Tor}_{q,*}^{Z/_2[\pi 2]} = 0$ for Type (i) if $q > o$ and $\mathrm{Tor}_{o,*}^{Z/_2[\pi 2]}$ is the module of coinvariants for both types of module. Finally, for Type (ii) $\mathrm{Tor}_{q,o}^{Z/_2[\pi 2]} \cong Z/_2$, generated by the class of $u_\alpha^{\otimes 2}\otimes e_q$, and $\mathrm{Tor}_{q,1}^{Z/_2[\pi 2]} = 0$.

$\mathrm{Cotor}_{K^*(\pi_p;Z/_p)}^{*,*}(K^*(U,V;Z/_p)^{\otimes p}, Z/_p)$ is obtained from the dual of the complex (1.6). Hence $\mathrm{Cotor}_{K^*(\pi_p;Z/_p)}^{o,*}$ is the submodule of π_p-invariants in $K^*(U,V;Z/_p)^{\otimes p}$ and for $q > 0$ there are natural isomorphisms

$$\phi_2 : \mathrm{Cotor}^{q,o}_{K^*(\pi_2;Z/_2)}(K^*(U,V;Z/_2)^{\otimes 2}, Z/_2) \stackrel{\cong}{\to} \frac{\ker\beta_2}{\mathrm{im}\beta_2} \subset \frac{K^*(U,V;Z/_2)}{\mathrm{im}\beta_2}$$

and $(p \neq 2)$

$$\phi_p : \mathrm{Cotor}^{q,*}_{K^*(\pi_p;Z/_p)}(K^*(U,V;Z/_p)^{\otimes p}, Z/_p) \stackrel{\cong}{\to} K^*(U,V;Z/_p)$$

given by

$$\phi_2(x^{\otimes 2} \otimes e_q) = x + \mathrm{im}\beta_2,$$

$$\phi_p(x^{\otimes p} \otimes e_q) = x.$$

The $\mathrm{Cotor}^{r,o}_{K^*(\pi_p;Z/_p)}(Z/_p, Z/_p)$-module action on $\mathrm{Cotor}^{q,\alpha}_{K^*(\pi_p;Z/_p)}(K^*(U,V;Z/_p)^{\otimes p}, Z/_p)$ is given by

$(1^{\otimes p} \otimes e_r).(x^{\otimes p} \otimes e_q) = x^{\otimes p} \otimes e_{q+r}$ and $(1^{\otimes p} \otimes e_r).z = o$ otherwise.

Now let $i : H \subset G$ be an inclusion of groups and let M be a right $Z/_p[G]$-module.

Define $j : M \to M \otimes_{Z/_p[H]} Z/_p[G]$ by $j(m) = m \otimes 1$,

$\mu : M \otimes_{Z/_p[H]} Z/_p[G] \to M$ by $\mu(m \otimes g) = m.g$,

$(m \in M,\ g \in G)$ and $\bar{\mu} : M \to M \otimes_{Z/_p[H]} Z/_p[G]$

by $\bar{\mu}(m) = \sum_{i=1}^{|G:H|} m.g_i \otimes g_i^{-1}$ where $\{g_i;\ 1 \leq i \leq |G:H|\}$ is a set of left coset representatives of H in G. Thus j is a $Z/_p[H]$-module homomorphism and $\mu, \bar{\mu}$ are $Z/_p[G]$-module homomorphisms. There is a commutative diagram

$$\begin{array}{ccc}
\mathrm{Tor}^{Z/_p[H]}(M,Z/_p) & \xrightarrow[\mathrm{Tor}^i(j,Z/_p)]{\cong} & \mathrm{Tor}^{Z/_p[G]}(M \otimes_{Z/_p[H]} Z/_p[G], Z/_p) \\
& \mathrm{Tor}^i(M,Z/_p) \searrow \quad \swarrow \mathrm{Tor}^{Z/_p[G]}(\mu,Z/_p) & \\
& \mathrm{Tor}^{Z/_p[G]}(M,Z/_p) & \\
& \Big\downarrow \mathrm{Tor}^{Z/_p[G]}(\bar{\mu},Z/_p) & \\
& \mathrm{Tor}^{Z/_p[G]}(M \otimes_{Z/_p[H]} Z/_p[G], Z/_p). &
\end{array}$$

If $p \nmid |G:H|$ then $\mathrm{Tor}^i(M,Z/_p)$ is a split epimorphism since $\mu(\bar{\mu}(m)) = \mu(\Sigma\, m.g_k \otimes g_k^{-1}) = |G:H|m$. We now use this fact to calculate the E_2-terms of the spectral sequences of Theorem 1.4 in the case when $G = \Sigma_p$ and $(X,Y) = (U,V)^p$. We deal separately with $\mathrm{Tor}_o^{Z/_p[\Sigma_p]}(M,Z/_p)$. The Σ_p-bar resolution commences as

$$Z/_p \xleftarrow{\varepsilon} Z/_p[\Sigma_p] \xleftarrow{d_o} Z/_p[\Sigma_p] \otimes \ker \varepsilon \leftarrow \ldots.$$

where d_o is induced by the multiplication and

$\mathrm{im}\, d_o = \{$submodule generated by $(g-1)$ as g runs through $\Sigma_p\}$.

Hence $\mathrm{Tor}_o^{Z/_p[\Sigma_p]}(M,Z/_p) = M/_{\mathrm{im}(1 \otimes d_o)} = M/_{\{m - m.g\}}$, the Σ_p- coinvariants of M.

Now let p be an odd prime. We study the epimorphism

$$\mathrm{Tor}_{*,*}^{Z/_p[\pi_p]}(M(u_\alpha),Z/_p) \xrightarrow{\mathrm{Tor}^i(M(u_\alpha),Z_p)} \mathrm{Tor}_{*,*}^{Z/_p[\Sigma_p]}(M(u_\alpha),Z/_p),$$

$(u_\alpha \in K_*(U,V;Z/_p))$. Since $M(u_\alpha) \cong Z/_p$ as a Σ_p-module when $\alpha \equiv 0 \pmod 2$ we have isomorphisms

$$\mathrm{Tor}_{q,\deg(u_\alpha)}^{Z/_p[H]}(M(u_\alpha),Z/_p) \cong \mathrm{Tor}_{q,o}^{Z/_p[H]}(Z/_p,Z/_p) \cong H_q(BH;Z/_p),$$

for $H = \pi_p$ or Σ_p and $\mathrm{Tor}^i(M(u_\alpha),Z/_p)$ corresponds under these isomorphisms to the map induced by $B(i):B\pi_p \to B\Sigma_p$. Hence, [D-L,p.51], $\mathrm{Tor}^i_{2j,\deg(u_\alpha)}(M(u_\alpha),Z/_p)$ $(j \geq o)$ is zero unless $2j$ is an even multiple of $(p-1)$. Now the homology Bockstein

$\beta_p:H_k(B\pi_p;Z/_p) \to H_{k-1}(B\pi_p;Z/_p)$ is zero when k is odd and an isomorphism when k is even. Also $H_*(B(i);Z/_{p^2})$ is a split epimorphism. Thus, from the Bockstein exact sequence, $\mathrm{Tor}^i_{2j-1}(M(u_\alpha),Z/_p)$ is non-trivial if and only if $\mathrm{Tor}^i_{2j}(M(u_\alpha),Z/_p)$ is. Hence it will follow from the $\mathrm{Tor}^{Z/_p[\pi_p]}(Z/_p,Z/_p)$ algebra structure that if

$$\mathrm{Tor}^{Z/_p[\pi_p]}_{2(p-1)}(Z/_p,Z/_p) \stackrel{\cong}{\to} \mathrm{Tor}^{Z/_p[\Sigma_p]}_{2(p-1)}(Z/_p,Z/_p)$$

then, if $\alpha \equiv 0 \pmod 2$,

$$\mathrm{Tor}^{Z/_p[\Sigma_p]}_{j,\deg u_\alpha}(M(u_\alpha),Z/_p) = \begin{cases} Z/_p \text{ generated by } u_\alpha^{\otimes p} \otimes e_j \\ \quad \text{if } j = 2k(p-1)-1 \\ \quad \text{or } j = 2k(p-1), \\ 0 \text{ otherwise.} \end{cases}$$

However, in §2.8, we will show by considering the transfer in representation theory that the dual group, $\mathrm{Cotor}^{2(p-1),o}_{K^*(\Sigma_p;Z/_p)}(Z/_p,Z/_p)$ is non-zero.

When $\alpha \equiv 1 \pmod 2$ $M(u_\alpha) \cong Z/_p$ as a $Z/_p[\pi_p]$- module but $M(u_\alpha)$ is the sign representation as a $Z/_p[\Sigma_p]$- module. By studying $\{E_s^{*,*}((S^1)^P;\ G{:}Z/_p)\}$ for $G = \pi_p$ and Σ_p we will show that if $\alpha \equiv 1 \pmod 2$ then

$$\mathrm{Tor}^{Z/_p[\Sigma_p]}_{j,\deg u_\alpha}(M(u_\alpha),Z/_p) = \begin{cases} Z/_p \text{ generated by } u_\alpha^{\otimes p} \otimes e_j \\ \quad \text{if } j = (2k+1)(p-1)-1 \text{ or} \\ \quad j = (2k+1)(p-1), \\ 0 \text{ otherwise.} \end{cases}$$

Let X be a Σ_p-space and let $E \to X$ be a π_p-vector bundle. If $g \in N(\pi_p)$, the normaliser of π_p, then $g^*E \to X$ is a π_p-vector bundle with π_p-action given by conjugating the old action by g. If E is a Σ_p-vector bundle then $g^*E \cong E$ as π_p-vector bundles. This yields a multiplicative endomorphism, g^*, of $K^*_{\pi_p}(-)$, which on representations of π_p is simply the homomorphism induced by the inner automorphism $h \longmapsto g.h.g^{-1}$. Now let 1 and ι be generators of $K^0(S^1;Z/_p)$ and $K^1(S^1;Z/_p)$ respectively. From the results of §2 on the transfer homomorphism we have that the restriction map embeds

$\{E_s^{**}((S^1)^p;\ \Sigma_p;Z/_p)\}$ as a direct summand in

$\{E_s^{*,*}((S^1)^p;\ \pi_p;Z/_p)\}$ and from §3 we see that the latter spectral sequence has only d_{2p-1} non-trivial, which kills $\mathrm{Cotor}^{j,*}_{K^*(\pi_p;Z/_p)}$ precisely when $j \notin (0,2,4,\dots,2(p-1)$. Let $y \in R(\pi_p)$ be the canonical one-dimensional representation so that if $\sigma = 1-y$ then $R(\pi_p) \otimes Z/_p \cong Z/_p[\sigma]/_{(\sigma^p)}$. From [Sn 2] and the results of §3 we have that $\iota^{\otimes p} \in K_{\Sigma_p}((S^1 \times (I,\partial I))^p)$ restricts to an element in $K^{-1}_{\pi_p}((S^1)^p)$ which is represented in the π_p-spectral sequence by $\iota^{\otimes p} \otimes [(p-1)/_2]!\, e_{p-1}$, since in $R(\pi_p)\otimes Z/_p$ $\prod_1^{(p-1)/2}(1-y^j) \equiv [p-1)/_2]!\sigma^{(p-1)/2}$ (mod higher powers of σ). For $g \in N(\pi_p)$, g^* operates on the spectral π_p-spectral sequence and fixes the image of the Σ_p-spectral sequence. However, we may realise $g^*(y) = y^k$ for any $1 < k \le p-1$, so if $g^*(y) = y^k$ then results of §3 on representatives in the π_p-spectral sequence show that

$$g^*(\iota^{\otimes p} \otimes e_{2j}) = g^*(\iota^{\otimes p}) \otimes g^*(e_{2j})$$
$$\equiv k^{(p-1)/2}\, \iota^{\otimes p} \otimes k^j\, e_{2j} \ (\mathrm{mod}\ p).$$

Hence $\mathrm{Cotor}^{2j,\deg u_\alpha}_{K^*(\Sigma_p;Z/_p)}(M(u_\alpha),Z/_p)$, $0 \leq j \leq p-1$, is non-zero onl[y] when $j = (p-1)/_2$. From the behaviour of the Bockstein we have $\mathrm{Cotor}^{2j+1,\deg u_\alpha}_{K^*(\Sigma_p;Z/_p)}(M(u_\alpha),Z/_p)$, $0 \leq j \leq (p-2)$, is non-zero only when $j = (p-3)/_2$. Finally since d_{2p-1} is the only non-zero differentia[l] which must annihilate $\iota^{\otimes p} \otimes e_{p-2}$, the module structure over the π_p-spectral sequence for a point and the Bockstein behaviour give th[e] periodicity of period $2(p-1)$. Dualising gives the results on $\mathrm{Tor}^{*,*}_{Z/_p[\Sigma_p]}(M(u_\alpha),Z/_p)$.

We summarise these calculations in the following proposition.

<u>Proposition 1.7</u>:

Let p be a prime.

(a)(i) For $H = \pi_p$ or Σ_p the quotient

$$\mathrm{Tor}^{Z/_p[H]}(K_*(U,V;Z/_p)^{\otimes p},Z/_p) \text{ of } K_*(U,V;Z/_p)^{\otimes p}$$

is isomorphic to the coinvariants of the H-action.

(ii) There are natural isomorphisms $(q > 0)$

$$\phi_2:\mathrm{Tor}^{Z/_2[\pi_2]}(K_*(U,V;Z/_2)^{\otimes 2},Z/_2) \xrightarrow{\cong} \frac{\ker \beta_2}{\mathrm{im}\ \beta_2} \subset \frac{K_*(U,V;Z/_2)}{\mathrm{im}\ \beta_2}$$

and $(p \neq 2)$

$$\phi_p:\mathrm{Tor}^{Z/_p[\pi_p]}(K_*(U,V;Z/_p)^{\otimes p},Z/_p) \xrightarrow{\cong} K_*(U,V;Z/_p)$$

given by $\phi_2(x^{\otimes 2} \otimes e_q)\ 2\ x + \mathrm{im}\ \beta_2$,

$$\phi_p(x^{\otimes p} \otimes e_q) = x.$$

For $p = 2$, $q > 0$ $\mathrm{Tor}_{q,1}$ is zero.

(b)(i) For $H = \pi_p$ or Σ_p subgroup

$$\mathrm{Cotor}^{o,*}_{K^*(H;Z/_p)}(K^*(U,V;Z/_p)^{\otimes p}, Z/_p) \quad \text{of} \quad K^*(U,V;Z/_p)^{\otimes p}$$

is isomorphic to the invariants of the H-action.

(ii) There are isomorphisms $(q > o)$

$$\phi_2 : \mathrm{Cotor}^{q,o}_{K^*(\pi_2;Z/_2)}(K^*(U,V;Z/_2)^{\otimes 2}, Z/_2) \overset{\cong}{\to} \frac{\ker\beta_2}{\mathrm{im}\beta_2} \subset \frac{K^*(U,V;Z/_2)}{\mathrm{im}\beta_2}$$

and $(p \neq 2)$

$$\phi_p : \mathrm{Cotor}^{q,*}_{K^*(\pi_p;Z/_p)}(K^*(U,V;Z/_p)^{\otimes p}; Z/_p) \overset{\cong}{\to} K^*(U,V;Z/_p)$$

given by $\quad \phi_2(x^{\otimes 2} \otimes e_q) = x + \mathrm{im}\beta_2,$

$$\phi_p(x^{\otimes p} \otimes e_q) = x\ .$$

For $p = 2$, $q > o$ the $\mathrm{Cotor}^{q,1}$ are zero.

(c) The inclusion $i : \pi_p \subset \Sigma_p$ induces isomorphisms

$$\mathrm{Tor}^{Z/_p[\pi_p]}_{j,\alpha}(K_*(U,V;Z/_p)^{\otimes p}, Z/_p) \overset{\cong}{\to} \mathrm{Tor}^{Z/_p[\Sigma_p]}_{j,\alpha}(K_*(U,V;Z/_p)^{\otimes p}, Z/_p)$$

and

$$\mathrm{Cotor}^{j,\alpha}_{K^*(\Sigma_p;Z/_p)}(K^*(U,V;Z/_p)^{\otimes p}, Z/_p) \overset{\cong}{\to} \mathrm{Cotor}^{j,\alpha}_{K^*(\pi_p;Z/_p)}(K^*(U,V;Z/_p)^{\otimes p} Z/_p)$$

$$\text{if} \begin{cases} \alpha \equiv 0 (\mathrm{mod}\ 2),\ 0 < j = 2k(p-1) - 1 \ \text{ or } \ 2k(p-1), \\ \alpha \equiv 1 (\mathrm{mod}\ 2),\ 0 < j = (2k+1)(p-1) - 1 \ \text{ or } \ (2k+1)(p-1). \end{cases}$$

For $0 < j$ the groups $\mathrm{Tor}^{Z/_p[\Sigma_p]}_{j,\alpha}$ and $\mathrm{Cotor}^{j,\alpha}_{K^*(\Sigma_p;Z/_p)}$ are zero except in the case mentioned above.

§2:

Let $f: X \to Y$ be a finite covering between compact spaces. If E is a vector bundle over X the direct image bundle [At 1, §2], $f_!(E)$ over Y, is the bundle whose fibre at $y \in Y$ is $\bigoplus_{f(x)=y} E_x$ where E_x is the fibre of E over $x \in X$. The direct image construction is functorial on vector bundles and commutes with direct sums to yield a homomorphism, $f_!: K^*(X) \to K^*(Y)$. This homomorphism will also be called the transfer homomorphism associated with f. The transfer can be defined for the reduced theory to give

$f_!: \tilde{K}^*(X) \to \tilde{K}^*(Y)$, which can also be produced as the homomorphism, g^*, induced by a map $g: S^{2N}Y \to S^{2N}X$ for large N [K-P, §§1,2].

If F is a vector bundle over Y and E is a vector bundle over X then

(2.1) $f_!(E \otimes f^*(F)) \cong f_!(E) \otimes F$, [At 1, §2].

Suppose now that X is a compact G-space and $Y \subset X$ is a G-invariant, closed subspace. Let $j: H \subset G$ be the inclusion of a subgroup of finite index. The map $f: (G \times X)/_H \to X$ and its restriction $f: (G \times Y)/_H \to Y$ given by $f[g,x] = x.g^{-1}$ ($g \in G, x \in X$, $[-] \equiv$ H-orbit equivalence class) are finite coverings. The group, G, acts on $(G \times X)/_H$ via left multiplication on the G factor in such a way that f is a G-map. However

$$K_G^*((G \times X)/_H, (G \times Y)/_H) \cong K_H^*(X,Y),$$

[Se, §2], so we obtain $j_! = f_!: K_H^*(X,Y) \to K_G^*(X,Y)$ since the direct image construction assigns to a G-vector bundle over $(G \times X)/_H$ a G-vector bundle over X. When $(X,Y) = (pt., \phi)$ $K_H^*(pt,\phi) = R(H)$, the complex representation ring of H, and $j_!$

is the familiar induced representation construction. These homomorphisms are compatible in the sense that if the finite covering is the canonical projection $f:X_{/H} \to X_{/G}$ where X is a free G-space then under the identifications $K_G^*(X) \cong K^*(X_{/G})$, $K_H^*(X) \cong K^*(X_{/H})$ [Se, §2.1) the homomorphisms $f_!$ and $j_!$ coincide.

Proposition 2.2

Let $j:H \subset G$ be an inclusion of groups. Let $j^!$ be the dual of the transfer, $j_!$. The homomorphism, $j^!$, induces a homomorphism of spectral sequences

$$E^r(j^!): E^r_{*,*}(X,Y;G;Z/_p) \to E^r_{*,*}(X,Y;H;Z/_p).$$

Also the homomorphism, $E^2(j^!)$, from

$$E^2_{*,*}(X,Y;G;Z/_p) = \mathrm{Tor}^{Z/_p[G]}_{*,*}(K_*(X,Y;Z/_p),Z/_p)$$

to $E^2_{*,*}(X,Y;H;Z/_p) \cong \mathrm{Tor}^{Z/_p[G]}_{*,*}(K_*(X,Y;Z/_p) \otimes_{Z/_p[H]} Z/_p[G],Z/_p)$

is that induced by the homomorphism

$$\bar{\mu}: K_*(X,Y;Z/_p) \to K_*(X,Y;Z/_p) \otimes_{Z/_p[H]} Z/_p[G]$$

of §1.

Proof:

We may assume that (X,Y) is a finite CW pair and under this assumption we are permitted to prove the dual assertion. The equivarient K-theory definition of $j_!$ clearly commutes with coboundaries and so induces a homomorphism of spectral sequences.

If $\{EH_n\}$ and $\{EG_n\}$ are respectively H and G-resolutions there is a map of resolutions between them induced by j. The spectral sequence obtained by applying

$K_*(-;Z/_p)$ to

$((^X/_Y) \wedge EG^+)/_H$ has E_2-term

$$\mathrm{Tor}_{*,*}^{Z/_p[G]}(K_*(X,Y;Z/_p) \otimes_{Z/_p[H]} Z/_p[G], Z/_p)$$

and the map of resolutions induces an isomorphism of spectral sequences which on the E_2-term is the "change of rings" isomorphism of §1. Hence, to complete the proof, it suffices to determine $j_!: K_H^*(Z \times (EG_n, EG_{n-1})) \to K_G^*(Z \times (EG_n, EG_{n-1}))$ where Z is a compact G-space.

The identification of the E_1-term is made by use of the relative homeomorphism of G-spaces

$$\phi: Z \times (DG_n, EG_{n-1}) \times G \to Z \times (EG_n, EG_{n-1})$$

given by $\phi(z,d,g) = (z.g, \phi_n(d,g))$, [Sn 1, §2].
This induces an isomorphism

$$K_H^*(Z \times (DG_n, EG_{n-1}) \times G) \cong K_H^*(Z \times (EG_n, EG_{n-1}))$$ where,

in the spaces on the left, G acts trivially on Z and DG_n. It is easy to see in the Milnor resolution that $K^*(DG_n, EG_{n-1})$ is free. Hence for $F = H$ or G we have

$$K_F^*(Z \times (DG_n, EG_{n-1}) \times G) \cong K_F^*(Z \times G) \otimes K^*(DG_n, EG_{n-1}).$$

Thus equation (2.1) implies that $j_!$ is a $K^*(DG_n, EG_{n-1})$-module homomorphism and it suffices to determine

$j_!: K_H^*(Z \times G) \to K_G^*(Z \times G)$ when G acts trivially on the space, Z. Now let $\{g_i H\}$ be the set of left cosets of H in G. By [Se, §2.1] there are isomorphisms

$$K_G^*(Z \times G) \cong K^*(Z) \quad \text{and} \quad K_H^*(Z \times G) \cong K^*(Z \times {}^G/_H)$$

under which $j_!$ becomes the transfer associated with the trivial covering $\tilde{j}: {}^G/_H \to (\text{pt.})$. Now $K^*(Z \times {}^G/_H)$ is the direct sum of $|G:H|$ copies of $K^*(Z)$. Call the i-th of these copies $K^*(Z)_{g_i}$, so $K^*(Z \times {}^G/_H) \cong \bigoplus_{i=1}^{|G:H|} K^*(Z)_{g_i}$ and $\tilde{j}_!: K^*(Z)_{g_i} \to K^*(Z)$ is just the identity. However $\psi: Z \times G \to Z \times G$ given by

$$\psi(z,g) = (z.g,g)$$

is a homeomorphism of G-spaces between the right-factor action and the diagonal action.
Since $K_H^*(W \times G; Z/_p) \cong K^*(W) \otimes_{Z/_p[H]} Z/_p[G]$ is an isomorphism using the diagonal action on $W \times G$ the transfer must be evaluated for this action. This is just

$$K_H^*(Z \times G) \xrightarrow{\psi^*} K_H^*(Z \times G) \xrightarrow{\tilde{j}_!} K^*(Z)$$

which is clearly g_i^* on the summand

$$K_H^*(Z \times g_i H) \cong K^*(Z). \text{ Finally the dual of}$$

$$\oplus\, g_i^*: \oplus\, K_H^*(- \times g_i H; Z/_p) \to K^*(-; Z/_p)$$

is the homomorphism, $\bar{\mu}$.

We now examine the spectral sequence $\{E_r^{**}(X,Y;\pi_2;Z/_2)\}$ in which the transfer homomorphism appears in a different manner, which is described in Proposition 2.4.

Let S_π^n be the n-sphere as a free π_2-space with the antipodal action and let $D_n \subset S_\pi^n$ be the n-disc obtained as the northern hemisphere.
Thus $\{\dots \subset S_\pi^{n-1} \subset D_n \subset S_\pi^n \subset \dots \subset S_\pi^\infty = E\pi_2\}$ is the Milnor π_2-resolution. Put $D_\pi^n = CS_\pi^{n-1}$, with the conewise action.

<u>Proposition 2.3</u>:

Let X be a compact π_2-space. There are isomorphisms $(m > o)$

$$K_{\pi_2}^*(X,Y;Z/_2) \cong K_{\pi_2}^*((X,Y) \times (D_\pi^{2m}, S_\pi^{2m-1}); Z/_2)$$

$$\cong K^*_{\pi_2}((X,Y) \times (E\pi_2, S^{2m-1}_\pi); Z/_2).$$

Proof:

The first isomorphism is multiplication by the Thom class of the representation given by $\mathbb{C}^m$ with the antipodal action, [Se, §3]. The second isomorphism results from the commutative diagram of π_2-maps

$$\begin{array}{ccc} S^{2m-1}_\pi & \subset & S^\infty_\pi \\ {\scriptstyle 1}\downarrow & & \downarrow \\ S^{2m-1}_\pi & \rightarrow & (\text{point}) \end{array} ,$$

the isomorphism $K^*_{\pi_2}(-;Z/_2) \cong K^*_{\pi_2}(- \times E\pi_2; Z/_2)$, remarked in §1, and the five lemma.

Proposition 2.4:

Let $i_1:\{1\} \subset \pi_2$ be the inclusion of the identity. Let X be a compact π_2-space. Under the isomorphism of Proposition 2.3 and the isomorphism

$$K^\alpha_{\pi_2}((X,Y) \times (S^1_\pi, S^0_\pi); Z/_2) \cong K^{\alpha+1}(X,Y;Z/_2)$$

the coboundary

$$\delta: K^\alpha_{\pi_2}((X,Y) \times (S^1_\pi, S^0_\pi); Z/_2) \rightarrow K^{\alpha+1}_{\pi_2}((X,Y) \times (E\pi_2, S^1_\pi); Z/_2)$$

corresponds to the transfer, $(i_1)_!$.

Proof:

The relative homeomorphism

$$\phi: (X,Y) \times (D_1, S^0_\pi) \times \pi_2 \rightarrow (X,Y) \times (S^1_\pi, S^0_\pi)$$

given by $\phi(x,d,g) = (x.g, \psi_1(d,g))$ induces isomorphisms

$$K^{\alpha-1}(X,Y;Z/_2) \otimes K((I,\partial I) \times (D_1, S^0_\pi))$$

$$\cong K^{\alpha-1}((X,Y) \times (I,\partial I) \times (D_1, S^0_\pi); Z/_2)$$

$\cong K^{\alpha}((X,Y) \times (D_1,S^o_\pi);Z/_2)$

$\cong K^{\alpha}_{\pi_2}((X,Y) \times (D_1,S^o_\pi) \times \pi_2;Z/_2)$

$\cong K^{\alpha}_{\pi_2}((X,Y) \times (S^1_\pi,S^o_\pi);Z/_2)$.

Let $w \in K^{\alpha-1}(X,Y;Z/_2)$ be represented by a complex of vector bundles [Se, §3; Sn 3, I §1 & II §1] (W,d_W) over $(X,Y) \times (I^{\alpha-1},\partial I^{\alpha-1}) \times (M,m_o)$ where $M = M_2 = \mathbb{R}P^2$, the Moore space. Let (B,d_B) be a complex, $0 \longrightarrow \mathbb{C} \xrightarrow{d_B} \mathbb{C} \longrightarrow 0$, over $(I,\partial I) \times (D_1,S^o_\pi)$ representing the Bott element in $K((I,\partial I) \times (D_1,S^o_\pi))$. If $(s,t) \in I \times D_1$, choose d_B to satisfy $1_{\mathbb{C}} = d_B(s,o) = d_B(s,1): \mathbb{C} \to \mathbb{C}$, then $w \otimes [B,d_B]$ gives rise to an element of $K^{\alpha}_{\pi_2}((X,Y) \times (S^1_\pi,S^o_\pi);Z/_2)$ represented by a π_2-complex over $(I^{\alpha},\partial I^{\alpha}) \times (X,Y) \times (S^1_\pi,S^o_\pi) \times (\mathbb{R}P^2,m_o)$ of the following form. Let D_1^+ and D_1^- be the northern and southern semi-circles in S^1_π and let $(B,1)$ be the complex $0 \to \mathbb{C} \xrightarrow{1} \mathbb{C} \to 0$ over $(I,\partial I) \times (D_1,S^o_\pi)$. Let $\tau:(X,Y) \to (X,Y)$ be the involution given by the π_2-action. Now form the complex which is

$$(W,d_W) \otimes (B,d_B^+) \oplus \tau^*(W,d_W) \otimes (B,1)$$

over $(I^{\alpha-1},\partial I^{\alpha-1}) \times (X,Y) \times (I,\partial I) \times (D_1^+,S^o_\pi) \times (\mathbb{R}P^2,m_o)$ and

$$(W,d_W) \otimes (B,1) \oplus \tau^*(W,d_W) \otimes (B,d_B^-)$$

over $(I^{\alpha-1},\partial I^{\alpha-1}) \times (X,Y) \times (I,\partial I) \times (D_1^-,S^o_\pi) \times (\mathbb{R}P^2,m_o)$ where d_B^+ and d_B^- are the differential, d_B, running respectively in the positive and negative senses of the parameter $t \in D_1$. The π_2-action on this complex is given in the obvious manner. In $K^{-1}(S^1_\pi,S^o_\pi;Z/_2) \cong K^{-1}(D_1^+,S^o_\pi;Z/_2) \oplus K^{-1}(D_1^-,S^o_\pi;Z/_2)$ the complex

(B,d_B^+) represents the generator, b, of the first factor. Hence the complex described above represents $i_!(w \otimes b)$. If $\bar{\delta}: K^*(S^1_\pi, S^0_\pi; Z/_2) \to K^*(D^2_\pi, S^1_\pi; Z/_2)$ is the coboundary then $0 \neq \bar{\delta}(b) = z$ which is in the image of the epimorphic forgetful map $(i_1)^*: K^*_{\pi_2}(D^2_\pi, S^1_\pi; Z/_2) \to K^*(D^2_\pi, S^1_\pi; Z/_2)$.

Put $i_1^*(a) = z$.

$$\begin{aligned} \text{Hence} \quad \delta((i_1)_!(w \otimes b)) &= (i_1)_!(\delta(w \otimes b)) \\ &= (i_1)_!(w \otimes \bar{\delta}(b)) \\ &= (i_1)_!(w \otimes (i_1)^*(a)) \\ &= (i_1)_!(w).a \in K^{\alpha-1}_{\pi_2}(X,Y;Z/_2). \end{aligned}$$

Finally the formula (2.1) shows that if $y \in R(\pi_2) \otimes Z/_2$ is the element produced from the non-trivial one dimensional complex representation of π_2 then $0 = (i_1)_!(w.(i_1)^*(1+y)) = (i_1)_!(w).(1+y$ so if $a = 1 \otimes p_1 + (1+y) \otimes p_2 \in R(\pi_2) \otimes Z/_2$, $(p_i \in Z/_2)$ then $a.(i_1)_!(w) = p_1.(i_1)_!(w)$. Thus $\delta((i_1)_!(w \otimes b)$ is either $(i_1)_!(w)$ or zero. However the case $(X,Y) = (pt.,\phi)$ will show [Proposition 2.6] that $\delta \neq 0$.

<u>Proposition 2.5</u>:

Let i_1 and (X,Y) be as in Proposition 2.4. Under the isomorphisms of Proposition 2.3 and the isomorphism $K^\alpha_{\pi_2}((X,Y) \times (S^2_\pi, S^1_\pi); Z/_2) \cong K^\alpha(X,Y;Z/_2)$ the restriction $j: K^\alpha_{\pi_2}((X,Y) \times (E\pi_2, S^1_\pi); Z/_2) \to K^\alpha_{\pi_2}((X,Y) \times (S^2_\pi, S^1_\pi); Z/_2)$ corresponds to the homomorphism, $(i_1)^*$.

<u>Proof</u>:

By Proposition 2.3 an arbitrary element of

$K^{\alpha}_{\pi_2}((X,Y) \times (E\pi_2, S^1_\pi); Z/_2)$ is of the form, $w \otimes T$, where $w \in K^{\alpha}_{\pi_2}(X,Y;Z/_2)$ and T corresponds to the Thom class in $K_{\pi_2}(D^2_\pi, S^1_\pi)$. Hence $j(w \otimes T)$ is represented by the restriction of $w \otimes T$ to $(X,Y) \times (D_2, S^1_\pi)$ which is just the product of $i_1^*(w)$ and the Bott element, $i_1^*(T) \in K(D_2, S^1_\pi)$.

We now determine the spectral sequences $E^r_{*,*}((\text{point},\phi);G;Z/_p)$ and $E_r^{*,*}((\text{point},\phi);G;Z/_p)$ when $G = \pi_p$ or Σ_p. Let

$$\begin{array}{ccc} \{1\} & \xrightarrow{i_1} & \pi_p \\ & {\scriptstyle i}\searrow \quad \swarrow{\scriptstyle i_2} & \\ & \Sigma_p & \end{array}$$

be the commutative diagram of canonical inclusions. Let $y \in R(\pi_p)$ denote the class of the one-dimensional complex representation of π_p whose character is $e^{i2\pi/p}$ on the canonical p-cycle. Thus $R(\pi_p) = Z[y]/_{(y^p-1)}$. Denote by σ the element $(1-y) \in R(\pi_p)$. Hence we have $0 = \sigma^p \in R(\pi_p) \otimes Z/_p$ and $\{1,\sigma,\ldots,\sigma^{p-1}\}$ form a $Z/_p$-basis for $R(\pi_p) \otimes Z/_p$. From [Sn 2] we know that the spectral sequence obtained by applying integral K^*-theory to the $B\pi_p$-filtration collapses and whose only non-zero groups are

$$E_2^{2q,o} = \operatorname{Cotor}^{2q,o}_{K^*(\pi_p)}(Z,Z) = \begin{cases} Z & , q = o, \\ Z/_p & , q > o. \end{cases}$$

From this discussion and Theorem 1.4 it is simple to deduce that the only possible behaviour for the $K^*(-;Z/_p)$-spectral sequence is the following one.

<u>Proposition 2.6:</u>

(a) (i) $E_2^{q,t}((\text{point},\phi);\pi_p;Z/_p) = \begin{cases} Z/_p & , t \equiv o \pmod 2, \\ 0 & , t \equiv 1 \pmod 2. \end{cases}$

(ii) $E_2^{2q,o}((\text{point},\phi);\pi_p;Z/_p)$ consists of permanent cycles.

(iii) Any element of the form

$$\sigma^q + \sum_{j>q} \sigma^j \otimes a_j \in R(\pi_p) \otimes Z/_p$$

is represented by the canonical generator of $E_2^{2q,o} \cong E_\infty^{2q,o}$.

(iv) The only non-zero differential is

$$d_{2p-1}: E_2^{2q+1,o} \cong E_{2p-1}^{2q+1,o} \xrightarrow{\cong} E_{2p-1}^{2(q+p),o} \cong E_2^{2(q+p),o}.$$

Dually

(b)(i) $E^2_{q,t}((\text{point},\phi);\pi_p;Z/_p) = \begin{cases} Z/_p, & t \equiv o \pmod 2 \\ 0, & t \equiv 1 \pmod 2 \end{cases}$.

(ii) If j is odd or $j < 2p$, $E^2_{j,o}((\text{point},\phi);\pi_p;Z/_p)$ consists of permanent cycles.

(iii) The only non-zero differential is

$$d^{2p-1}: E^2_{2(q+p),o} \cong E^{2p-1}_{2(q+p),o} \xrightarrow{\cong} E^{2p-1}_{2q+1,o} \cong E^2_{2q+1,o}.$$

<u>Remark 2.7</u>:

For π_2 Proposition 2.6 also follows from Propositions 2.4, 2.5 since $(i_1)_!(1) = \sigma \in R(\pi_2) \otimes Z/_2$.

<u>Proposition 2.8</u>:

(a) $\text{Cotor}^{2(p-1),o}_{K^*(\Sigma_p;Z/_p)}(Z/_p,Z/_p) \cong Z/_p \cong \text{Tor}^{Z/_p[\Sigma_p]}_{2(p-1),o}(Z/_p,Z/_p)$.

(b)(i) $E_2^{q,t}((\text{point},\phi);\Sigma_p;Z/_p) = \begin{cases} Z/_p, & q = 2k(p-1)-1, 2k(p-1) \text{ and } t \equiv o \pmod 2, \\ 0 & \text{otherwise.} \end{cases}$

(ii) $E_2^{2(p-1),o}((\text{point},\phi);\Sigma_p;Z/_p)$ consists of permanent cycles.

(iii) The element $i_!(1) \in R(\Sigma_p) \otimes Z/_p$ is represented by a generator of $E_2^{2(p-1),o}$.

(iv) The only non-zero differential is

$$d_{2p-1}: E_2^{2k(p-1)-1,o} \cong E_{2p-1}^{2k(p-1)-1,o} \xrightarrow{\cong} E_{2p-1}^{2(k+1)(p-1),o} .$$

Dually

(c) (i) $E^2_{q,t}((\text{point},\phi);\Sigma_p,Z/_p) = \begin{cases} Z/_p, q = 2k(p-1)-1, 2k(p-1) \\ \quad \text{and} \quad t \equiv o(\text{mod } 2), \\ o \quad \text{otherwise.} \end{cases}$

(ii) $E^2_{2(p-1),o}((\text{point},\phi);\Sigma_p;Z/_p)$ consists of permanent cycles.

(iii) The only non-zero differential is

$$d^{2p-1}: E^2_{2(k+1)(p-1),o} = E^{2p-1}_{2(k+1)(p-1),o} \xrightarrow{\cong} E^{2p-1}_{2k(p-1)-1,o} .$$

Proof:

(a) and (b)(iii):

Let $N(\pi_p)$ be the normaliser of π_p in Σ_p. $N(\pi_p)$ acts transitively on $\{g \in \pi_p | g \neq 1\}$ and on $R(\pi_p) \otimes Z/_p$. However $N(\pi_p)$, acting as inner automorphisms, acts trivially on the image of the restriction, $i_2^*:R(\Sigma_p) \otimes Z/_p \to R(\pi_p) \otimes Z/_p$. Now $0 \neq (i_1)_!(1) \in R(\pi_p)$ is $\sum_{j=o}^{p-1} y^j$ and is characterised by the fact that its character is zero on π_p except at the identity. Hence $(i_1)_!(1)$ is invariant under the action of $N(\pi_p)$. However $(1 - y^j)^k = (1 - y)^k(1 + y + y^2 + \ldots + y^{j-1})^k$

$$= j^k \sigma^k + \sum_{q>k} a_q.\sigma^q \in R(\pi_p) \otimes Z/_p$$

and $(1 - y^j)^k$ is the image of σ^k under the action of an element of $N(\pi_p)$. Thus the only $N(\pi_p)$-invariant elements in $R(\pi_p) \otimes Z/_p$ are multiples of σ^{p-1}. Hence for some $o \neq a \in Z/_p$ $(i_1)_!(1) = a.\sigma^{p-1}$ and by Propositions 2.2 and 2.6(a)(iii)

$0 \neq (i_2)_!(i_1)_!(1) = i_!(1) \in R(\Sigma_p) \otimes Z/_p$ is represented by a non-zero element in $\mathrm{Cotor}^{2(p-1),o}_{K^*(\Sigma_p;Z/_p)}(Z/_p,Z/_p)$.

(b)(ii),(iv) and (c):

These parts now follow from Propositions 1.7, 2.2 and 2.6.

Let $i:H \subset G$ and $\mathrm{Tor}^{Z/_p[G]}(\bar{\mu}\circ\mu, Z/_p)$ be the homomorphisms of §1. For $F = H$ or G let $B(F) \to Z/_p$ be the left bar resolution for $Z/_p[F]$. Let $\{g_i;\ 1 \leq i \leq |G:H|\}$ be the set of left coset representatives of H in G used in the definition of $\bar{\mu}$. Let

$$h_{-1}: Z/_p[G] \otimes_{Z/_p[H]} Z/_p \to Z/_p$$

be the left $Z/_p[H]$-module homomorphism defined by

$h_{-1}(g_i^{-1} \otimes 1) = 1 \in Z/_p$, $(1 \leq i \leq |G:H|)$.

We will need the following simple description of $\mathrm{Tor}^{Z/_p[G]}(\bar{\mu}\circ\mu, Z/_p)$.

Proposition 2.9:

Let $C \xrightarrow{\varepsilon} Z/_p$ be a free, left $Z/_p[H]$-module resolution of $Z/_p$. Let $\{h_n: Z/_p[G] \otimes_{Z/_p[H]} C_n \to C_n;\ n \geq o\}$ be a chain morphism of left $Z/_p[H]$-module homomorphisms such that $h_{-1}\circ(1 \otimes \varepsilon) = \varepsilon\circ(h_o)$. The homology of the complex $M \otimes_{Z/_p[H]} C$ is $\mathrm{Tor}^{Z/_p[H]}(M,Z/_p)$ and in terms of this complex $\mathrm{Tor}^{Z/_p[G]}(\bar{\mu}\circ\mu, Z/_p)$ is induced by

$$\tilde{\mu}: M \otimes_{Z/_p[H]} C \to M \otimes_{Z/_p[H]} C,$$

$$\tilde{\mu}(m \otimes c_n) = \sum_{i=1}^{|G:H|} m.g_i \otimes h_n(g_i^{-1} \otimes c_n),$$

$$(m \in M,\ c_n \in C_n).$$

<u>Proof</u>:

It suffices to prove this for $C = B(H)$. In terms of $\mathrm{Tor}^{Z/_p[G]}(M \otimes_{Z/_p[H]} Z/_p[G], Z/_p)$, which is the homology of $M \otimes_{Z/_p[H]} Z/_p[G] \otimes_{Z/_p[G]} B(G) \cong M \otimes_{Z/_p[H]} B(G)$, the homomorphism is induced by

$$\mu_1 : M \otimes_{Z/_p[H]} B(G) \to M \otimes_{Z/_p[H]} B(G) ,$$

$\mu_1(m \otimes x) = \sum_i m.g_i \otimes g_i^{-1}.x$. However to find the appropriate chain map on $M \otimes_{Z/_p[H]} B(H)$ we observe that there is a commutative diagram

$$\begin{array}{ccc} M \otimes_{Z/_p[H]} Z/_p[G] \otimes_{Z/_p[H]} B(H) & \xrightarrow{1 \otimes 1 \otimes B(i)} & M \otimes_{Z/_p[H]} B(G) \\ \downarrow{\scriptstyle \mu_2} & & \downarrow{\scriptstyle \mu_1} \\ M \otimes_{Z/_p[H]} Z/_p[G] \otimes_{Z/_p[H]} B(H) & \xrightarrow[1 \otimes 1 \otimes B(i)]{} & M \otimes_{Z/_p[H]} B(G) \end{array}$$

where $\mu_2(m \otimes g \otimes y) = \sum_i m.g_i \otimes g_i^{-1}.g \otimes y$. Now $1 \otimes B(i)$ is a $Z/_p[H]$-chain homomorphism covering h_{-1}, so if $\bar{h}: B(G) \to B(H)$ is a $Z/_p[H]$-homomorphism covering the identity on $Z/_p$ then $\bar{h} \circ (1 \otimes B(i))$ is chain homotopic to h. If $j: M \to M \otimes_{Z/_p[H]} Z/_p[G]$ is given by $j(m) = m \otimes 1$ the required homomorphism on $M \otimes_{Z/_p[H]} B(H)$ is

$$(1 \otimes \bar{h}) \circ \mu_1 \circ (1 \otimes 1 \otimes B(i)) \circ (j \otimes 1) \simeq (1 \otimes h) \circ \mu_2 \circ (j \otimes 1).$$

Now we apply Proposition 2.9 to some calculations for $i: \pi_2 \wr \pi_2 \subset \Sigma_4$ and $p = 2$. Let $D \xrightarrow{\varepsilon} Z/_2$ be the free, left $Z/_2[\pi_2]$-module resolution of §1.5, which is in fact $B(\pi_2)$.

Let $(D\int D, d\int d)$ be the complex of left $Z/_2[\pi_2\int\pi_2]$-modules given by $(D\int D)_q = \bigoplus_{j+k+\ell=q} D_j \otimes D_k \otimes D_\ell$ with $d\int d = d\otimes 1\otimes 1 + 1\otimes d\otimes 1 + 1\otimes 1\otimes d$ where d is the differential, $d:D \to D$, of §1.5. If π_2 acts on $D \otimes D$ via the switching map the left $(\pi_2\int\pi_2)$-action on $D\int D$ is

$[g_o \cdot (g_1,g_2)] \cdot [a_o \otimes a_1 \otimes a_2] = g_o(a_o) \otimes g_o(g_1(a_1) \otimes g_2(a_2))$,

$(g_t \in \pi_2,\ a_t \in D,\ 0 \leq t \leq 2)$. Let $\varepsilon\int\varepsilon = \varepsilon\otimes\varepsilon\otimes\varepsilon : D_o\otimes D_o\otimes D_o \to Z/_2$ and write $D_i \otimes D_j \otimes D_k = D_i\int D_j \otimes D_k$,

$$e_i \otimes e_j \otimes e_k = e_i\int e_j \otimes e_k$$

where $e_i \in D_i$ is the canonical basis element. The complex

(2.10) $\ldots \to (D\int D)_1 \xrightarrow{d\int d} (D\int D)_o \xrightarrow{\varepsilon\int\varepsilon} Z/_2 \to 0$ is a free, left $Z/_2[\pi_2\int\pi_2]$-resolution of the trivial module, $Z/_2$, since it is the complex realised by applying $K_*(-;Z/_2)$ to the wreath product of two Milnor π_2-resolutions. If $\tau = (13)(24)$, $\tau_1 = (1,2)$, $\tau_2 = (3,4)$ and $\nu = (123)$ are generators of Σ_4 then $\pi_2\int\pi_2$ is generated by τ, τ_1 and τ_2. These following relations hold

(2.11)
- (i) $\tau_1 \cdot \tau_2 = \tau_2 \cdot \tau_1,\ \tau\tau_1 = \tau_2 \cdot \tau,\ \tau^2 = \tau_1^2 = \tau_2^2 = 1 = \nu^3$.
- (ii) $\nu^2 \cdot \tau_1 = \tau_1 \cdot \nu\ ,\ \nu \cdot \tau_1 = \tau_1 \cdot \nu^2$.
- (iii) $\nu \cdot \tau_2 = \tau \cdot \tau_1 \cdot \nu^2\ ,\ \nu^2 \cdot \tau_2 = \tau \cdot \tau_2 \cdot \nu$.
- (iv) $\nu \cdot \tau = \tau \cdot \tau_1 \cdot \tau_2 \cdot \nu\ ,\ \nu^2 \cdot \tau = \tau_1 \cdot \tau_2 \cdot \nu^2$.

Proposition 2.12:

For the $Z/_2[\pi_2\int\pi_2]$-resolution $D\int D \to Z/_2$ the homomorphism $h: Z/_2[\Sigma_4] \otimes_{Z/_2[\pi_2\int\pi_2]} D\int D \to D\int D$ of Proposition 2.9 may be chosen to satisfy the following conditions.

(a) $h|D\int D$ = identity,

$h_o(\nu\otimes e_o\int e_o\otimes e_o) = h(\nu^2\otimes e_o\int e_o\otimes e_o) = e_o\int e_o\otimes e_o.$

(b) $h_1(\nu\otimes e_o\int e_1\otimes e_o) = h_1(\nu^2\otimes e_o\int e_1\otimes e_o) = e_o\int e_1\otimes e_o.$

(c)(i) $h_1(\nu^2\otimes e_1\int e_o\otimes e_o) = \tau_2.[e_o\int e_1\otimes e_o + e_o\int e_o\otimes e_1]$

(ii) $h_1(\nu\otimes e_1\int e_o\otimes e_o) = \tau_2.[e_o\int e_1\otimes e_o + e_o\int e_o\otimes e_1] + \tau_1\tau_2(e_1\int e_o\otimes e_o)$

(d)(i) $h_1(\nu^2\otimes e_o\int e_o\otimes e_1) = \tau.[e_1\int e_o\otimes e_o + e_o\int e_o\otimes e_1]$

(ii) $h_1(\nu\otimes e_o\int e_o\otimes e_1) = \tau.[e_1\int e_o\otimes e_o + e_o\int e_1\otimes e_o].$

(e) $h_2(\nu\otimes e_o\int e_2\otimes e_o) = h_2(\nu^2\otimes e_o\int e_2\otimes e_o) = e_o\int e_2\otimes e_o.$

(f)(i) $h_2(\nu^2\otimes e_o\int e_o\otimes e_2) = e_o\int e_2\otimes e_o + \tau[e_1\int e_o\otimes e_1] + \tau_1[e_2\int e_o\otimes e_o]$

(ii) $h_2(\nu\otimes e_o\int e_o\otimes e_2) = e_o\int e_o\otimes e_2 + \tau[e_1\int e_1\otimes e_o] + \tau_2[e_2\int e_o\otimes e_o].$

(g)(i) $h_2(\nu^2\otimes e_1\int e_1\otimes e_o) = \tau_2[e_o\int e_1\otimes e_1] + \tau_1\tau_2[e_2\int e_o\otimes e_o + e_1\int e_1\otimes e_o]$

(ii) $h_2(\nu\otimes e_1\int e_1\otimes e_o) = \tau_2[e_o\int e_1\otimes e_1].$

(h)(i) $h_2(\nu^2\otimes e_o\int e_1\otimes e_1) = \tau[e_1\int e_o\otimes e_1]$

(ii) $h_2(\nu\otimes e_o\int e_1\otimes e_1) = \tau[e_1\int e_o\otimes e_1 + e_o\int e_o\otimes e_2 + e_o\int e_2\otimes e_o].$

(i)(i) $h_2(\nu^2\otimes e_2\int e_o\otimes e_o) = e_o\int e_2\otimes e_o + e_o\int e_o\otimes e_2 + e_o\int e_1\otimes e_1$

(ii) $h_2(\nu\otimes e_2\int e_o\otimes e_o) = \tau_2[e_1\int e_o\otimes e_1 + e_1\int e_1\otimes e_o] + e_2\int e_o\otimes e_o.$

(j)(i) $h_2(\nu^2\otimes e_1\int e_o\otimes e_1) = \tau[e_1\int e_1\otimes e_o] + \tau_2[e_o\int e_2\otimes e_o] + e_o\int e_o\otimes e_2$

(ii) $h_2(\nu\otimes e_1\int e_o\otimes e_1) = e_2\int e_o{}^2 + \tau_2[e_1\int e_o\otimes e_1 + e_1\int e_1\otimes e_o]$

$+ \tau\tau_1[e_o\int e_o\otimes e_2] + \tau[e_o\int e_2\otimes e_o].$

Proof:

(a) Clearly $h|D\int D$ = identity is a chain map, and h_o covers the correct $h_{-1}: Z/_2[\Sigma_4] \otimes_{Z/_2[\pi_2\int\pi_2]} Z/_2 \to Z/_2$. Noticing that $\tau.e_i\int e_j\otimes e_k = \tau[e_i\int e_k\otimes e_j]$ one computes that the following expressions are the common values of $(d\int d)h(x)$ and $h(d\int d)(x)$.

Case	$(d/d)h(x) = h(d/d)(x)$
(b)	$(1+\tau_1)[e_o/e_o\otimes e_o]$.
(c)(i)	$(1+\tau_1\tau_2)[e_o/e_o\otimes e_o]$.
c(ii)	$(1+\tau_1\tau_2\tau)[e_o/e_o\otimes e_o]$.
d(i)	$(1+\tau_1\tau_2)[e_o/e_o\otimes e_o]$.
d(ii)	$(1+\tau\tau_1)[e_o/e_o\otimes e_o]$.
(e)	$(1+\tau_1)[e_o/e_1\otimes e_o]$.
f(i)	$\tau_1[e_o/e_1\otimes e_o] + \tau[e_o/e_o\otimes e_1] + (\tau+\tau_1)[e_1/e_o\otimes e_o]$.
f(ii)	$\tau_2[e_o/e_o\otimes e_1] + \tau[e_o/e_1\otimes e_o] + (\tau+\tau_2)[e_1/e_o\otimes e_o]$.
g(i)	$(1+\tau_2+\tau_1\tau_2)[e_o/e_1\otimes e_o] + (\tau_1\tau_2\tau+\tau_2)[e_1/e_o\otimes e_o]$ $+ (\tau_2+\tau_1\tau_2 + \tau_1\tau_2\tau)[e_o/e_o\otimes e_1]$.
g(ii)	$(\tau_2 + \tau_1\tau_2)[e_o/e_o\otimes e_1] + (1+\tau_2)[e_o/e_1\otimes e_o]$.
h(i)	$(\tau + \tau\tau_2)[e_1/e_o\otimes e_o] + \tau[e_o/e_o\otimes e_1] + e_o/e_1\otimes e_o$.
h(ii)	$(\tau + \tau\tau_2)[e_1/e_o\otimes e_o] + (1+\tau+\tau\tau_1)[e_o/e_1\otimes e_o]$ $+ \tau\tau_2[e_o/e_o\otimes e_1]$.
i(i)	$(\tau_1 + \tau_2)[e_o/e_o\otimes e_1 + e_o/e_1\otimes e_o]$.
i(ii)	$(\tau_2 + \tau\tau_1)[e_o/e_o\otimes e_1 + e_o/e_1\otimes e_o]$.
j(i)	$(\tau + \tau\tau_1)[e_1/e_o\otimes e_o] + \tau_2[e_o/e_o\otimes e_1]$ $+ (\tau_2 + \tau + \tau_1\tau_2)[e_o/e_1\otimes e_o]$.
j(ii)	$(\tau + \tau_1\tau_2)[e_1/e_o\otimes e_o] + (\tau + \tau_2)[e_o/e_1\otimes e_o]$ $+ (\tau_2 + \tau\tau_1\tau_2)[e_o/e_o\otimes e_1]$.

Proposition 2.13:

Let $w \in K_\alpha(U,V;Z/_2)$ and put $\beta = \beta_2 w$.

In $K_*(U,V;Z/_2)^{\otimes 4} \otimes_{Z/_2[\pi_2/\pi_2]} D/D$ the elements

$$z_1 = (w \otimes \beta)^{\otimes 2} \otimes e_1 \int e_o \otimes e_o + (w^{\otimes^2}) \otimes (\beta^{\otimes^2}) \otimes e_o \int e_1 \otimes e_o$$

and

$$z_2 = w^{\otimes^4} \otimes e_2 \int e_o \otimes e_o + [w\otimes\beta\otimes w\otimes\beta + \beta\otimes w\otimes\beta\otimes w] \otimes e_1 \int e_1 \otimes e_o$$

$$+ (w\otimes\beta)^{\otimes^2} \otimes e_o \int e_1 \otimes e_1 + [w\otimes\beta\otimes\beta\otimes w] \otimes e_o \int e_2 \otimes e_o$$

$$+ (\beta\otimes w)^{\otimes^2} \otimes e_o \int e_o \otimes e_2$$

satisfy the following conditions.

(i) $(1 \otimes d\int d)(z_i) = 0$, $1 \leq i \leq 2$.

(ii) The homology class of z_1 is in the kernel of $\mathrm{Tor}^{Z/_2[\Sigma_4]}(\bar{\mu}\circ\mu, Z/_2)$.

Proof:

Recall that the Bockstein on $K_*(U,V;Z/_2)^{\otimes^2}$ is $(\beta_2 \otimes 1 + 1 \otimes \beta_2)$.

(i) This is straightforward.

(ii) This follows from Propositions 2.9 and 2.12 by direct computation.

Remark 2.14:

The result of Proposition 2.13(ii) will later be used to show the decomposability of the operation $Q(-.\beta_2(-))$ on $K_*(-;Z/_2)$ of an infinite loopspace. One may obtain the decomposability of $Q((-)^2)$ on $K_*(-; Z/_2)$ by evaluating $\mathrm{Tor}(\bar{\mu}\circ\mu, Z/_2)(z_2)$, using Propositions 2.9 and 2.12. However we will obtain this decomposability result otherwise, as a corollary of which one will see that $\mathrm{Tor}(\bar{\mu}\circ\mu, Z/_2)(z_2)$ is at least a permanent cycle in the $\{E^r((U,V)^4; \pi_2\int\pi_2; Z/_2\}$ spectral sequence which does not survive.

§3:

We now determine the bahaviour of the spectral sequences $\{E_r((U,V)^P;G;Z/_p)\}$ and $\{E^r((U,V)^P;G;Z/_p)\}$ when $G = \pi_p$ or Σ_p. It suffices to assume U compact and to determine the $\{E_r((U,V)^P;G;Z/_p)\}$ spectral sequences. We use the notation of Proposition 1.7.

Proposition 3.1:

Let p be a prime and $G = \pi_p$ or Σ_p. For $1 \leq i \leq p$ let $x_i \in K^{\alpha_i}(U,V;Z/_p)$. The element

$$\sum_{g\in G} g^*(x_1 \otimes x_2 \otimes \ldots \otimes x_p) \otimes e_o \in \mathrm{Cotor}^{o,\alpha}_{K^*(G;Z/_p)}(K^*(U,V;Z/_p)^{\otimes P}, Z/_p)$$

$(\alpha \equiv \sum_i \alpha_i \pmod 2)$ is an infinite cycle representing the image of $x_1 \otimes x_2 \otimes \ldots \otimes x_p$ under $j_!:K^*((U,V);Z/_p)^{\otimes P} \to K^*_G((U,V)^P;Z/_p)$.

Proof:

Apply Proposition 2.2 to the transfer between the spectral sequences $\{E_r((U,V)^P;1;Z/_p)\}$ and $\{E_r((U,V)^P;G;Z/_p)\}$.

Proposition 3.1 leaves in doubt only the question of which elements of the form $x^{\otimes P} \otimes e_j \in E_2^{j,*}$ (j mod 2(p-1)) are permanent cycles. For, as in [Sn 2,§1.1], the elements of §3.1 and the $x^{\otimes P} \otimes e_o$ (deg $x \equiv 0 \pmod 2$ for Σ_p) generate $E_2^{o,*}$. Once these permanent cycles are determined the module structure of the spectral sequence for $(U,V)^P$ over that for (point, ϕ) [c.f.§1] and Proposition 2.6 will determine the spectral sequence.

Proposition 3.2: Let p be an odd prime and $x \in K^{\alpha}(U,V;Z/_p)$. The element $x^{\otimes P} \otimes e_j \in E_2^{j,\alpha}((U,V)^P;\Sigma_p;Z/_p)$ of Proposition 1.7(c) is a permanent cycle when j is even. Equivalently the element $x^{\otimes P} \otimes e_o \in E_2^{o,\alpha}((U,V)^P;\pi_p;Z/_p)$ is a permanent cycle.

Proof:

(i) If the appropriate $x^{\otimes p} \otimes e_{2k}$ is a permanent cycle in the Σ_p-spectral sequence its image under i_2^*, $x^{\otimes p} \otimes e_{2k}$, is a permanent cycle in the π_p-spectral sequence. However the module structure over the π_p-spectral sequence for a point shows that $x^{\otimes p} \otimes e_{2j}$ is a permanent cycle for all $j \geq 0$.

(ii) If $x^{\otimes p} \otimes e_o$ is a permanent cycle in the π_p-spectral sequence the module structure implies that $x^{\otimes p} \otimes e_{2j}$ $(j \geq 0)$ is a permanent cycle. However, by §2, the transfer induces a map of spectral sequences and shows that the Σ_p-spectral sequence embeds as a direct summand of the π_p-spectral sequence. Hence the appropriate $x^{\otimes p} \otimes e_{2j}$ is a permanent cycle in the Σ_p-spectral sequence.

(iii) We give two proofs of the statement about permanent cycles of the form $x^{\otimes p} \otimes e_{2j}$. The proof which takes place in the π_p-spectral sequence is to be found in Appendix I, where the more extensive Proposition 3.3 below is proved. Here we sketch a proof which takes place in the Σ_p-spectral sequence. The two proofs are given because they emphasise different ways in which the odd prime case differs from the case $p = 2$.

The following proof uses the computation of $\mathrm{Cotor}_{K^*}(\Sigma_p;Z/_p)$ given in Proposition 1.7(c). However that proposition was proved using the knowledge of $\{E_s^{**}((S^1)^p;\Sigma_p;Z/_p)\}$. Hence in order to avoid a circular argument either one uses Appendix I and treats the following sketch as one for amusement only or one imitates the geometric proof of Proposition 1.7(c) purely algebraically.

From Proposition 1.7(c) and degree considerations the only possibly non-zero differentials on the elements $x^{\otimes p} \otimes e_{2j}$ in question are d_{p-1} and d_{2p-3} for if these are both zero the module structure and Propositions 2.6 and 3.1 imply that after the

E_{2p}-term the differentials must be zero because there is nothing left for $x^{\otimes p} \otimes e_o$ (deg $x \equiv 0 (\mathrm{mod}\ 2)$) or $x^{\otimes p} \otimes e_{p-1}$ (deg $x \equiv 1 (\mathrm{mod}\ 2$ to hit.

Let $b \in K(D^2,S^1)$ be the Bott element and let $b^{\boxtimes p} \in K^o_{\Sigma_p}((D^2,S^1)^p)$ be the external p-th power of b [At 3;Sn 2]. In $\{E_s^{*,*}((D^2,S^1)^p;\Sigma_p;Z/_p)\}$ $b^{\otimes p} \otimes e_o$ is a permanent cycle representing $b^{\boxtimes p}$. Hence in $\{E_s^{*,*}((((U,V) \times (D^2,S^1))^p;\Sigma_p;Z/_p)\}$ $d_k((x.b)^{\otimes p} \otimes e_j) = d_k(x^{\otimes p} \otimes e_j).(b^{\otimes p} \otimes e_o)$, from the multiplicative structure. The stability of d_{p-1} or d_{2p-3} under multiplication by b implies that d_{p-1} induces an additive stable operation $\phi_1: \tilde{K}^\alpha(-;Z/_p) \to \tilde{K}^{\alpha+1}(-;Z/_p)$ using the natural isomorphisms of Proposition 1.7. Here an operation is called stable if it commutes with the Bott periodicity isomorphism and hence is not necessarily stable in the sense of [M]. However, from the computations of [M] we have the following result. For each $\alpha \in Z/_2$, $\phi_1: \tilde{K}^\alpha(-;Z/_p) \to \tilde{K}^{\alpha-1}(-;Z/_p)$ is a linear combination of the $\{\psi^i \circ \beta_p\}$, where ψ^i is the Adams operation. From the results of [M] it suffices to show that ϕ_1 vanishes on all spaces of the form $M \wedge S^N \wedge CP^n$ (where M is the Moore space, M_p) in order to show $\phi_1 = 0$. However the derivation properties of d_{p-1} and the results of [Sn 2] on permanent cycles in these spectral sequences for torsion free spaces like $S^N \wedge CP^n$ show that $\phi_1 = \lambda.\beta_p$ for some $\lambda \in Z/_p$ and is determined by its behaviour on $K^*(M,m_o;Z/_p)$. Finally using the results of [D-L] on $H^*((M,m_o)^p \times_{\Sigma_p} E\Sigma_p;Z/_p)$, applying the Atiyah-Hirzebruch spectral sequence and inspecting the first non-zero differential shows that $\lambda = 0$.

Now consider $\phi_o: \tilde{K}^\alpha(-;Z/_p) \to \tilde{K}^\alpha(-;Z/_p)$, the additive stable operation constructed using d_{2p-3}. From [M] this is a linear combination of the $\{\psi^i\}$ and it suffices to consider the effect on spaces $S^N \wedge CP^n$. Hence the results of [Sn 2] show that $\phi_o = 0$.

Let S^1_π be the circle with π_p-action given by the generator of π_p actine as the rotation, $(-.e^{2\pi i/_p})$. Let S^{2t-1}_π be the join of t copies of S^1_π, then

$$K^\alpha(S^{2t-1}_\pi;Z/_p) \cong Z/_p, \quad (\alpha \equiv 0 \text{ or } 1 \pmod 2),$$

with generators $\iota \in K^o(S^{2t-1}_\pi;Z/_p)$ and $\xi \in K^1(S^{2t-1}_\pi;Z/_p)$, $(t \geq 1)$. The following result is proved in Appendix §A.1.

<u>Proposition 3.3:</u>

Let p be an odd prime. Let $x \in K^\alpha(U,V;Z/_p)$.

$(i)_t$ There is $z(x) \in K^{\alpha+1}(U,V;Z/_p)$ such that $[\xi \otimes z(x)^{\otimes p} + \iota \otimes x^{\otimes p}] \otimes e_o$ is a permanent cycle in $E_2^{o,\alpha}(S^{2t+1}_\pi \times (U,V)^p;\ \pi_p;Z/_p)$.

$(ii)_t$ $(\iota \otimes x^{\otimes p}) \otimes e_o \in E_2^{o,\alpha}(S^{2t+1}_\pi \times (U,V)^p;\pi_p;Z/_p)$ is a permanent cycle.

(iii) $x^{\otimes p} \otimes e_o \in E_2^{o,\alpha}((U,V)^p;\pi_p;Z/_p)$.

The problem of determining the permanent cycles in $\{E_r((U,V)^2;\pi_2;Z/_2)\}$, $(r \geq 2)$, is much more delicate than that for odd primes and requires a detailed analysis which will largely be relegated to Appendix II. However, in order to state the results, we first recall the definition of the triple

Massey product [Sn 3] and define the quadratic construction in K-theory. These constructions are defined in integral K-theory.

§3.4: Massey triple products.

If $x_i \in K_G(X_i,Y_i)$ are elements satisfying $0 = x_1.x_2 \in K_G((X_1,Y_1) \times (X_2,Y_2))$ and $0 = x_2.x_3 \in K_G((X_2,Y_2) \times (X_3,Y_3))$ the *external Massey triple product* is defined and denoted by

$$< x_1,x_2,x_3 > \quad \in \frac{K_G^{-1}(\prod_{i=1}^{3} (X_i,Y_i))}{\{x_1.K_G^{-1}((X_2,Y_2) \times (X_3,Y_3)) + K_G^{-1}((X_1,Y_1) \times (X_2,Y_2)).x_3\}}.$$

It is constructed by the following Toda bracket construction. Let (E_i,d_i) be a complex of G-vector bundles representing x_i such that there exist homotopies, $H_{i,i+1;s}$ $(i = 1,2)$, of differentials on the underlying family of G-vector bundles of $E_i \otimes E_{i+1}$ through complexes over $(X_i,Y_i) \times (X_{i+1},Y_{i+1})$ to an exact complex and starting with $H_{i,i+1;o} = d_i \otimes d_{i+1}$. An element of $<x_1,x_2,x_3>$ is represented by the complex of G-vector bundles over $(I,\partial I) \times \prod_{i=1}^{3} (X_i,Y_i)$ whose underlying family of G-vector bundles is that of $I \times (E_1 \otimes E_2 \otimes E_3)$ and whose differential at (t,x_1,x_2,x_3), $(t \in I,\ x_i \in X_i)$, is given by

$$d_1(x_1) \otimes H_{2,3;1-2t}(x_2,x_3) \qquad (0 \leq t \leq \tfrac{1}{2})$$

$$H_{1,2;2t-1}(x_1,x_2) \otimes d_3(x_3) \qquad (\tfrac{1}{2} \leq t \leq 1).$$

For example, if there exists an integer, n, such that

$0 = n.x_1 \in K_G(X_1,Y_1)$ and $0 = n.x_3 \in K_G(X_3,Y_3)$ then $\langle x_1,n,x_3\rangle$ is a subset of $K_G^{-1}((X_1,Y_1) \times (X_3,Y_3))$. Also there exist internal Massey triple products.

We shall use Massey triple products to define mod 2 K-theory classes in $\ker \beta_2$ and also to define their squares under the exotic mod 2 multiplication. On these elements we will also want to perform the quadratic construction.

§3.5: Quadratic construction.

Let (X,Y) be a pair of π_2-spaces and let $x \in K_{\pi_2}(X,Y)$ belong to $\ker i_1^*$. The quadratic construction, quad(x), is an element of $\dfrac{K^{-1}(X,Y)}{\{(1+\tau^*).K^{-1}(X,Y)\}}$ where $1 \neq \tau \in \pi_2$.

It is constructed in the following manner. Let (E,d) be a complex of π_2-vector bundles over (X,Y) and let H_s be a homotopy on the underlying family of vector bundles of E through complexes over (X,Y) from $H_o = d$ to an exact complex. An element of quad(x) is represented by the complex of vector bundles over $(I,\partial I) \times (X,Y)$ whose underlying family of vector bundles is that of $I \times E$ and whose differential at (t,x), $(t \in I, x \in X)$, is given by

$$\begin{cases} H_{1-2t}(x) & (0 \leq t \leq \tfrac{1}{2}) \\ \tau(H_{2t-1}(\tau.x)) & (\tfrac{1}{2} \leq t \leq 1) . \end{cases}$$

The set, quad(x), is the totality of elements representable in this manner. The indeterminacy arising from choices of H_s or (E,d) is contained in $(1+\tau^*).K^{-1}(X,Y)$ [c.f.Sn 3,I §4]. If H_s is an equivariant homotopy then quad(x) contains zero. The quadratic construction is a natural operation

$$\text{quad}(-):\ker(i_1)^* \to \frac{K^*(-)}{\{(1+\tau^*).K^*(-)\}}$$

of odd degree.

Proposition 3.6:

Let $0 \neq a \in K^0(M_2,m_o) \cong Z/_2$. If $w \in K^\alpha(U,V)$ satisfies $4w = 0$ and $x \in \langle 2w,2,a\rangle \subset K^{\alpha-1}(U,V;Z/_2)$ then there exists $w_1 \in \langle 2w,2,2w\rangle$ such that $x^{\otimes 2} \in \langle w_1,2,a\rangle \subset K^0((U,V)^2;Z/_2)$. Also if $w^{\otimes 2} \in K_{\pi_2}((U,V)^2)$ is the external square of w [At 3; Sn 2] then $\langle (5+3y)w^{\otimes 2},2,a\rangle \subset K^{-1}_{\pi_2}((U,V)^2;Z/_2)$ is defined and contains an element $x_1 \in \ker(i_1^*)$ such that $x^{\otimes 2} \in \text{quad}(x_1)$.

Proof:

This will be given using explicit constructions with complexes of π_2-vector bundles. The proof is straightforward but very long and for this reason has been relegated to Appendix II.

Proposition 3.7:

For $\alpha = 0$ or 1 let $x \in \ker \beta_2 \subset K^\alpha(U,V;Z/_2)$. The element $x^{\otimes 2} \otimes e_\alpha \in E_2^{\alpha,0}((U,V)^2;\pi_2;Z/_2)$ is a permanent cycle.

Proof:

If $x \in \ker \beta_2$ is in $\text{im}\{K^\alpha(U,V) \to K^\alpha(U,V;Z/_2)\}$ this result follows from [Sn 2,§2]. The cofibration $M_2 \to S^2 \xrightarrow{(2.-)} S^2$ is a geometric resolution of $\tilde{K}^*(M_2) \cong Z/_2$ in the sense of

[Sn 3,I & II §4]. Also if β is the Bockstein associated with the exact sequence $0 \to Z \to Z \to Z/_2 \to 0$ then $\beta(x) = 2w$ for some $w \in K^{\alpha+1}(U,V)$ satisfying $4w = 0$. The results of [Sn 3,II §3.2] show that $\beta^{-1}(-) = \langle -,2,a\rangle$ so $x \in \langle 2w,2,a\rangle \subset K^*(U,V;Z/_2)$. If $\alpha = -1$ then by Proposition 3.6 $w^{\boxtimes 2} \in K^*_{\pi_2}((U,V)^2)$ satisfies $0 = (4w)^{\boxtimes 2} = (4^{\boxtimes 2})(w^{\boxtimes 2}) = 2[(5+3y)w^{\boxtimes 2}]$ and $\langle (5+3y)w^{\boxtimes 2},2,a\rangle \subset K^{-1}_{\pi_2}((U,V)^2;Z/_2)$ is defined.

However the proof of [Sn 2,§2] shows that if $z \in K^{-1}_{\pi_2}((U,V)^2;Z/_2) \cap \ker i_1^*$ then z may be pulled back and mapped into

$$E_1^{1,*}((U,V)^2;\pi_2;Z/_2) \cong K^*(U,V;Z/_2)^{\otimes 2}$$

(this isomorphism assumes the use of the Milnor π_2-resolution) to give quad(z). Hence by Proposition 3.6 there exists an element of $\langle (5+3y)w^{\boxtimes 2},2,a\rangle$ represented by $x^{\otimes 2} \otimes e_1$.

Now suppose $\alpha = 0$, then there is $w \in K^{-1}(U,V)$ such that $\beta(x) = 2w$, $4w = 0$ and $x \in \langle 2w,2,a\rangle \subset K^{-2}(U,V;Z/_2)$. Let I_π be the unit interval with the involution $\tau(s) = 1-s$. Thus $\langle (5+3y)w^{\boxtimes 2},2,a\rangle \subset K_{\pi_2}((I,\partial I) \times (I,\partial I)^2 \times (U,V)^2;Z/_2)$ where $(I,\partial I)^2$ has the action which interchanges the factors. Hence Proposition 3.6 implies there is an element, $x_1 \in \langle (5+3y)w^{\boxtimes 2},2,a\rangle$ such that $x^{\otimes 2} \in \text{quad}(x_1)$. However the constructions of [Sn 2,§2] show that quad(x_1) is obtained by pulling x_1 back to

$$(3.7.1) \quad K^*_{\pi_2}((I_\pi,\partial I_\pi) \times (I,\partial I) \times (I,\partial I)^2 \times (U,V)^2;Z/_2) \cong K^*_{\pi_2}((U,V)^2;Z/_2)$$

and mapping it into $K^*((U,V)^2;Z/_2) \cong E_1^{2,*}((U,V)^2;\pi_2;Z/_2)$. The isomorphism (3.7.1) is the Thom isomorphism, since

$(I,\partial I)^2 \cong (I_\pi,\partial I_\pi) \times (I,\partial I)$. However Proposition 2.3 implies that

$$K^*_{\pi_2}((I_\pi,\partial I_\pi) \times (I,\partial I) \times (I,\partial I)^2 \times (U,V)^2;Z/_2)$$

$$\cong K^*_{\pi_2}((U,V)^2 \times (E\pi_2,S^1_\pi);Z/_2).$$

Using this isomorphism the map

$$K^*_{\pi_2}((U,V)^2;Z/_2) \to E_1^{2,*}((U,V)^2;\pi_2;Z/_2)$$

is, by [Sn 2,§2], the homomorphism of Proposition 2.5, the forgetful map $(i_1)_*$. Hence $x^{\otimes 2} \otimes e_o$ and $x^{\otimes 2} \otimes e_2$ are permanent cycles.

We may now state the behaviour of the Rothenberg-Steenrod spectral sequences. We use the notation of Proposition 1.7.

<u>Theorem 3.8</u>:

Let p be a prime.

(a) In the spectral sequences $\{E^r((U,V)^P;G;Z/_p);r \geq 2\}$ $(G = \pi_p$ or $\Sigma_p)$ the only differential is d_{2p-1} which is a shift operator of the following form. Let $x \in K_\alpha(U,V;Z/_p)$.

(i) <u>$p \neq 2$</u>

$$d_{2p-1}(x^{\otimes P} \otimes e_{2q}) = x^{\otimes P} \otimes e_{2(q-p)+1}$$

and d_{2p-1} is zero otherwise.

(ii) <u>$p = 2$</u>

$$d_3(x^{\otimes 2} \otimes e_{2q}) = x^{\otimes 2} \otimes e_{2q-3} \quad \text{if} \quad \alpha \equiv 0 (\text{mod } 2),$$

$$d_3(x^{\otimes 2} \otimes e_{2q+1}) = x^{\otimes 2} \otimes e_{2(q-1)} \quad \text{if} \quad \alpha \equiv 1 (\text{mod } 2),$$

$(x \in \ker \beta_2 - \text{im } \beta_2)$ and d_3 is zero otherwise.

Dually

(b) In the spectral sequences $\{E_r((U,V)^P;G;Z/_p);r \geq 2\}$ $(G = \pi_p$ or $\Sigma_p)$ the only differential is d_{2p-1} which is a shift operator of the following form. Let $x \in K^{\alpha}(U,V;Z/_p)$.

(i) $\underline{p \neq 2}$

$d_{2p-1}(x^{\otimes p} \otimes e_{2q+1}) = x^{\otimes p} \otimes e_{2(p+q)}$ and d_{2p-1} is zero otherwise.

(ii) $\underline{p = 2}$

$d_3(x^{\otimes 2} \otimes e_{2q+1}) = x^{\otimes 2} \otimes e_{2(q+2)}$ if $\alpha \equiv 0 \pmod 2$,

$d_3(x^{\otimes 2} \otimes e_{2q}) = x^{\otimes 2} \otimes e_{2q+3}$ if $\alpha \equiv 1 \pmod 2$,

($x \in \ker \beta_2 - \mathrm{im}\, \beta_2$) and d_3 is zero otherwise.

Proof:

(b) The fact that d_{2p-1} is a shift operator follows from Propositions 2.6 and 2.8 and the module structure of $\{E_r((U,V)^p;G;Z/_p)\}$ over $\{E_r(\mathrm{point},\phi;G;Z/_p)\}$, [c.f.§1.7]. The existence of permanent cycles as described in Propositions 3.3 and 3.7 show where and when the shifting starts. After d_{2p-1} $E_r^{q,*} = 0$ for $q \geq 2p$ so there are no further differentials.

Part (a) is dual to (b).

From Theorem 3.8 we obtain the following exact sequences. We present only the $K_*(-;Z/_p)$-versions. Let $J\pi_p = \tilde{K}_*(B\pi_p;Z/_p)$ and let M_G denote the quotient module of coinvariants of the $Z/_p[G]$-module, M.

Theorem 3.9:

There exist natural exact sequences

(i) For $p \neq 2$,

$$0 \to [K_*(U,V;Z/_p)^{\otimes p}]_{\pi_p} \xrightarrow{(i_1)_*} K_*^{\pi_p}((U,V)^p;Z/_p) \xrightarrow{\bar{a}_p} J\pi_p \otimes K_*(U,V;Z/_p) \to 0$$

(ii)

$$0 \to \{[K_*(U,V;Z/_2)^{\otimes 2}]_{\pi_2}\}_{/N} \xrightarrow{i_*} K_*^{\pi_2}((U,V)^2;Z/_p) \xrightarrow{a_2} \ker \beta_2/_{\mathrm{im}\beta_2} \to 0$$

where $\ker \beta_2/_{\mathrm{im}\beta_2} \subset K_*(U,V;Z/_2)/_{\mathrm{im}\beta_2}$

and $N = d_3(E^3_{3,o})$, the submodule generated by elements of the form $x^{\otimes 2}$ for $x \in \ker \beta_2$ and $\deg x \equiv 1 (\mathrm{mod}\ 2)$.

(iii) For $p \neq 2$

$$0 \to [K_*(U,V;Z/_p)^{\otimes p}]_{\Sigma_p} \xrightarrow{i_*} K_*^{\Sigma_p}((U,V)^p;Z/_p) \xrightarrow{a_p} K_*(U,V;Z/_p) \to 0.$$

All the homomorphisms are of degree zero.

<u>Proof</u>:

Parts (ii) and (iii) follow from the identifications of $E^2((U,V)^p;G;Z/_p)$ in Proposition 1.7 and the fact that E^∞ has only two non-zero groups in each total degree. For $p \neq 2$ the non-zero groups are $E^\infty_{o,*} \cong E^2_{o,*} \cong [K_*(U,V;Z/_p)^{\otimes p}]_{\Sigma p}$ and $E^\infty_{2p-2,o} \cong K_o$, $E^\infty_{p-1,1} \cong K_1$. For $p = 2$, the non-zero groups are $E^\infty_{o,*} \cong \{[K_*(U,V;Z/_2)^{\otimes 2}]_{\pi_2}\}_{/N}$,

$$E^\infty_{1,o} \cong \{x + \mathrm{im}\ \beta_2 \in \ker \beta_2/_{\mathrm{im}\ \beta_2} \mid \deg x \equiv 1 (\mathrm{mod}\ 2)\} \subset E^2_{1,o}$$

and $E^\infty_{2,o} \cong \{x + \mathrm{im}\ \beta_2 \in \ker \beta_2/_{\mathrm{im}\ \beta_2} \mid \deg x \equiv o (\mathrm{mod}\ 2)\} \subset E^2_{2,o}$.

Part (i) is proved in [H 1, Theorem 3], by extending the secondary edge homomorphism

$$K_*(U,V;Z/_p) \cong E^2_{2,*} \to K_*^{\pi_p}((U,V)^p;Z/_p)/_{E^\infty_{o,*}}$$

using the $K_*(B\pi_p;Z/_p)$-comodule structure of $K_*^{\pi_p}(-;Z/_p)$.

Let $(M)^\alpha$ denote the part of M of degree α.

<u>Proposition 3.10</u>:

Let U be a compact space. There exist the following natural isomorphisms.

(i) $K^0(U,V;Z/_2) \otimes K^1(U,V;Z/_2) \oplus (\ker\beta_{2}{/_{\mathrm{im}\beta_2}})^1 \cong K^1_{\pi_2}((U,V)^2;Z/_2)$

(ii) $[K^*(U,V;Z/_2)^{\otimes 2}]^0/_{Q_2} \cong \mathrm{im}\, i_! = \ker(\sigma.-) \subset K^0_{\pi_2}((U,V)^2;Z/_2)$

where Q_2 is the submodule generated by elements of the form $\{(1+\tau^*)(x_1 \otimes x_2) \mid \tau \in \pi_2\}$ and $\{x^{\otimes 2} \mid \deg x \equiv 1 (\mathrm{mod}\ 2),\ x \in \ker \beta_2\}$.

(iii) $K^*(U,V;Z/_p)^{\otimes p}/_{Q_p} \cong \mathrm{im}\, i_! \subset K^*_{\Sigma_p}((U,V)^p;Z/_p)$ where Q_p is the submodule generated by elements of the form $\{(1-g^*)(x_1 \otimes \ldots \otimes x_p) \mid g \in \Sigma_p\}$. Also $(\mathrm{im}\, i_!)^1 \oplus K^1(U,V;Z/_p) \cong K^1_{\Sigma_p}((U,V)^P;Z/_p)$.

<u>Proof</u>:

(i) We use the dual of the exact sequence of Theorem 3.9(ii).

Let $a_2^!$ be the dual of a_2. Since $\ker \beta_2{/_{\mathrm{im}\beta_2}}$ in $K_*(-;Z/_2)$ is dual to $\ker \beta_2{/_{\mathrm{im}\,\beta_2}}$ in $K^*(-;Z/_2)$ we obtain a homomorphism

$$q' : K^0(U,V;Z/_2) \otimes K^1(U,V;Z/_2) \oplus (\ker\beta_2{/_{\mathrm{im}\beta_2}})^1 \to K^1_{\pi_2}((U,V)^2;Z/_2)$$

given by $q'(x_o \otimes x_1,\ x+\mathrm{im}\beta_2) = i_!(x_o \otimes x_1) + a_2^!(x+\mathrm{im}\beta_2)$.

For $z \in K^1_{\pi_2}((U,V)^2;Z/_2)$ there exists a unique

$z' \in K^0(U,V;Z/_2) \otimes K^1(U,V;Z/_2)$ such that $i^*(i_!(z')) = i^*(z)$ and then $z + i_!(z') \in \mathrm{im}\, a_2^!$.

(ii) By (2.1) we have $i_!(i^*(z)) = z.\sigma$,

$(\sigma = 1 - y \in R(\pi_2) \otimes Z/_2,\ z \in K^*_{\pi_2}((U,V)^2;Z/_2))$.

Thus if $x \in \ker \beta_2$ then $i_!(x^{\otimes 2}) = a_2^!(x + \mathrm{im}\, \beta_2)$.

Also, by (2.1), $\sigma.i_!(x_1 \otimes x_2) = i_!((x_1 \otimes x_2)i^*(\sigma)) = 0$

so $\operatorname{im} i_! \subset \operatorname{Ker}(\sigma.-)$. Now if $\deg x \equiv 1 (\bmod 2)$ and $x \in \ker \beta_2 - \operatorname{im} \beta_2$ then $x^{\otimes 2} \otimes e_1$ is a permanent cycle and by Proposition 2.4 $0 = \delta(x^{\otimes 2} \otimes e_1)$ implies $i_!(x^{\otimes 2}) = 0$. Hence the factorisation $\tilde{i}_! : [K^*(U,V;Z/_2)^{\otimes 2}]^{o}/\Omega_2 \to \operatorname{Ker}(\sigma.-)$ exists. If $i^*(z) = \Sigma x_i^{\otimes 2} + (1+\tau^*)(w)$, $(x_i \in \ker \beta_2 - \operatorname{im} \beta_2)$ then $\sigma.z = a_2'(\Sigma x_i)$ so $\operatorname{im} i_! = \ker(\sigma.-)$. If $i_!(\Sigma x' \otimes x'') = 0$ then $i^* i_!(\Sigma x' \otimes x'') = 0$, but $i^* i_! = 1 + \tau^*$ so by the analysis used to compute the homology of (1.6) we have $\Sigma x' \otimes x'' = (1+\tau^*)(w) + \Sigma x_i^{\otimes 2}$, $(x_i \in \ker \beta_2 - \operatorname{im} \beta_2)$. Thus $0 = i_!(\Sigma x' \otimes x'') = a_2'(\Sigma_j x_j)$ where the sum is taken over those x_j of degree zero. Hence $\tilde{i}_!$ is monic.

(iii) This is similar to (ii), using Proposition 2.8(iii) to show $i_!(x^{\otimes p})$ is a non-zero multiple of $a_p'(x)$ and noting that $i_!(w) = i_!(g^* w)$, $(g \in \Sigma_p)$. The sum decomposition uses the inclusion of $K^1 \cong E_\infty^{p-1,1}$.

Proposition 3.11:

Under the hypothesis of Proposition 3.10 if $\alpha \not\equiv 1 (\bmod 2)$ the subspace $\operatorname{im} i_! \subset K^\alpha_{\Sigma_p}((U,V)^p;Z/_p)$ is dual to $K_\alpha^{\Sigma_p}((U,V)^p;Z/_p)/\operatorname{Ind}_p(U,V)$ where $\operatorname{Ind}_p(U,V)$ is given by

$$\operatorname{Ind}_p(U,V) = \begin{cases} \text{submodule generated by } \{i_*(x^{\otimes p})\} & \text{if } p \neq 2 \\ \text{submodule generated by } \{i_*(x^{\otimes 2}) \mid \beta_2 x = 0\} & \text{if } p=2. \end{cases}$$

Proof:

(i) $p \neq 2$: Let $\langle\, , \rangle : K_\alpha(-;Z/_p) \otimes K^\alpha(-;Z/_p) \to Z/_p$ be the non-singular pairing. If $z \in K_\alpha^{\Sigma_p}((U,V)^p;Z/_p)$ satisfies $0 = \langle z, i_!(w)\rangle$ for all $w \in K^*(U,V;Z/_p)^{\otimes p}$ we must show $z \in \operatorname{ind}_p(U,V)$. Putting $w = x^{\otimes p}$ we have

$$0 = \langle z, i_!(x^{\otimes p})\rangle$$

$$= \langle z, \lambda\, a_p'(x)\rangle\ , \qquad (\lambda \neq 0 \in Z/_p),$$

$$= \lambda\, \langle a_p(z), x\rangle$$

so $z = i_*(z')$. However

$$0 = \langle z, i_!(x_1 \otimes x_2 \otimes \ldots \otimes x_p)\rangle$$

$$= \langle i_*(z'), i_!(x_1 \otimes \ldots \otimes x_p)\rangle$$

$$= \langle z', i^* i_!(x_1 \otimes \ldots \otimes x_p)\rangle$$

$$= \sum_{g \in \Sigma_p} \langle z', g^*(x_1 \otimes \ldots \otimes x_p)\rangle$$

for all $x_1 \otimes x_2 \otimes \ldots \otimes x_p$ so $z \in \mathrm{Ind}_p(U,V)$, since $\Sigma\, g^*(z') = 0$.

<u>(ii) p = 2:</u>

As in (i) if $0 = \langle z, i_!(w)\rangle$ for all $w \in K^*(U,V;Z/_2)^{\otimes 2}$ then $z = i_*(z')$ and $z' = \Sigma\, z''^{\otimes 2}$. Finally

$$\langle i_*(z''^{\otimes 2}),\ i_!(x_1 \otimes x_2)\rangle$$

$$= \langle z'' \otimes z'',\ i^* i_!(x_1 \otimes x_2)\rangle$$

$$= \langle z'' \otimes z'',\ x_1 \otimes x_2 + x_2 \otimes x_1 + \beta_2 x_2 \otimes \beta_2 x_1\rangle$$

$$= \langle z'',\ \beta_2 x_2\rangle \langle z'',\ \beta_2 x_1\rangle$$

$$= \langle \beta_2 z'',\ x_2\rangle \langle \beta_2 z'',\ x_1\rangle$$

so $z' = \Sigma\, z''^{\otimes 2} \in \mathrm{Ind}_2(U,V)$.

§4.1:

We now use the results of Propositions 3.10 and 3.11 to construct a natural operation

$$q: K_\alpha(U,V;Z/_p) \to K_\alpha^{\Sigma_p}((U,V)^p;Z/_p)_{/\mathrm{Ind}_p(U,V)} .$$

In order to construct q it suffices to assume U is compact. Using the isomorphism $K_*(-;Z/_p) \cong \mathrm{Hom}(K^*(-;Z/_p),Z/_p)$ and the duality result it suffices to define the value of $q(f)$ on $\mathrm{im}\, i_!$ when $(\alpha \equiv 0(\mathrm{mod}\ 2))$ $(f \in K_\alpha(U,V;Z/_2)$.

Define $q(f): K^\alpha_{\Sigma_p}((U,V)^p;Z/_p) \to Z/_p$, $(\alpha \equiv 0(\mathrm{mod}\ 2))$

to be any functional which fits into the commutative diagram

$$\begin{array}{ccc} [K^*(U,V;Z/_p)^{\otimes p}{}_{/Q_p}]^\alpha & \xrightarrow{f^{\otimes p}} & Z/_p \\ \searrow i_! & & \nearrow q(f) \\ & K^\alpha_{\Sigma_p}((U,V)^p;Z/_p) & \end{array} .$$

Notice that $f^{\otimes p}(Q_p) = 0$. If $p = 2$ and $x \in \ker\beta_2$, $(\deg x \equiv 1(\mathrm{mod}\ 2))$, then $f^{\otimes 2}(x^{\otimes 2}) = 0$ and, since $\beta_2(f) = 0$, $f^{\otimes 2}(x_1 \otimes x_2) = f^{\otimes 2}(\tau^*(x_1 \otimes x_2))$. For $\alpha \equiv 1(\mathrm{mod}\ 2)$, $(Q_p)^1 = 0$ and we define q(f) as the composition of f with projection onto $E_\infty^{p-1,1}$. For example, $p = 2$, $\alpha \equiv 1(\mathrm{mod}\ 2)$ we define the following operation which has no indeterminacy. If $f: K^1(U,V;Z/_2) \to Z/_2$ satisfies $\beta_2(f) = 0$, $q(f)$ is the composition

$$\begin{array}{c} K^1_{\Sigma_2}((U,V)^2;Z/_2) \cong K^0(U,V;Z/_2) \otimes K^1(U,V;Z/_2) \oplus (\ker\beta_{2/\mathrm{im}\beta_2})^1 \\ \downarrow \text{projection} \\ (\ker\beta_{2/\mathrm{im}\,\beta_2})^1 \\ \downarrow f \\ Z/_2 \end{array}$$

<u>Remark 4.2</u>:

From Proposition 1 of [H1] we see that any "reasonable operation", q, must have some indeterminacy as follows. By a "reasonable operation", q, we mean an operation of degree zero from a subset of $K_*(U,V;Z/_p)$ to a quotient $K_*^{\Sigma_p}((U,V)^p;Z/_p)$ such that $\operatorname{im} q$ and $\operatorname{im}(i_*)$ generate $K_*^{\Sigma_p}((U,V)^p;Z/_p)$.

Now let $\mu_{p+1}:(QS^o)_o \to (QS^o)_o$ be the $(p+1)$-st power map. We have [H2]

$$K_*((QS^o)_o;Z/_p) \cong Z/_p[\tilde\theta_p,\tilde\theta_{p^2},\ldots],$$

(see also §6). The comultiplication, Δ, is given by the following formulae (the notation is that of [H2])

$$\Delta(\tilde\theta_p) = \tilde\theta_p \otimes 1 + 1 \otimes \tilde\theta_p \,,$$

$$\Delta(\tilde\sigma_{p^2}) = \tilde\sigma_{p^2} \otimes \tilde\sigma_{p^2} \quad \text{where}$$

$$\tilde\sigma_{p^2} = 1 + p(\tilde\theta_p)^p + p^2\,\tilde\theta_{p^2} \quad \text{so}$$

$$(4.2.1)\ \Delta(\tilde\theta_{p^2}) = \tilde\theta_{p^2}\otimes 1 + 1\otimes\tilde\theta_{p^2} + \tilde\theta_p{}^p\otimes\tilde\theta_p{}^p - \sum_{i=1}^{p-1} \binom{p}{i}/_p\ \tilde\theta_p^i\otimes\tilde\theta_p^{p-i}\,.$$

Hence $(\mu_{p+1})_*$ fixes $\tilde\theta_p$ but not $\tilde\theta_{p^2}$. Now take p odd and let X be the $(2p-1)$-skeleton of $(QS^o)_o$ then the results of [H2, §3] show that $\tilde K_o(X;Z/_p) \cong Z/_p$, generated by u, and $\tilde K_1(X;Z/_p) \cong Z/_p$ generated by $\beta_p u$. Hence for q to be "reasonable", in Theorem 3.9(iii)

$$a_p(q(u)) = \lambda.u\,, \qquad (0 \neq \lambda \in Z/_p).$$

Let $f:X \to X$ be the restriction of a cellular version of μ_{p+1} and let $j:X \to (QS^o)_o = Y$ be the inclusion. There is a commutative diagram

(4.2.2)
$$\begin{array}{ccccc}
X^p \times_{\Sigma_p} E\Sigma_p & \xrightarrow{j^p \times_{\Sigma_p} 1} & Y^p \times_{\Sigma_p} E\Sigma_p & \xrightarrow{\nu} & Y \\
\Big\downarrow {\scriptstyle f^p \times_{\Sigma_p} 1} & & & & \Big\downarrow {\scriptstyle \mu_{p+1}} \\
X^p \times_{\Sigma_p} E\Sigma_p & \xrightarrow[j^p \times_{\Sigma_p} 1]{} & Y^p \times_{\Sigma_p} E\Sigma_p & \xrightarrow[\nu]{} & Y
\end{array}$$

Also $j_*(u) = \tilde{\theta}_p$ and $\tilde{\theta}_{p^2}$ is the image under $\nu_*(j^p \times_{\Sigma_p} 1)_*$ of some element, z, which satisfies $a_p(z) = u$. Hence if $q(u)$ were an element then

$\nu_*(j^p \times_{\Sigma_p} 1)_*(q(u)) = \lambda . \tilde{\theta}_{p^2} + g(\tilde{\theta}_p)$ where $g \in Z/_p[t]$ is a polynomial, $0 \neq \lambda \in Z/_p$. The diagram (4.2.1) and $f_* u = u$ imply

$$\begin{aligned}
& \lambda(\mu_{p+1})_*(\tilde{\theta}_{p^2}) \\
= \; & (\mu_{p+1})_* \nu_*(j^p \times_{\Sigma_p} 1)_*[q(u) - i_*(g(u))] \\
= \; & \nu_*(j^p \times_{\Sigma_p} 1)_*(f^p \times_{\Sigma_p} 1)_*[q(u) - i_*(g(u))] \\
= \; & \nu_*(j^p \times_{\Sigma_p} 1)_*[q(f(u))] - \nu_* i_*(g(\tilde{\theta}_p)) \\
= \; & \nu_*(j^p \times_{\Sigma_p} 1)_*[q(u) - i_*(g(\tilde{\theta}_p))] \\
= \; & \lambda . \tilde{\theta}_{p^2}
\end{aligned}$$

contradicting $(\mu_{p+1})_*(\tilde{\theta}_{p^2}) \neq \tilde{\theta}_{p^2}$.

We have given the example for $p \neq 2$. However, after we have defined the Dyer-Lashof operations in the next section it will be clear that if there existed a mod 2 operation without indeterminacy, q, in degree zero then there would be a corresponding indeterminacy free Dyer-Lashof operation and from the results of §6, working with $(QS^O)_o$ instead of its 3-skeleton, we again see that a "reasonable" operation, q, would imply that $(\mu_3)_*(\tilde\theta_4) = \tilde\theta_4 \in K_o((QS^O)_o;Z/_2)$. We cannot use the 3-skeleton, X, because $\beta_2 u \neq 0$ and q is defined only on $\ker \beta_2$.

<u>Proposition 4.3</u>:

Let $x,y \in K_\alpha(U,V;Z/_p)$ $(p \neq 2)$ or $x,y \in \ker \beta_2 \subset K_\alpha(U,V;Z/_2)$.

If $\alpha \equiv 0 (\mathrm{mod}\ 2)$ then

$$q(x+y) = q(x) + q(y) + \sum_{i=1}^{p-1}\left\{-\binom{p}{i}\Big/p\right\} i_*(x^{\otimes i} \otimes y^{\otimes p-i})$$ and if

$\alpha \equiv 1 (\mathrm{mod}\ 2)$

$$q(x+y) = q(x) + q(y) .$$

<u>Proof</u>:

For $(1 \leq j \leq p)$, let x_j be homogeneous elements of $K^*(U,V;Z/_p)$. By definition $q(x+y)$ is specified by the equation

$$\langle q(x+y), i_!(x_1 \otimes \ldots \otimes x_p)\rangle = \langle (x+y)^{\otimes p}, x_1 \otimes \ldots \otimes x_p\rangle,$$

and we have

$$\langle (x+y)^{\otimes p}, x_1 \otimes \ldots \otimes x_p\rangle$$

$$= \langle q(x), i_!(x_1 \otimes \ldots \otimes x_p)\rangle + \langle q(y), i_!(x_1 \otimes \ldots \otimes x_p\rangle$$

$$+ \sum_T \prod_i \langle x, x_i\rangle \prod_j \langle y, x_j\rangle$$

where the sum is taken over subsets, T, of $\{1,2,\ldots,p\}$ with $i \in T$, $j \notin T$ and $t = |T|$.

However $\langle i_*(x^{\otimes s} \otimes y^{\otimes p-s}), i_!(x_1 \otimes \ldots \otimes x_p)\rangle$

$$= \langle x^{\otimes s} \otimes y^{\otimes p-s}, i^* i_!(x_1 \otimes \ldots \otimes x_p)\rangle$$

$$= \langle x^{\otimes s} \otimes y^{\otimes p-s}, \sum_{g \in \Sigma_p} g^*(x_1 \otimes \ldots \otimes x_p)\rangle$$

$$= \sum_{|T|=s} s!(p-s)! \prod_{i \in T} \langle x,x_i\rangle \prod_{j \notin T} \langle y,x_j\rangle$$

which proves the formula except when $\alpha \equiv 1 \pmod 2$. However, in the case $\alpha \equiv 1 \pmod 2$ the construction of the composite, $q(f)$, in §4.1 is clearly linear in f.

Remark 4.4:

The operations of [H1], defined for $p \neq 2$, had larger indeterminacy than q, namely $im(i_*)$. Since they were the quotient by $im(i_*)$ of the operations, q, of §4.1 they were additive, by Proposition 4.3. This deviation from additivity is the necessary price that one pays for the more efficient indeterminacy, as the following example shows. Suppose that q is a "reasonable operation" in the sense of Remark 4.2. Let p be odd and put X equal to the $(2p-2)$-skeleton of $B\Sigma_p$. Let $K_*^j(-;Z/_p)$ be the j-th skeletal filtration of $K_*(-;Z/_p)$, the image of $K_*(-;Z/_p)$ of the j-skeleton. There is a map $j: B\Sigma_p \to (QS^0)_p$ such that if $x \in K_0(X;Z/_p)$ is the generator then $j_*(x) = \theta_p$. Here we use the notation of [H1], $\tilde{\theta}_{p^k} = \theta_{p^k} \cdot \theta_1^{-p^k}$ where $\theta_1^{p^k} = * \in K_0((QS^0)_{p^k};Z/_p)$ is the translation operator. As in Remark 4.2, $a_p(q(x)) = \lambda x$, $(0 \neq \lambda \in Z/_p)$. There is a map [c.f.§4.2]

$$j_1: X^p \times_{\Sigma_p} E\Sigma_p \to (QS^0)^p_{2p-2} \times_{\Sigma_p} E\Sigma_p \xrightarrow{\nu} (QS^0),$$

such that $(j_1)_*(q(x)) = \lambda\, \theta_{p^2} + \mu\, \theta_p^p + a$ where the element, a, comes from the $2(p^2-p-1)$-skeleton. Now let $k_1, k_2: X \to X \times X$ be the axial inclusions and $\Delta: X \to X \times X$ the diagonal. Since $j_*(x) = \theta_p$ and $\tilde{\theta}_p$ is primitive we have

$$\begin{aligned} j_*(\Delta_*(x)) &= \Delta_* j_*(x) \\ &= \Delta_*(\theta_p) \\ &= \theta_p \otimes \theta_1^p + \theta_1^p \otimes \theta_p \qquad (4.4.1) \\ &= [(k_1)_* + (k_2)_*](\theta_p) \\ &= j_*[(k_1)_* + (k_2)_*](x). \end{aligned}$$

Since all maps preserve the skeletal filtration

$$\begin{aligned} &\Delta_*(\theta_{p^2}) - (k_1)_*(\theta_{p^2}) - (k_2)_*(\theta_{p^2}) \qquad (4.4.2) \\ &\equiv (j_1)_*[\Delta_* - (k_1)_* - (k_2)_*](\lambda^{-1}q(x)) \\ &\quad (\mathrm{mod}\ K_0^{2p^2-2p-2}((QS^0)_{p^2} \times (QS^0)_{p^2}; Z/_p) \\ &\qquad + (j_1)_*(\mathrm{Ind}_p(X \times X))). \end{aligned}$$

The use of $(j_1)_*(\mathrm{ind}_p(X \times X))$ in (4.4.2) is justified by the fact that $\mathrm{Ind}_p(-)$ is a functor. Converting the formula (4.2.1) we have

$$\begin{aligned} \Delta_*(\theta_{p^2}) &= \theta_{p^2} \otimes \theta_1^{p^2} + \theta_1^{p^2} \otimes \theta_{p^2} + \theta_p^p \otimes \theta_p^p \\ &\quad - \sum_{i=1}^{p-1} \binom{p}{i}/_p\, (\theta_p^i\, \theta_1^{p(p-i)} \otimes \theta_p^{p-i}\, \theta_1^{pi}). \end{aligned}$$

Since $[\Delta_* - (k_1)_* - (k_2)_*](\theta_p^p) = 0$ we have

$$\begin{aligned} (4.4.3) \quad &[\Delta_* - (k_1)_* - (k_2)_*](\theta_{p^2}) \\ &\equiv - \sum_{i=1}^{p-1} \binom{p}{i}/_p\, (\theta_p^i\, \theta_1^{p(p-i)} \otimes \theta_p^{p-i}\, \theta_1^{pi}) \\ &\quad (\mathrm{mod}\ K_0^{2p^2-2p-2}((QS^0)_{p^2} \times (QS^0)_{p^2}; Z/_p) + \mathrm{Ind}_p((QS^0)^2)), \end{aligned}$$

and this is non-zero. But if q were additive we would have $q((k_1)_* + (k_2)_*) = [(k_1)_* + (k_2)_*]q$ and since $q\,\Delta_* = \Delta_* q$ by (4.4.1) $(\Delta_* - (k_1)_* - (k_2)_*(q(x))$ would be zero in $\mathrm{Ind}_p(X)$. This contradicts the non-vanishing in equation (4.4.3) by virtue of (4.4.2).

We have shown that "reasonable operations", q, cannot be additive for $p \neq 2$, but for $p = 2$ the space X has $\beta_2 x \neq 0$ and the operation q is not defined. However a more careful analysis of $K_0((QS^0)_0; Z/_2)$, [c.f.§6, and H2] shows that one may work with $(QS^0)_0$ instead of X to show that a "reasonable operation", q, cannot be additive in degree zero when $p = 2$.

Proposition 4.5: (Cartan Formula).

Let $x \in K_\alpha(U,V;Z/_p)$ and $y \in K_\beta(U',V';Z/_p)$, both belong to $\ker \beta_2$ if $p = 2$. Let Σ_p and $\Sigma_p \times \Sigma_p$ respectively act on $[(U,V) \times (U',V')]^p$ and $(U,V)^p \times (U',V')^p$ in the canonical manner. Let Δ denote the equivariant shuffle map between these products.

(i) $p \neq 2$, $\alpha \equiv \beta \pmod 2$

$$\Delta_* q(x \otimes y) = \begin{cases} \lambda q(x) \otimes q(y)\,, & (0 \neq \lambda \in Z/_p),\ \text{if } \alpha \equiv 1 \pmod 2. \\ q(x) \otimes i_*(y^{\otimes p}) + i_*(x^{\otimes p}) \otimes q(y), & \text{if } \alpha \equiv 0 \pmod 2. \end{cases}$$

in $$K_*^{\Sigma_p}((U,V)^p; Z/_p) \otimes K_*^{\Sigma_p}((U',V')^p; Z/_p) \Big/ \{\mathrm{Ind}_p(U,V) \otimes \mathrm{Ind}_p(U',V')\}.$$

(ii) $p = 2$, $\alpha \equiv \beta \pmod 2$

$$\Delta_* q(x \otimes y) = \begin{cases} q(x) \otimes i_*(y^{\otimes 2}) + i_*(x^{\otimes 2}) \otimes q(y), & \text{if } \alpha \equiv 0 \pmod 2 \\ q(x) \otimes q(y), & \text{if } \alpha \equiv 1 \pmod 2 \end{cases}$$

in $K_*^{\Sigma_2}((U,V)^2;Z/_p) \otimes K_*^{\Sigma_2}((U',V')^2;Z/_2) \Big/ \{\mathrm{Ind}_2(U,V) \otimes \mathrm{Ind}_2(U',V')\}$.

(iii) $\alpha + \beta \equiv 1 \pmod 2)$

$$\Delta_* q(x \otimes y) = q(x) \otimes i_*(y^{\otimes p}) + i_*(x^{\otimes p}) \otimes q(y)$$

in $K_*^{\Sigma_p}((U,V)^p;Z/_p) \otimes K_*^{\Sigma_p}((U',V')^p;Z/_p)$.

[Recall if $\deg z \equiv 1 \pmod 2$ and $\beta_2 z = 0$ then $i_*(z^{\otimes 2}) = 0$ and for $p \neq 2$ $i_*(z^{\otimes p}) = 0$].

Proof:

Assume U and U' compact.

(i) The dual of $\mathrm{Ind}_p(U,V) \otimes \mathrm{Ind}_p(U',V')$ is

$K^*_{\Sigma_p}((U,V)^p;Z/_p) \otimes \mathrm{im}(i_!) + \mathrm{im}(i_!) \otimes K^*_{\Sigma_p}((U',V')^p;Z/_p)$.

For $z \in K^*_{\Sigma_p}((U,V)^p;Z/_p)$ and $w \in K^*(U',V';Z/_p)^{\otimes p}$

$$\begin{aligned}
&\langle \Delta_* q(x \otimes y), z \otimes i_!(w)\rangle \\
= &\langle q(x \otimes y), \Delta^*(z \otimes i_!(w))\rangle \\
= &\langle q(x \otimes y), \pi_1^*(z) i_!(1 \otimes w)\rangle, \quad ((\pi_1 : (U \times U')^p \to U^p) = (\text{proj.})) \\
= &\langle q(x \otimes y), i_!(i^*(z) \otimes w)\rangle, \quad \text{by (2.1)}, \\
= &\langle x^{\otimes p} \otimes y^{\otimes p}, i^*(z) \otimes w\rangle \\
= &\langle x^{\otimes p}, i^*(z)\rangle \langle y^{\otimes p}, w\rangle .
\end{aligned}$$

Also $\langle q(x) \otimes i_*(y^{\otimes p}), z \otimes i_!(w)\rangle$

$$\begin{aligned}
= &\langle q(x), z\rangle \langle y^{\otimes p}, i^* i_!(w)\rangle \\
= &\ 0
\end{aligned}$$

and

$$\begin{aligned}
&\langle i_*(x^{\otimes p}) \otimes q(y), z \otimes i_!(w)\rangle \\
= &\langle x^{\otimes p}, i^* z\rangle \langle y^{\otimes p}, w\rangle .
\end{aligned}$$

These equations and their transpositions prove (i) when $\alpha \equiv 0 \pmod 2)$.

The case $\alpha \equiv 1 \pmod 2$ is similar to (ii) below.

(ii) Write

$$z_i, z_i', z_i'' \ldots \text{ for elements in } K^i_{\Sigma_2}(-;Z/_2);$$

$$w_i, w_i', w_i'' \ldots \text{ for elements in } (K^*(-;Z/_2)^{\otimes 2})^i$$

and $v_i, v_i', v_i'' \ldots$ for elements in $K^i(-;Z/_2)$.

In terms of the direct sum decomposition of Proposition 3.10(i) an element of $K^1_{\Sigma_2}((U,V)^2;Z/_2)$ is written as $i_!(w_1) + v_1$ where $v_1 \in \ker \beta_2$. The internal product is given by

$$[i_!(w_1) + v_1][i_!(w_1') + v_1']$$

$$= \quad i_!((v_1 v_1')^{\otimes 2}) + i_!([i^* i_!(w)]w')$$

since $i_!(w_1)i_!(w_1') = i_!([i^* i_!(w_1)]w_1')$ by (2.1)

and $i_!(w_1)v_1' = 0 = i_!(w_1')v_1$, $v_1 . v_1' = i_![(v_1 v_1')^{\otimes 2}]$,

from the multiplicative structure of the spectral sequence, $\{E_r((U,V)^2;\Sigma_2;Z/_2)\}$. Hence if $\alpha \equiv \beta \pmod 2$

$$\langle \Delta_* q(x \otimes y),\ z_0 \otimes i_!(w_0) + i_!(w_0') \otimes z_0' + (i_!(w_1)+v_1) \otimes (i_!(w_1')+v_1')\rangle$$

$$\doteq \quad \langle q(x \otimes y), i_!\{i^*(z_0) \otimes w_0 + w_0' \otimes i^*(z_0') + (v_1 \otimes v_1')^{\otimes 2} + i^* i_!(w_1) \otimes w_1'\}\rangle$$

$$= \begin{cases} \langle x^{\otimes 2}, i^* z_0\rangle\langle y^{\otimes 2} \otimes w_0\rangle + \langle x^{\otimes 2}, w_0'\rangle\langle y^{\otimes 2}, i^* z_0'\rangle, & \text{if } \alpha \equiv 0 \pmod 2 \\ \langle x, v_1\rangle\langle y, v_1'\rangle & \text{if } \alpha \equiv 1 \pmod 2. \end{cases}$$

However, as in the proof of (i), the expression when $\alpha \equiv 0 \pmod 2$ is the pairing of

$q(x) \otimes i_*(y^{\otimes 2}) + i_*(x^{\otimes 2}) \otimes q(y)$ with the element on the right. When $\alpha \equiv 1 \pmod 2$ we have

$$\langle q(x) \otimes q(y), z_0 \otimes i_!(w_0) + i_!(w_0') \otimes z_0' + (i_!(w_1)+v_1) \otimes (i_!(w_1')+v_1')\rangle$$

$$= \quad \langle q(x), i_!(w_1)+v_1\rangle \ \langle q(y), i_!(w_1') + v_1\rangle$$

$= \langle x,v_1\rangle \langle y,v_1'\rangle$.

Since $\mathrm{Ind}_2 \otimes \mathrm{Ind}_2$ is dual to

$$K^*_{\Sigma_2} \otimes [K^1_{\Sigma_2} + i_!(K^0(-;Z/_2)^{\otimes 2})] + [K^1_{\Sigma_2} + i_!(K^0(-;Z/_2)^{\otimes 2})]\otimes K^*_{\Sigma_2}$$

the proof of (ii) is complete.

(iii) Write any element of

$$(i^{*-1})(x^{\otimes 2}) \subset K^0_{\Sigma_2}((U,V)^2;Z/_2)$$

as $x^{\otimes 2} \otimes e_0$, so any element of $K^0_{\Sigma_2}((U,V)^2;Z/_2)$ can be written as $i_!(w_0) + v_0^{\otimes 2} \otimes e_0$. With the notation of part (ii) we have

$$\begin{aligned}&\langle \Delta_* q(x \otimes y),(i_!(w_0) + v_0^{\otimes 2} \otimes e_0) \otimes (i_!(w_1) + v_1)\\ &\qquad + (i_!(w_1') + v_1') \otimes (i_!(w_0') + v_0'^{\otimes 2} \otimes e_0)\rangle\\ = \;&\langle q(x \otimes y), i_!\{i^* i_!(w_0) \otimes w_1 + v_0^{\otimes 2} \otimes w_1\} + v_0 \otimes v_1\\ &\qquad + i_!\{i^* i_!(w_1') \otimes w_0' + w_1' \otimes v_0'^{\otimes 2}\} + v_1' \otimes v_0'\rangle\\ = \;&\langle x,v_0\rangle \langle y,v_1\rangle + \langle x,v_1'\rangle \langle y,v_0'\rangle ,\end{aligned}$$

since, from the multiplication in the spectral sequence, we see that $i_!(w)(z^{\otimes 2} \otimes e_0) = i_!((wz)^{\otimes 2})$ is a well-defined element and for $\ker \beta_2/_{\mathrm{im}\,\beta_2} \subset K^1_{\Sigma_2}((U,V)^2;Z/_2)/_{\mathrm{im}\,\beta_2}$

$$(\ker \beta_2/_{\mathrm{im}\,\beta_2})(\mathrm{im}(i_!)) \subset i_![i^*(\ker \beta_2/_{\mathrm{im}\,\beta_2}).K^*_{\Sigma_2}] = 0.$$

Now suppose $\alpha \equiv 1 \pmod 2$ then

$$\begin{aligned}&\langle q(x) \otimes i_*(y^{\otimes 2}), (i_!(w_0) + v_0^{\otimes 2} \otimes e_0) \otimes (i_!(w_1) + v_1)\\ &\qquad + (i_!(w_1') + v_1') \otimes (i_!(w_0') + v_0'^{\otimes 2} \otimes e_0)\rangle\\ &= \langle x,v_1'\rangle \langle i_*(y^{\otimes 2}), i_!(w_0') + v_0'^{\otimes 2} \otimes e_0\rangle\\ &\doteq \langle x,v_1'\rangle \langle y^{\otimes 2}, i^* i_!(w_0') + v_0'^{\otimes 2}\rangle\\ &= \langle x,v_1'\rangle \langle y,v_0'\rangle .\end{aligned}$$

The case $p \neq 2$ is similar.

Proposition 4.6:

Let p be a prime which does not divide $k \in Z$. If ψ^k is the Adams operation in $K^*(-;Z/_p)$ and $(\psi^k)_*$ is the dual operation then, for q as in (§4.1)

$$q(\psi_*^k(x)) = \psi_*^k(q(x)) .$$

Proof:

The operation, ψ^k, is stable [M] and so commutes with the stable map inducing the transfer. Hence ψ_*^k commutes with q in degree zero. Similarly ψ^k operates in the spectral sequence $\{E_r((U,V)^p;G;Z/_p)\}$. Hence ψ_*^k commutes with q in degree one.

Proposition 4.7:

Let p be an odd prime. If $w \in K_{\alpha+1}(U,V;Z/_p)$ then $\beta_p(q(w))$ is an element. There are $\gamma_i \in Z/_p$ (i=0 or 1), independent of (U,V) such that

$$\beta_p q(w) = \begin{cases} \gamma_1 q(\beta_p w) - i_*(w^{\otimes p-1} \otimes \beta_p w) & \text{if } \deg w \equiv 0 \\ \gamma_o\ i_*((\beta_p w)^{\otimes p}) & \text{if } \deg w \equiv 1. \end{cases}$$

Proof:

Since

$$\beta_p(i_*(x^{\otimes p})) = i_*(\sum_{i=o}^{p-1} x^{\otimes i} \otimes \beta_p x \otimes x^{\otimes p-i-1})$$

$$= p\ i_*(x^{\otimes p-1} \otimes \beta_p x)$$

$$= 0$$

we have $\beta_p(q(w))$ is an element. Now β_p induces an endomorphism of the spectral sequence $\{E_r((U,V)^p;\Sigma_p;Z/_p)\}$ which kills $E_\infty^{2p-2,o}$ and maps $E_\infty^{p-1,1}$ to $E_\infty^{2p-2,o} \subset K^o$. Put $x_o^{\otimes p} \otimes e_o$ for an element of $(i^{*-1})(x_o^{\otimes p}) \subset K_{\Sigma_p}^o((U,V)^p;Z/_p)$

then $\beta_p(x_o^{\otimes p} \otimes e_o)$ is an element. From the spectral sequence it is clear that $\beta_p(x_1^{\otimes p} \otimes e_{p-1}) = z(x_1)^{\otimes p} \otimes e_{2p-2}$ and

$$\beta_p(x_\alpha^{\otimes p} \otimes e_o) = i_!(z(x_\alpha)^{\otimes p} - x_\alpha^{\otimes p-1} \otimes \beta_p x_\alpha)$$

since $i^*\beta_p(x_\alpha^{\otimes p} \otimes e_o)$

$$= \quad \beta_p(x_\alpha^{\otimes p})$$

$$= \quad [\sum_{i=o}^{p-1} x_\alpha^{\otimes i} \otimes \beta_p x_\alpha \otimes x_\alpha^{p-i-1}]$$

$$= \quad [- i^* i_!(x_\alpha^{\otimes p-1} \otimes \beta_p x_\alpha)], \quad ((p-1)! + 1 \equiv 0(\text{mod } p)).$$

This construction defines a natural homomorphism, $z(-)$, of degree one. Also if x_β is an integral class then

$z(x_\alpha x_\beta) = z(x_\alpha)x_\beta$ since the integral class $x_\beta^{\otimes p}$ is in $x_\beta^{\otimes p} \otimes e_o$ and β_p is a derivation. Since it suffices to test $z(-)$ on spaces, $M_p \wedge S^N \wedge \mathbb{C}P^n$, because it is stable in the sense of [Proposition 3.2, proof] the multiplicative property implies that $z = \mu_i . \beta_p : K^i \to K^{i+1}, (\mu_i \in Z/_p)$.

We prove the formula when deg $w \equiv 0(\text{mod } 2)$, the other degree is similar and simpler.

An arbitrary generator of $K^1_{\Sigma_p}((U,V)^P;Z/_p)$ may be written as $i_!(x_1 \otimes \ldots \otimes x_p) + v_1^{\otimes p} \otimes e_{p-1}$. If $(1 \leq j \leq p)$ $n_j = \sum_{k<j} \deg x_k$ we have

$$< \beta_p q(w), i_!(x_1 \otimes \ldots \otimes x_p) + v_1^{\otimes p} \otimes e_{p-1} >$$

$$= < q(w), i_!(\sum_1^p (-1)^{n_j} x_1 \otimes \ldots \otimes \beta_p x_j \otimes \ldots \otimes x_p) + \mu_1(\beta_p v_1)^{\otimes p} \otimes e_{2p-2} >$$

$$= \sum_j (-1)^{n_j} (\prod_{k \neq j} < w, x_k >) < w, \beta_p x_j > + \gamma_1 < w, \beta_p v_1 >,$$

when $\gamma_1 = \lambda\mu_1$ and $1/\lambda e_{2p-2} = i_!(1) \in R(\Sigma_p) \otimes Z/_p$.

Also $< \gamma_1 q(\beta_p w) - i_*(w^{\otimes p-1} \otimes \beta_p w), i_!(x_1 \otimes \ldots) + v_1^{\otimes p} \otimes e_{p-1} >$

$= \gamma_1 < \beta_p w, v_1 > - < w^{\otimes p-1} \otimes \beta_p w, i^* i_! (x_1 \otimes \dots) >$

$= \gamma_1 < \beta_p w, v_1 > - (p-1)! \sum_j (\prod_{k \neq j} < w, x_k > < w, \beta_p x_j >).$

The result follows since $1 + (p-1)! \equiv 0 (\text{mod } p)$ and if n_j is odd then $(\prod_k < w, x_k >) < w, \beta_p x_j > = 0$.

Remark 4.8:

We now examine two features of the mod 2 operation which are connected with iterability of q, mod 2. The first difficulty is that $q(x)$ is defined only for $x \in \ker \beta_2$. Hence one must investigate $\beta_2(q(x))$. Also we should remark here the importance of having indeterminacy equal to Ind_2, having a larger indeterminacy would mean that there might often be some elements in $q(x)$ on which q was defined and some on which it was not. However, in contrast to the odd primary case, $\beta_2(q(x))$ can involve second order Bocksteins.

The second feature of iterability is that $q(i_*(x \otimes x))$ and $q(i_*(x \otimes \beta_2 x))$ are always defined in $K_*^{\Sigma_2}((X^2 \times_{\Sigma_2} E\Sigma_2)^2; Z/_2)_{/\text{Ind}_2}$ and non-zero modulo $\text{im}(i_*)$. Since elements of this form, a composition of q following an i_*-map, are difficult to control under further iterations of q we would like these to be expressible as something in the image of an i_*-map when mapped into $K_*^G(X^4 : Z/_2)$ for some larger group, G. Although the Cartan formula does not apply, since $\beta_2 x \neq 0$ in general, we show that these elements are decomposable in $K_*^{\Sigma_4}(X^4; Z/_2)$.

Let $\beta_2 = B_1, B_2, \dots, B_n, \dots$ be the family of Bocksteins in $K^*(-;Z/_2)$-theory. Thus B_i is the i-th differential in the spectral sequence [A-T,II §11]

$$E_1^* = \tilde{K}^*(X;Z/_2) \Longrightarrow (\tilde{K}^*(X)/_{\mathrm{Tors}}) \otimes Z/_2 .$$

Denote the dual Bocksteins by the same symbols.

We have $B_2\colon \ker \beta_2/_{\mathrm{im}\,\beta_2} \to \ker \beta_2/_{\mathrm{im}\,\beta_2}$.

<u>Proposition 4.9</u>:

Let $x \in \ker \beta_2 \subset K_\alpha(U,V;Z/_2)$.

(i) If $\alpha \equiv 0 \pmod 2$ the element $\beta_2(q(x))$ satisfies

$$B_1(q(x)) = \beta_2(q(x)) = q(B_2(x)) \in K_1^{\Sigma_2}((U,V)^2;Z/_2).$$

(ii) If $\alpha \equiv 1 \pmod 2$ then

$$B_1(q(x)) = \beta_2(q(x)) = i_*[(B_2x)^{\otimes 2}] \in K_0^{\Sigma_2}((U,V)^2;Z/_2).$$

[Here $i_*[(B_2x)^{\otimes 2}]$ means $i_*(w^{\otimes 2})$ for any $w \in B_2(x)$.]

<u>Proof</u>:

To prove these equations we compute the effect of β_2 on $K^*_{\Sigma_2}((U,V)^2;Z/_2)$ and use the pairing. Now $\beta_2 i_! = i_! \beta_2$ since $i_!$ is induced by a stable map. For this reason, or since β_2 induces an endomorphism of spectral sequences which annihilates $E_\infty^{2,0}((U,V)^2;\Sigma_2;Z/_2)$, we see that if $z \in \ker \beta_2 \subset K^0(U,V;Z/_2)$ then $\beta_2 i_!(z^{\otimes 2}) = 0$. Write $z_0^{\otimes 2} \otimes e_0$ for any element of $(i^*)^{-1}(z_0^{\otimes 2})$ and $z_1^{\otimes 2} \otimes e_1$ for the element represented by this class in $E_\infty^{1,0}((U,V)^2;\Sigma_2;Z/_2)$, $(z_i \in K^i(U,V;Z/_2) \cap \ker \beta_2)$.

In Proposition 4.10 we show that

$$\beta_2(z_0^{\otimes 2} \otimes e_0) = B_2(z_0)^{\otimes 2} \otimes e_1$$

and $\beta_2(z_1^{\otimes 2} \otimes e_1) = i_!(B_2(z_1)^{\otimes 2})$.

Hence if $\alpha \equiv 0 \pmod 2$ we have

$$<\beta_2 q(x),\ i_!(v \otimes v') + v_1^{\otimes 2} \otimes e_1$$

$$= \ <q(x),\ i_!(\beta_2 v \otimes v' + v \otimes \beta_2 v' + B_2(v_1)^{\otimes 2})>$$

$$= \ <x,\beta_2 v> <x,v'> + <x,v> <x,\beta_2 v'> + <x,B_2(v_1)>$$

$$= \ <B_2(x),v_1> \quad , \text{ since } \beta_2 x = 0,$$

$$= \ <qB_2(x),\ i_!(v \otimes v') + v_1^{\otimes 2} \otimes e_1> .$$

If $\alpha \equiv 1 \pmod 2$ then

$$<\beta_2 q(x),\ i_!(v \otimes v' + v_o^{\otimes 2}) + v_o'^{\otimes 2} \otimes e_o>$$

$$= <q(x),\ i_!(\beta_2 v \otimes v' + x \otimes \beta_2 v') + B_2(v_o')^{\otimes 2} \otimes e_1>$$

$$= <x,B_2(v_o')>$$

$$= <B_2(x),v_o'>$$

$$= <i_*[B_2(x)^{\otimes 2}],\ i_!(v \otimes v' + v_o^{\otimes 2}) + v_o'^{\otimes 2} \otimes e_o> .$$

<u>Proposition 4.10:</u>

(i) Let $z_1^{\otimes 2} \otimes e_1 \in K^1_{\Sigma_2}((U,V)^2;Z/_2)$ be the element represented by this class in $E_\infty^{1,o}((U,V)^2;\Sigma_2;Z/_2)$. Then

$\beta_2(z_1^{\otimes 2} \otimes e_1) = i_!(B_2(z_1)^{\otimes 2}) \in K^o_{\Sigma_2}((U,V)^2;Z/_2)$.

(ii) Let $z_o^{\otimes 2} \otimes e_o \in i^{*-1}(z_o^{\otimes 2}) \subset K^o_{\Sigma_2}((U,V)^2;Z/_2)$

then the element $\beta_2(z_o^{\otimes 2} \otimes e_o)$ is given by

$\beta_2(z_o^{\otimes 2} \otimes e_o) = B_2(z_o)^{\otimes 2} \otimes e_1 \in K^1_{\Sigma_2}((U,V)^2;Z/_2)$.

<u>Proof:</u>

The elements z_o and z_1 are in $\ker \beta_2$. From [Sn 2,§2] if z_o or z_1 are integral classes then the elements represented by $z_o^{\otimes 2} \otimes e_o$ and $z_1^{\otimes 2} \otimes e_1$ may be chosen as integral classes and the identities are trivial. We now prove the other cases.

(i) In the notation of Proposition 3.6 and Appendix II $\beta_2 z_1 = 0$ implies $z_1 \in \langle 2w,2,a\rangle = \beta^{-1}(2w)$ where $\beta: K^*(-;Z/_2) \to K^*(-;Z)$ is the Bockstein for the sequence $0 \to Z \to Z \to Z/_2 \to 0$ and $w \in K^o(U,V)$. In Appendix II it is shown that $z_1^{\otimes 2} \otimes e_1 \in \langle(5+3y)w^{\boxtimes 2}, 2,a\rangle \subset K^1_{\Sigma_2}((U,V)^2;Z/_2)$ so

(4.10.1) $\beta(z_1^{\otimes 2} \otimes e_1) = (5+3y)w^{\boxtimes 2} \in K^o_{\Sigma_2}((U,V)^2)$ by [Sn 3,II §3.5]. Since

$$\sigma = (1-y) \equiv (5+3y) \qquad \text{(mod 2)}$$

and the mod 2 reduction of $\sigma.w^{\boxtimes 2}$ is $i_!(w^{\otimes 2})$ it remains only to observe that

$$B_i(z_1) = p_2(\beta(z_1)_{/2^{i-1}}) \quad \text{where} \quad p_2: K^*(-;Z) \to K^*(-;Z/_2)$$

is reduction mod 2.

(ii) The proof of part (ii) is slightly more circuitous than that of (i) because the representives constructed in Appendix II were for $\sigma(z_o^{\otimes 2} \otimes e_o) = i_!(z_o^{\otimes 2})$.

Suppose $z_o \in \langle 2w,2,a\rangle \subset K((U,V) \times (I,\partial I) \times (I,\partial I);Z/_2)$

$$\cong K^o(U,V;Z/_2)$$

then $w \in K^o((U,V) \times (I,\partial I)) \cong K^{-1}(U,V)$. Write I_τ for the unit interval with the Σ_2-action, $\tau(u) = 1-u$. Thus if I^2 has the Σ_2-action which interchanges the factors then $I^2 \cong I \times I_\tau$.

In Appendix II it is shown that there exists

$$z \in \langle 5+3y)w^{\boxtimes 2},2,a\rangle \subset K_{\Sigma_2}((U,V)^2 \times (I,\partial I) \times (I_\tau,\partial I_\tau) \times (I,\partial I);Z/_2)$$

with the property that

$z_0^{\otimes 2} \in \mathrm{quad}(z) \subset K((U,V)^2 \times (I,\partial I)^4;Z/_2) \cong K^0((U,V)^2;Z/_2)$.

Consider the exact sequence, derived from the inclusion

$\partial I_\tau = \Sigma_2 \subset I_\tau$,

$$\begin{array}{ccc}
 & \vdots & \\
 & \downarrow & \\
 & K^{-1}_{\Sigma_2}((U,V)^2\times(I,\partial I)\times(I_\tau,\partial I_\tau)\times \partial I_\tau\times(I,\partial I);Z/_2) \cong K^0((U,V)^2;Z/_2 & \\
(4.10.2) & \downarrow \delta \qquad\qquad\qquad\qquad\qquad\qquad\qquad\qquad i_! \downarrow & \\
 & K_{\Sigma_2}((U,V)^2\times(I,\partial I)\times(I_\tau,\partial I_\tau)\times(I_\tau,\partial I_\tau)\times(I,\partial I);Z/_2)\cong K^0_{\Sigma_2}((U,V)^2 Z/ & \\
 & \downarrow j & \\
 & K_{\Sigma_2}((U,V)^2\times(I,\partial I)\times(I_\tau,\partial I_\tau)\times(I,\partial I);Z/_2) & \\
 & \downarrow & \\
 & \vdots &
\end{array}$$

where δ was identified with $i_!$ in Proposition 2.4.

From [Sn 2,§2] we know that $\mathrm{quad}(z)$ is in fact $i^*(j^{-1}(z))$. Hence if $\Lambda(\mathbb{C}y) \in K_{\Sigma_2}((I_\tau,\partial I_\tau)^2)$ is the Thom class [Se,§3] there exist elements $v \in K^0_{\Sigma_2}((U,V)^2\times(I,\partial I)\times(I,\partial I);Z/_2)$ and $v' \in K^0((U,V)^2;Z/_2) \cong K^0((U,V)^2\times(I,\partial I)^2;Z/_2)$ satisfying

$j([v + i_!(v')] \otimes \Lambda(\mathbb{C}y)) = z$.

Consider the commutative diagram

$$\begin{array}{ccc}
 & \vdots & \\
 & \delta \downarrow & \\
(4.10.3) & K_{\Sigma_2}((U,V)^2\times(I,\partial I)\times(I_\tau,\partial I_\tau)^2) \cong K_{\Sigma_2}((U,V)^2\times(I,\partial I)) & \\
 & \downarrow j \qquad\qquad\qquad\qquad\qquad \downarrow (-.\sigma) & \\
 & K_{\Sigma_2}((U,V)^2\times(I,\partial I)\times(I_\tau,\partial I_\tau)) \overset{\gamma}{\to} K_{\Sigma_2}((U,V)^2\times(I,\partial I)). &
\end{array}$$

In (4.10.3) we have

$$j([\beta(v) + i_!(\beta(v'))] \otimes \Lambda(\mathbb{C}y))$$
$$= (5+3y)w^{\boxtimes 2}$$
$$= 4(1+y)w^{\boxtimes 2} + (1-y)w^{\boxtimes 2}$$
$$= 4w^{\boxtimes 2}\, i_!(1) + \sigma w^{\boxtimes 2}$$
$$= i_! i^*(4w^{\boxtimes 2}) + \sigma w^{\boxtimes 2} \quad , \quad \text{by (2.1)},$$
$$= i_!(4w^{\otimes 2}) + \sigma . w^{\boxtimes 2}$$
$$= \sigma . w^{\boxtimes 2} \quad , \quad \text{since} \quad 4w = 0.$$

For $\beta_2(z_o^{\otimes 2} \otimes e_o)$ we require to find the mod 2 reduction

$$p_2([\beta(v)+i_!(\beta(v'))]\otimes\Lambda(\mathbb{C}y)) \in K_{\Sigma_2}((U,V)^2\times(I,\partial I)\times(I_\tau,\partial I_\tau)^2;Z/_2)$$
$$= K^{-1}_{\Sigma_2}((U,V)^2;Z/_2).$$

Suppose that in the integral spectral sequence $\beta(v) + i_!(\beta(v'))$ is represented by an element

$v'' \in E_2^{q,*}((U,V)^2;\Sigma_2;Z)$ then $\sigma(\beta(v) + i_!(\beta(v')))$ is represented by the translation of v'' to $E_2^{q+2,*}((U,V)^2;\Sigma_2;Z)$. But $\sigma(\beta(v) + i_!(\beta(v')))$ is $(\sigma.w^{\boxtimes 2}|(U,V)^2\times(I,\partial I)) = \gamma(\sigma w^{\boxtimes 2})$, where $(I,\partial I) \subset (I,\partial I)\times(I_\tau,\partial I_\tau)$ corresponds to the diagonal in $(I^2,\partial I^2)$. Since, by [Sn 2,§2], $\gamma(w^{\boxtimes 2})$ is represented by $w^{\otimes 2} \otimes e_1 \in E_2^{1,*}((U,V)^2;\Sigma_2;Z)$ then $\gamma(\sigma w^{\boxtimes 2})$ is represented by $w^{\otimes 2} \otimes e_3 \in E_2^{3,*}((U,V)^2;\Sigma_2;Z)$ and since translation is an isomorphism (multiplication by $\mathrm{Cotor}^{2,o}_{K^*(\Sigma_2)}(Z,Z)$)
$E_2^{1,*}((U,V)^2;\Sigma_2;Z) \xrightarrow{\cong} E_2^{3,*}((U,V)^2;\Sigma_2;Z)$ then $\beta(v) + i_!(\beta(v'))$ is represented by

$$w^{\otimes 2} \otimes e_1 \in E_2^{1,o}((U,V)^2;\Sigma_2;Z)$$

and $p_2(\beta(v) + i_!(\beta(v'))) = w^{\otimes 2} \otimes e_1 \in E_\infty^{1,o}((U,V)^2;\Sigma_2;Z/_2)$.

Proposition 4.11:

Let $x \in \ker \beta_2 \subset K_\alpha(U,V;Z/_2)$.

[I(i)] If these exists

$$z_\alpha \in q(x) \subset K_\alpha^{\Sigma_2}((U,V)^2;Z/_2) \quad \text{such that}$$

$$B_i(z_\alpha) \subset \ker \beta_2 \subset K_{\alpha+1}^{\Sigma_2}((U,V)^2;Z/_2)$$

is defined then $B_i(x)$ is defined and contains zero, so $B_{i+1}(x)$ is defined.

[II(i)] For $z'_{\alpha+1} \subset K^{\alpha+1}(U,V;Z/_2)$ suppose that $B_{i+1}(z'_{\alpha+1}) \subset K^\alpha(U,V;Z/_2)$ is defined. Then, in the notation of Proposition 4.10

$$\langle B_i z_1, z_0'^{\otimes 2} \otimes e_0\rangle = \langle i_*(B_{i+1}x)^{\otimes 2}, z_0'^{\otimes 2} \otimes e_0\rangle$$

$$\langle B_i z_0, z_1'^{\otimes 2} \otimes e_1\rangle = \langle q(B_{i+1}(x)), z_1'^{\otimes 2} \otimes e_1\rangle \tag{4.11.1}$$

[$i_*(B_{i+1}x)^{\otimes 2}$ is interpreted as $i_*(s^{\otimes 2})$ for any $s \in B_{i+1}(x)$ and under this interpretation both sides of (4.11.1) are elements of $Z/_2$].

Proof:

In Proposition 4.9 [II(l)] is proved in a sharper form than (4.11.1).

[I(i)] and [II(i)] when $\alpha \equiv 0 \pmod 2$

If z_1' is integral then $z_1'^{\otimes 2} \otimes e_1$ is also integral and both sides of (4.11.1) are zero. Otherwise $z_1' \in \langle 2w,2,a\rangle$ and

$$\begin{aligned} \beta(z_1'^{\otimes 2} \otimes e_1) &= (5+3y)w^{\boxtimes 2} \\ &= \sigma . w^{\boxtimes 2} \in K_{\Sigma_2}^0((U,V)^2) , \end{aligned}$$

as in Proposition 4.10 (proof). From this equation it is easy to see, by induction, that a necessary and sufficient condition for $B_i(-) = p_2(\beta(-)_{/2^{i-1}})$ to be defined on $z_1'^{\otimes 2} \otimes e_1$ is that $B_{i+1}(z_1')$ is defined. For $B_{i+1}(z_1')$ is defined if and only if $w = 2^{i-1}w'$ in which case $p_2(w') \in B_{i+1}(z_1')$.

However

$$\sigma(2w'')^{\boxtimes 2} = \sigma(3+y)(w'')^{\boxtimes 2}$$
$$= \sigma[2(1+y)+\sigma](w'')^{\boxtimes 2}$$
$$= \sigma^2(w'')^{\boxtimes 2}$$
$$= (-2\sigma)(w'')^{\boxtimes 2} \in K^0_{\Sigma_2}((U,V)^2) \;,$$

so by induction we see that $w = 2^{i-1}w'$ if and only if $\sigma w^{\boxtimes 2} = \pm 2^{i-1}\sigma(w')^{\boxtimes 2}$ and $i_!(w'^{\otimes 2}) \in B_i(z_1'^{\otimes 2} \otimes e_1)$.

In this case

$$\langle B_i z_o, z_1'^{\otimes 2} \otimes e_1\rangle = \langle z_o, B_i[z_1'^{\otimes 2} \otimes e_1]\rangle$$
$$= \langle q(x), i_!(w'^{\otimes 2})\rangle$$
$$(4.11.2) \qquad = \langle x, w'\rangle$$
$$= \langle x, B_{i+1}z_1'\rangle$$
$$= \langle B_{i+1}(x), z_1'\rangle$$
$$= \langle q(B_{i+1}x), z_1'^{\otimes 2} \otimes e_1\rangle \;.$$

To show [I(i)] it suffices to observe that there exists an element, z_o, such that $B_i(z_o)$ is defined if and only if $B_j(z_o)$ is defined and contains zero for $j < i$.

Hence by (4.11.2) if $B_i(z_o)$ is defined then $(j < i)$

$\langle x, B_{j+1}z_1'\rangle = 0$ for all z_1' such that $B_i(z_1')$ is defined.

Hence by induction $B_i(x)$ is defined and contains zero.

$\ulcorner I(i) \urcorner$ and $\ulcorner II(i) \urcorner$ when $\alpha \equiv 1 \pmod 2$

Again the identities are trivial when z_o' is integral. Otherwise, in the notation of Proposition 4.10 (proof),

$$z_o' \in \langle 2w, 2, a \rangle \quad \text{and}$$

$$\beta(z_o'^{\otimes 2} \otimes e_o) = \ulcorner \beta(v) + i_! \beta(v') \urcorner \otimes \Lambda(\mathbb{C}y) \in K_{\Sigma_2}((U,V)^2 \times (I,\partial I) \times (I_\tau, \partial I_\tau)^2) .$$

The image of this under the restriction, $\gamma \circ j$ of (4.10.3), in $K_{\Sigma_2}((U,V)^2 \times (I,\partial I)$ is $\sigma \ulcorner w^{\otimes 2} | (U,V)^2 \times (I,\partial I) \urcorner$, which is represented by $w^{\otimes 2} \otimes e_3$ in the integral spectral sequence $\{E_r((U,V)^2; \Sigma_2; Z)\}$. Hence we may argue as in the case $\alpha \equiv 0 \pmod 2)$ that $B_{i+1}(z_o')$ is defined if and only if $w = 2^{i-1}w'$ which in turn is if and only if $B_i(z_o'^{\otimes 2} \otimes e_o)$ is defined. In this case, arguing as in Proposition 4.10(ii)(proof) we see that $(w'^{\otimes 2} \otimes e_1) \in B_i(z_o'^{\otimes 2} \otimes e_o)$.

Hence

$$\begin{aligned}
\langle B_i(z_1), z_o'^{\otimes 2} \otimes e_o \rangle &= \langle z_1, B_i(z_o'^{\otimes 2} \otimes e_o) \rangle \\
&= \langle q(x), w'^{\otimes 2} \otimes e_1 \rangle \\
&= \langle x, w' \rangle \\
&= \langle x, B_{i+1} z_o' \rangle \\
&= \langle B_{i+1}, x, z_o' \rangle \\
&= \langle i_*(B_{i+1}x)^{\otimes 2}, z_o'^{\otimes 2} \otimes e_o \rangle .
\end{aligned} \tag{4.10.3}$$

Now $\ulcorner I(i) \urcorner$ follows from (4.10.3) .

Remark 4.12:

From (4.11.1) it appears as if there is a sense in which

$B_i q(-) = i_*(B_{i+1}(-)^{\otimes 2})$ in degree one and

$B_i q(-) = q(B_{i+1}(-))$ in degree zero. However in order to prove a more accurate form of (4.11.1) of this type it would be necessary to pair $q(x)$ with all the elements in $K^*_{\Sigma_2}((U,V)^2;Z/_2)$ on which B_i is defined. At this point the difficulty occurs that there exist elements in $\mathrm{im}(i_!)$ for which $B_i(i_!(s)$ is defined without $B_i(s)$ being defined and then the crucial inclusion

$$B_i(i_!(s)) \supset i_!(B_i(s))$$

is unavailable.

We now show that if $x \in K_\alpha(U,V;Z/_2)$ then the images of $q(i_*(x\otimes x))$ and $q(i_*(x\otimes\beta_2 x))$ under the homomorphism

$$\begin{array}{ccc} K_*^{\Sigma_2}(((U,V)^2\times_{\Sigma_2} E\Sigma_2)^2;Z/_2) & \xrightarrow{\cong} & K_*^{\Sigma_2\int\Sigma_2}((U,V)^4;Z/_2) \\ & & \downarrow \\ & & K_*^{\Sigma_4}((U,V)^4;Z/_2) \end{array}$$

are decomposable.

Let

$$\begin{array}{ccc} & \{1\} & \\ {}^{j_1}\swarrow & & \searrow^{j} \\ \Sigma_2\int\Sigma_2 & \xrightarrow[j_2]{} & \Sigma_4 \end{array}$$

be the commutative diagram of canonical inclusions. We will show that

$$(j_2)_*(q(i_*x^{\otimes 2})) \subset \mathrm{im}(j_2\circ i)_*$$

and $$(j_2)_*(q(i_*(x\otimes\beta_2 x))) \subset \mathrm{im}(j_2\circ i)_*$$

in the following manner. Firstly we determine

$\ker\beta_2 \subset K_*^{\Sigma_2}((U,V)^2;Z/_2)$ and

$\ker\beta_2 \subset K^*_{\Sigma_2}((U,V)^2;Z/_2)$ and describe the generators of

$K_*^{\Sigma_2 \int \Sigma_2}((U,V)^4;Z/_2)$ and $K^*_{\Sigma_2 \int \Sigma_2}((U,V)^4;Z/_2)$ in terms of elements of $K_*(U,V;Z/_2)$ and $K^*(U,V;Z/_2)$.

We then examine the spectral sequences

$\{E_r((U,V)^4;\Sigma_2 \int \Sigma_2;Z/_2)\}$ and

$\{E^r((U,V)^4;\Sigma_2 \int \Sigma_2;Z/_2)\}$. Using the wreath product of the Milnor resolution [§1.1] these spectral sequences have E_1-terms obtained from the wreath product resolutions used in (2.10). We investigate the filtrations in which the known generators lie and obtain sufficient information about the representatives of $q(i_*(x^{\otimes 2}))$ and $q(i_*(x \otimes \beta_2 x))$ to show that any representative vanishes under the homomorphism $\mathrm{Tor}^{Z/_2[\Sigma_4]}(\bar{\mu}\circ\mu, Z/_2)$ of Proposition 2.13, which implies that these elements are in a lower filtration in $K_*^{\Sigma_4}((U,V)^4;Z/_2)$.

<u>Proposition 4.13</u>:

(i) Let U be compact.

In the notation of Proposition 4.10

(a) $\ker \beta_2/_{\mathrm{im}\,\beta_2} \subset K^1_{\Sigma_2}((U,V)^2;Z/_2)/_{\mathrm{im}\,\beta_2}$

is generated by cosets of elements of the form

$\{i_!(z_1 \otimes z_0) \mid z_i \in \ker \beta_2 \subset K^i(U,V;Z/_2)\}$,

$\{z_1^{\otimes 2} \otimes e_1 \mid z_1 \in \ker \beta_2 \subset K^1(U,V;Z/_2)$ and $0 \in B_2(z_1)\}$

and $\{i_!(z_\alpha \otimes \beta_2 z_\alpha) \mid z_\alpha \in K^\alpha(U,V;Z/_2)\}$.

(b) $\ker \beta_2/_{\mathrm{im}\,\beta_2} \subset K^0_{\Sigma_2}((U,V)^2;Z/_2)/_{\mathrm{im}\,\beta_2}$

is generated by cosets of elements of the form

$\{i_!(z_\alpha \otimes z'_\alpha) \mid z_\alpha, z'_\alpha \in \ker \beta_2 \subset K^\alpha(U,V;Z/_2)\}$,

$\{z_o^{\otimes 2} \otimes e_o \mid z_o \in \ker \beta_2 \subset K^o(U,V;Z/_2)$ and $0 \in B_2(z_o)\}$

and $\{i_!(z_\alpha \otimes z_\alpha) \mid z_\alpha \in K^\alpha(U,V;Z/_2)\}$.

Dually

(ii)(a) $\ker \beta_2{}_{/\mathrm{im}\,\beta_2} \subset K_1^{\Sigma_2}((U,V)^2;Z/_2)_{/\mathrm{im}\,\beta_2}$ is generated

by cosets of elements of the form

$\{i_*(z_1 \otimes z_o) \mid z_i \in \ker \beta_2 \subset K_i(U,V;Z/_2)\}$,

$\{q(z_1) \mid z_1 \in \ker \beta_2 \subset K_1(U,V;Z/_2)$ and $0 \in B_2(z_1)\}$

and $\{i_*(z_\alpha \otimes \beta_2 z_\alpha) \mid z_\alpha \in K_\alpha(U,V;Z/_2)\}$.

(b) $\ker \beta_2{}_{/\mathrm{im}\,\beta_2} \subset K_o^{\Sigma_2}((U,V)^2;Z/_2)_{/\mathrm{im}\,\beta_2}$ is generated

by cosets of elements of the form

$\{i_!(z_\alpha \otimes z'_\alpha) \mid z_\alpha, z'_\alpha \in \ker \beta_2 \subset K_\alpha(U,V;Z/_2)\}$,

$\{q(z_o) \mid z_o \in \ker \beta_2 \subset K_o(U,V;Z/_2)$ and $0 \in B_2(z_o)\}$

and $\{i_*(z_\alpha \otimes z_\alpha) \mid z_\alpha \in K_\alpha(U,V;Z/_2)\}$.

Proof:

It suffices to prove (i). The proofs are straightforward, so we will only give (i)(b).

Notice, for example,

$$\begin{aligned} \beta_2 i_!(z_\alpha \otimes \beta_2 z_\alpha) &= i_!(\beta_2 z_\alpha \otimes \beta_2 z_\alpha) \\ &= i_! i^* i_!(z_\alpha \otimes z_\alpha) \\ &= 0 . \end{aligned}$$

Similarly all the other generators listed in $\ker \beta_2$.

(i)(b):

Choose a basis for $K^*(U,V;Z/_2)$ consisting of elements

$v_1,\dots,v_k \in \operatorname{im}\beta_2$,

$x_1,\dots,x_k$ such that $\beta_2 x_i = v_i$,

$y_1,\dots,y_t \in \ker\beta_2$ such that $0 \notin B_2 y_i$, $1 \le i \le t$,

$w_1,\dots,w_t \in \ker\beta_2$ such that $w_i \in B_2 y_i$

and $u_1,\dots,u_s \in \ker\beta_2$ such that $0 \in B_2 u_i$.

Since $i_!(a \otimes b) \equiv i_!(\tau^*(a \otimes b))$ it suffices to show that if

$$i_!(\sum_j \lambda_j x_j \otimes a_j) + z_o^{\otimes 2} \otimes e_o$$

is in $\ker\beta_2 \subset K^o_{\Sigma_2}((U,V)^2;Z/_2)$ then $0 \in B_2 z_o$ and $i_!(\sum_j \lambda_j x_j \otimes a_j)$ is a linear combination of the elements listed in (i)(b).

However

$$0 = \beta_2(i_!(\sum_j \lambda_j x_j \otimes a_j) + z_o^{\otimes 2} \otimes e_o)$$

$$= i_!(\sum_j \lambda_j[\beta_2 x_j \otimes a_j + x_j \otimes \beta_2 a_j]) + B_2(z_o)^{\otimes 2} \otimes e_1$$

so $0 \in B_2(z_o)$ and

$$0 = i^* i_!(\sum_j \lambda_j[\beta_2 x_j \otimes a_j + x_j \otimes \beta_2 a_j])$$

$$= (1 + \tau^*)(\sum_j \lambda_j[v_j \otimes a_j + x_j \otimes \beta_2 a_j]) .$$

However $(1 + \tau^*)$ is the differential used in the calculation of $\operatorname{Tor}^{Z/_2[\Sigma_2]}(M^{\otimes 2},Z/_2)$ in (§1) and hence the general form of the $(1 + \tau^*)$-cycle

$$(\sum_j \lambda_j[v_j \otimes a_j + x_j \otimes \beta_2 a_j])$$

is known in terms of the chosen basis. It is then simple to verify that $i_!(\sum_j \lambda_j x_j \otimes a_j) \pmod{\operatorname{im}\beta_2}$ is generated by the elements of (i)(b).

Now $(X^2 \times_{\Sigma_2} E\Sigma_2)^2 \times_{\Sigma_2} E\Sigma_2 \cong X^4 \times_{\Sigma_2 \int \Sigma_2} E\Sigma_2 \int \Sigma_2$

so Proposition 4.13 implies that a system of generators for $K^*_{\Sigma_2 \int \Sigma_2}((U,V)^4;Z/_2)$ and $K_*^{\Sigma_2 \int \Sigma_2}((U,V)^4;Z/_2)$ is given by the types of elements tabulated in the following lists.

For $K^*_{\Sigma_2 \int \Sigma_2}(-;Z/_2)$ it is permissible to work with the non-compact space, $U^2 \times_{\Sigma_2} E\Sigma_2$, since we may obtain the same results by working with $K_*(-;Z/_2)$ and dualising if U is compact. Henceforth U will be assumed compact in the statements of the $K^*(-;Z/_2)$ results.

Corollary 4.14:

The table 4.14.1 gives systems of generators for $K^1_{\Sigma_2 \int \Sigma_2}((U,V)^4;Z/_2)$ and $K_1^{\Sigma_2 \int \Sigma_2}((U,V)^4;Z/_2)$. (In this table

$z_i, z_i', z_i'' \in K^1(U,V;Z/_2)$ and

$y_i, y_i', y_i'' \in K_1(U,V;Z/_2)$.) .

4.15: $(j_2)_* q(i_*(y_\alpha \otimes \beta_2 y_\alpha))$.

Consider now the $K^1_{\Sigma_2 \int \Sigma_2}$ generators in Table 4.14.1.

In the spectral sequence

$\{E_r((U,V)^4;\Sigma_2 \int \Sigma_2;Z/_2)\}$ we have

$E_1 \cong K^*(U,V;Z/_2)^{\otimes 4} \otimes D \int D$ where

$D \int D$ denotes the resolution of (2.10) and its dual. Since elements (1)-(3) in the table belong to $\ker i^*$ they are represented in

$$E_1^{j,*} \cong K^*(-;Z/_2)^{\otimes 4} \otimes D \int D$$

Table 4.14.1

	$K^1_{\Sigma_2\int\Sigma_2}$	$K_1^{\Sigma_2\int\Sigma_2}$
(1)	$(i_!(z_1\otimes z_o))^{\otimes 2}\otimes e_1$ $z_i\in\ker\beta_2$	$q(i_*(y_1\otimes y_o))$ $y_i\in\ker\beta_2$
(2)	$(z_1^{\otimes 2}\otimes e_1)^{\otimes 2}\otimes e_1$ $z_1\in\ker\beta_2$ $0\in B_2 z_1$	$q(q(y_1))$ $y_1\in\ker\beta_2$ $0\in B_2 y_1$
(3)	$\lceil i_!(z_\alpha\otimes\beta_2 z_\alpha)\rceil^{\otimes 2}\otimes e_1$	$q(i_*(y_\alpha\otimes\beta_2 y_\alpha))$
(4)	$i_![(z_1^{\otimes 2}\otimes e_1)\otimes i_!(z'_\alpha\otimes z''_\alpha)]$ $z_1\in\ker\beta_2$	$i_*[q(y_1)\otimes i_*(y'_\alpha\otimes y''_\alpha)]$ $y_1\in\ker\beta_2$
(5)	$i_![(z_1^{\otimes 2}\otimes e_1)\otimes(z_o^{\otimes 2}\otimes e_o)]$ $z_i\in\ker\beta_2$	$i_*[q(y_1)\otimes i_*(y_o^{\otimes 2})]$ $y_i\in\ker\beta_2$
(6)	$i_![z_1^{\otimes 2}\otimes e_1)\otimes i_!(z_o^{\otimes 2})]$ $z_i\in\ker\beta_2$	$i_*[q(y_1)\otimes q(y_o)]$ $y_i\in\ker\beta_2$
(7)	$j_!(z_\alpha\otimes z_\beta\otimes z_\gamma\otimes z_\delta)$ $\alpha+\beta+\gamma+\delta\equiv 1 \pmod 2$	$j_*(y_\alpha\otimes y_\beta\otimes y_\gamma\otimes y_\delta)$ $\alpha+\beta+\gamma+\delta\equiv 1 \pmod 2$
(8)	$i_![i_!(z_1\otimes z_o)\otimes(z_o'^{\otimes 2}\otimes e_o)]$ $z'_o\in\ker\beta_2$	$j_*(y_1\otimes y_o\otimes y'_o\otimes y'_o)$ $y'_o\in\ker\beta_2$

for some $j \geq 1$. If represented in $E_1^{1,*}$ then the representatives of (1) and (2) must be of the forms

$$(z_1\otimes z_o)^{\otimes 2}\otimes e_1\int e_o\otimes e_o + z'\otimes e_o\int e_1\otimes e_o + z''\otimes e_o\int e_o\otimes e_1$$

and

$$z_1^{\otimes 4}\otimes e_1 \int e_o \otimes e_o + w'\otimes e_o \int e_1 \otimes e_o + w''\otimes e_o \int e_o \otimes e_1$$

respectively. Element (3) pulls back to

$$K^1((U,V)^4 \times_{\Sigma_2 \int \Sigma_2} (E\Sigma_2, S^o_\pi) \int E\Sigma_2 \times E\Sigma_2; Z/_2)$$

and if it pulled back to

$$K^1((U,V)^4 \times_{\Sigma_2 \int \Sigma_2} (E\Sigma_2 \int \Sigma_2, (E\Sigma_2 \int \Sigma_2)_1); Z/_2)$$

the common images of these elements would be zero modulo $\operatorname{im}(i^* i_!)$ in

$$K^*((U,V)^4 \times_{\Sigma_2 \int \Sigma_2} (S^1_\pi, S^o_\pi) \int S^o_\pi \times S^o_\pi; Z/_2) \cong K^*(U,V;Z/_2)^{\otimes 4}.$$

However, this image is $(z_\alpha \otimes \beta_2 z_\alpha)^{\otimes 2}$, so element (3) is represented in E_1^1 by an element of the form

$$(z_\alpha \otimes \beta_2 z_\alpha)^{\otimes 2} \otimes e_1 \int e_o \otimes e_o + u'\otimes e_o \int e_1 \otimes e_o + u''\otimes e_o \int e_o \otimes e_1.$$

Representatives of elements (4)-(8) are given in the following manner. If

$$a\otimes b\otimes e_i \otimes e_j \in E_1^{*,*}((U,V)^4; \Sigma_2 \times \Sigma_2; Z/_2)$$

then the transfer induces

$$E_1^{*,*}((U,V)^4; \Sigma_2 \times \Sigma_2; Z/_2) \xrightarrow{i_!} E_1^{*,*}((U,V)^4; \Sigma_2 \int \Sigma_2; Z/_2)$$

which, by Proposition 2.2, satisfies

$$i_!(a\otimes b\otimes e_i \otimes e_j) = a\otimes b\otimes e_o \int e_1 \otimes e_j + \tau^*(a\otimes b)\otimes e_o \int e_j \otimes e_i .$$

This formula gives the representatives of elements (4)-(8) in terms of representatives of their inverse images in $K^*_{\Sigma_2}((U,V)^2;Z/_2)^{\otimes 2}$. In particular elements (4)-(6) are represented in $E_1^{1,*}$ and (7)-(8) in $E_1^{o,*}$.

Dually in $K_*^{\Sigma_2 \int \Sigma_2}$ we obtain the following results. Element (3) is represented in $E_1^{1,*}$ by

$(y_\alpha \otimes \beta_2 y_\alpha)^{\otimes 2} \otimes e_1 \int e_o \otimes e_o + y' \otimes e_o \int e_1 \otimes e_o + y'' \otimes e_o \int e_o \otimes e_1$.

Elements (1) and (2) are represented in $E^1_{j,*}$ for some $j \geq 1$ and if $j = 1$ the representatives are

$(y_1 \otimes y_o)^{\otimes 2} \otimes e_1 \int e_o \otimes e_o + s' \otimes e_o \int e_1 \otimes e_o + s'' \otimes e_o \int e_o \otimes e_1$

and

$y_1^{\otimes 4} \otimes e_1 \int e_o \otimes e_o + t' \otimes e_o \int e_1 \otimes e_o + t'' \otimes e_o \int e_o \otimes e_1$,

respectively. The other representatives are calculated from the representatives of their inverse images in $E^1((U,V)^4;\Sigma_2 \times \Sigma_2;Z/_2)$. Thus by Proposition 2.13 $q(i_*(y_\alpha \otimes \beta_2 y_\alpha))$ is represented in $E^1_{1,*}$ by a permanent cycle of the form

$$[(y_\alpha \otimes \beta_2 y_\alpha)^{\otimes 2} \otimes e_1 \int e_o \otimes e_o + (y_\alpha \otimes y_\alpha \otimes \beta_2 y_\alpha \otimes \beta_2 y_\alpha) \otimes e_o \int e_1 \otimes e_o]$$

$$+ a$$

where a is an E^1-cycle. However, inspection of the possibilities shows that modulo $\mathrm{im}(d_1)$ a must be a permanent cycle representing an element in the image of i_* generated by elements of the form (4) and (5).

Thus $j_2^!(j_2)_* q_*(i_*(y_\alpha \otimes \beta_2 y_\alpha))$ by Proposition 2.13(ii) is represented by $j_2^!(j_2)_*(a)$. Since $j_2^!$ is a monomorphism this implies that

$(j_2)_* q(i_*(y_\alpha \otimes \beta_2 y_\alpha))$

$$= (j_2)_* i_*(\Sigma\, q(y_1) \otimes i_*(y'_\alpha \otimes y''_\alpha)) + j_*(\Sigma\, y_\alpha \otimes y_\beta \otimes y_\gamma \otimes y_\delta)$$

$$= (j_2 \circ i)_*(\Sigma\, q(y_1) \otimes i_*(y'_\alpha \otimes y''_\alpha)) + j_*(\Sigma\, y_\alpha \otimes y_\beta \otimes y_\gamma \otimes y_\delta) .$$

Table 4.16.1

	$K^{0}_{\Sigma_2\wr\Sigma_2}$	$K_{0}^{\Sigma_2\wr\Sigma_2}$
(1)	$i_!\lceil(z_0^{\otimes 2})^{\otimes 2}\rceil=(j_1)_!z_0^{\otimes 4}$ $z_0\in\ker\beta_2$	$q(q(y_0))$ $y_0\in\ker\beta_2$ $0\in B_2y_0$
(2)	$i_!\lceil(z_0^{\otimes 2}\otimes e_0)^{\otimes 2}\rceil$ $z_0\in\ker\beta_2$ $0\in B_2z_0$	$q(i_*(y_\alpha^{\otimes 2}))$ $y_\alpha\in\ker\beta_2$
(3)	$i_!\lceil i_!(z_\alpha\otimes z_\alpha)^{\otimes 2}\rceil$ $z_\alpha\notin\ker\beta_2$	$q(i_*(y_\alpha\otimes y_\alpha))$ $y_\alpha\notin\ker\beta_2$
(4)	$\lceil i_!(z_0^{\otimes 2})\rceil^{\otimes 2}\otimes e_0$ $z_0\in\ker\beta_2$	$i_*(q(y_0)^{\otimes 2})$ $y_0\in\ker\beta_2$
(5)	$\lceil i_!(z_\alpha\otimes z_\alpha)\rceil^{\otimes 2}\otimes e_0\, z_\alpha\notin\ker\beta_2$	$(j_1)_*(y_\alpha^{\otimes 4})$ $y_\alpha\notin\ker\beta_2$
(6)	$[z_0^{\otimes 2}\otimes e_0]^{\otimes 2}\otimes e_0$ $z_0\in\ker\beta_2$ $0\in B_2z_0$	$(j_1)_*(y_0^{\otimes 4})$ $y_0\in\ker\beta_2$
(7)	$i_!(\lceil z_0^{\otimes 2}\otimes e_0\rceil\otimes i_!(z_\alpha\otimes z_\beta))$ $z_0\in\ker\beta_2$	$(j_1)_*(y_0\otimes y_0\otimes y_\alpha\otimes y_\beta)$ $y_0\in\ker\beta_2$
(8)	$i_!(\lceil z_0^{\otimes 2}\otimes e_0\otimes\lceil z_0'^{\otimes 2}\otimes e_0\rceil)$ $z_0,z_0'\in\ker\beta_2$	$(j_1)_*(y_0^{\otimes 2}\otimes y_0'^{\otimes 2})$ $y_0,y_0'\in\ker\beta_2$
(9)	$(j_1)_!(z_\alpha\otimes z_\beta\otimes z_\gamma\otimes z_\delta)$	$(j_1)_*(y_\alpha\otimes y_\beta\otimes y_\gamma\otimes y_\delta)$
(10)	$i_!([z_1^{\otimes 2}\otimes e_1]\otimes i_!(z_\alpha\otimes z_\beta)$ $z_1\in\ker\beta_2$	$i_*(q(y_1)\otimes i_*(y_\alpha\otimes y_\beta))$ $y_1\in\ker\beta_2$
(11)	$i_!([z_1^{\otimes 2}\otimes e_1]\otimes\lceil z_1'^{\otimes 2}\otimes e_1]$ $z_1,z_1'\in\ker\beta_2$	$i_*(q(y_1)\otimes q(y_1'))$ $y_1,y_1'\in\ker\beta_2$
(12)		$i_*(q(y_0)\otimes q(y_0'))$ $y_0,y_0'\in\ker\beta_2$
(13)		$i_*(q(y_0)\otimes i_*(y_\alpha\otimes y_\beta)$ $y_0\in\ker\beta_2$

Corollary 4.16:

The table 4.16.1 gives systems of generators for $K^{o}_{\Sigma_2 \int \Sigma_2}((U,V)^4;Z/_2)$ and

$K_{o}^{\Sigma_2 \int \Sigma_2}((U,V)^4;Z/_2)$. (In this table

$z_i, z'_i, z''_i \in K^i(U,V;Z/_2)$ and

$y_i, y'_i, y''_i \in K_i(U,V;Z/_2)$).

4.17: Representatives of elements in $K^{o}_{\Sigma_2 \int \Sigma_2}$ and $K_{o}^{\Sigma_2 \int \Sigma_2}$.

In $E^1((U,V)^4;\Sigma_2 \int \Sigma_2;Z/_2)$ the elements (5)-(9) of $K^{\Sigma_2 \int \Sigma_2}((U,V)^4;Z/_2)$ are represented in $E^1_{o,*}$. The duals of these elements are represented in $E^{o,*}_1((U,V)^4;\Sigma_2 \int \Sigma_2;Z/_2)$. Let $\sigma \in R(\Sigma_2) \otimes Z/_2 = R(1 \int \Sigma_2) \otimes Z/_2$. The elements $(j_1)_*(y^{\otimes 4}) = i_*(i_*(y^{\otimes 2})^{\otimes 2})$ of (5) and (6) are represented in $E^1_{o,*}$ by $y^{\otimes 4} \otimes e_o \int e_o \otimes e_o$. To find the representative of $q(i_*(y^{\otimes 2}))$ when $\beta_2 y = 0$ first observe that a representative of $(j_1)_*(y^{\otimes 4})$ is detected by a non-trivial pairing with some permanent cycle $a \otimes e_o \int e_o \otimes e_o \in E^{o,*}_1$. Hence, since multiplication by σ translates representative up two dimensions then $q(i_*(y^{\otimes 2}))$ is detected in $E^{2,*}_1$ and so has a representative in $E^1_{j,*}$ $(j = 1$ or $2)$. However, the comodule structure of $K_*^{\Sigma_2 \int \Sigma_2}(-;Z/_2)$ over $K_*^{1 \int \Sigma_2}(-;Z/_2)$ and the corresponding comodule structure in $\{E^r_{*,*}((U,V)^4;\Sigma_2 \int \Sigma_2;Z/_2)\}$ shows that $q(i_*(y_\alpha^{\otimes 2}))$ is represented in $E^1_{2,*}$ and $(\beta_2 y_\alpha \neq 0)$ the pairings show that the representative is of the form

$y_\alpha^{\otimes 4} \otimes e_2 \int e_o \otimes e_o$ + (terms not involving $e_2 \int e_o \otimes e_o$) .

Also element (10) is represented in $E^1_{1,*}$ by

$y_1^{\otimes 2} \otimes y_\alpha \otimes y_\beta \otimes e_o \int e_1 \otimes e_o$ and element (11)

by $y_1^{\otimes 2} \otimes y_1'^{\otimes 2} \otimes e_o \int e_1 \otimes e_1$. Element (4) is represented in $E^1_{4,o}$ by $y_o^{\otimes 4} \otimes e_o \int e_2 \otimes e_2$. By considering the pairings, the module and the comodule structures, as for $q(i_*(y^{\otimes 2}))$, we find that element (1) is represented in $E^1_{j,*}$ for some $j \geq 6$. Finally elements (12) and (13) are represented by

$y_o^{\otimes 2} \otimes y_o'^{\otimes 2} \otimes e_o \int e_2 \otimes e_2$ and

$y_o \otimes y_o \otimes y_\alpha \otimes y_\beta \otimes e_o \int e_2 \otimes e_o$.

We are now in a position to prove the following result.

<u>Lemma 4.17.1</u>:

If $y_\alpha \notin \ker \beta_2$ then $(j_2)_* q(i_*(y_\alpha^{\otimes 2}))$ is decomposable in terms of the images under $(j_2)_*$ of the elements of Table 4.16.1 (5)-(11) and (13).

<u>Proof</u>:

From the previous discussion we have that $q(i_*(y_\alpha^{\otimes 2}))$ is represented by a permanent cycle in $E^1_{2,*}((U,V)^4;\Sigma_2 \int \Sigma_2;Z/2)$ of the form

$$y = y_\alpha^{\otimes 4} \otimes \{e_2 \int e_o \otimes e_o + \lambda_1 e_o \int e_o \otimes e_2 + \lambda_2 e_1 \int e_1 \otimes e_o + \lambda_3 e_1 \int e_o \otimes e_1\} + b$$

where b involves only terms of the form

$y_\alpha^{\otimes 4} \otimes e_i \int e_j \otimes e_k$ with

$(i,j,k) \notin \{(2,0,0);(0,0,2);(1,1,0);(1,0,1)\}$

or of the form $u_1 \otimes u_2 \otimes u_3 \otimes u_4 \otimes e_i/e_j \otimes e_k$ with some $u_i \neq y_\alpha$. Since y is a cycle comparing coefficients of $y_\alpha^{\otimes 4} \otimes e_o/e_o \otimes e_1$ in $(1 \otimes d/d)(y)$ shows $\lambda_2 = \lambda_3$. Now, using Proposition (2.12), it is clear that the coefficient of $y_\alpha^{\otimes 4} \otimes e_2/e_o \otimes e_o$ in the representative for $(j_2)^!(j_2)_*(y)$ is contributed to only by the term

$$y_\alpha^{\otimes 4} \otimes \{e_2/e_o \otimes e_o + \lambda_1 e_o/e_o \otimes e_2 + \lambda_2 e_1/e_1 \otimes e_o + \lambda_3 e_1/e_o \otimes e_1\}$$

and is given by $(1 + 1 + \lambda_1 + \lambda_1 + \lambda_2 + \lambda_3)$ which is zero mod 2. The same reasoning applied to calculating the coefficient of

$$y_\beta^{\otimes 4} \otimes e_2/e_o \otimes e_o \quad , \qquad (y_\beta \neq y_\alpha \text{ and } y_\beta \notin \ker \beta_2)$$

yields a coefficient $(\lambda_1' + \lambda_1' + \lambda_2' + \lambda_3')$, since there is no $y_\beta^{\otimes 4} \otimes e_2/e_o \otimes e_o$ term, and again $\lambda_2' = \lambda_3'$. Thus $(j_2)^!(j_2)_*(y)$ is a permanent cycle which has no term of the form $u^{\otimes 4} \otimes e_2/e_o \otimes e_o$ unless $\beta_2 u = 0$. Since $(j_2)_*(j_2)^!(j_2)_* = (j_2)_*$ we have $(j_2)_*(y)$ is a linear combination of $(j_2)_*$-images of representatives of elements of types (5)-(11),(13) and possibly (2). However, by the Cartan formula of Proposition 4.5(ii), images under $(j_2)_*$ of elements of type (2) are decomposable in terms of images under $(j_2)_*$ of elements of types (5)-(11), (13).

Summarising (4.15) and (4.17) we have the following result.

<u>Proposition 4.18</u>:

Let $y_\alpha \in K_\alpha(U,V;Z/_2)$.

(i) $0 = \beta_2 i_*(y_\alpha^{\otimes 2}) \in K_1^{\Sigma_2}((U,V)^2;Z/_2)$ and

$qi_*(y_\alpha^{\otimes 2}) \in K_0^{\Sigma_2 \int \Sigma_2}((U,V)^4;Z/_2)$ is defined.

There exist elements $\{q(u_{\beta,q}) \in K_\beta^{\Sigma_2}((U,V)^2;Z/_2)\}$ and $\{v_r \in K_*(U,V;Z/_2)\}$ such that $(j_2)_*(q(i_*(y_\alpha^{\otimes 2})))$

$$= \sum (j_2)_* i_* [q(u_{1,q}) \otimes q(u_{1,r}) + q(u_{o,q}) \otimes i_*(v_r \otimes v_s)]$$

$$+ \sum j_*(v_q \otimes v_r \otimes v_s \otimes v_t).$$

in $K_0^{\Sigma_4}((U,V)^4;Z/_2)$.

(ii) $0 = \beta_2(i_*(y_\alpha \otimes \beta_2 y_\alpha)) \in K_0^{\Sigma_2}((U,V)^2;Z/_2)$

and $q(i_*(y_\alpha \otimes \beta_2 y_\alpha)) \in K^{\Sigma_2 \int \Sigma_2}((U,V)^4;Z/_2)$ is defined. There exist elements

$$\{q(w_r) \in K_1^{\Sigma_2}((U,V)^2;Z/_2\}$$

and $\{x_r \in K_*(U,V;Z/_2)\}$ such that $(j_2)_* q(i_*(y_\alpha \otimes \beta_2 y_\alpha))$

$$= \sum (j_2)_* i_* [q(w_r) \otimes i_*(x_2 \otimes x_t)]$$

$$+ \sum j_*(x_q \otimes x_r \otimes x_s \otimes x_t)$$

in $K_1^{\Sigma_4}((U,V)^4;Z/_2)$.

§5: Dyer-Lashof operations in K-theory

Throughout this ection X will be an infinite loopspace. In [D - L, §1] structure maps $\mu_p: X^p \times_{\Sigma_p} E\Sigma_p \to X$ are constructed. These H^∞-space structure maps satisfy certain coherence conditions. In the language of [Ma, §1.4] the $\{\mu_p\}$ are given by the action of the operad $C_\infty = \{C_\infty(j): j \geq 1\}$ on X. Recall $C_\infty(j) = E\Sigma_j$. The composition

$$X^p \to X^p \times_{\Sigma_p} E\Sigma_p \to X$$

is the p-fold product. In $K_*(X;Z/_p)$ let $\mathrm{Ind}_p(X)$ be the submodule generated by elements of the form

$$\{x^p \mid x \in K_\alpha(X;Z/_p)\} \quad \text{if } p \neq 2$$

and $\{x^2 \mid x \in \ker \beta_2 \subset K_\alpha(X;Z/_2)\}$ if $p = 2$.

Theorem 5.1:

Let X be an H^∞-space and p a prime. There exist operations

$$Q: K_\alpha(X;Z/_p) \to K_\alpha(X;Z/_p)/\mathrm{Ind}_p(X), \quad \text{if } p \neq 2,$$

$$Q: \ker \beta_2 \to K_\alpha(X;Z/_2)/\mathrm{Ind}_2(X), \quad \text{if } p = 2$$

$$\ker \beta_2 \cap K_\alpha(X;Z/_2)$$

satisfying the following conditions.

(i) Q is natural for H^∞-maps.

(ii) Let $x,y \in K_\alpha(X;Z/_p)$.

$$Q(x+y) = Q(x) + Q(y) + \sum_1^{p-1} \left\{-\binom{p}{i}/_p\right\} x^i . y^{p-i}$$

if $\alpha \equiv 0 \pmod 2$

$$Q(x+y) = Q(x) + Q(y) \quad \text{if} \quad \alpha \equiv 1 \pmod 2).$$

(iii) Cartan formula

Let $x \in K_\alpha(X;Z/_p)$ and $y \in K_\beta(X;Z/_p)$.

(a) If $\alpha + \beta \equiv 1 (\mathrm{mod}\ 2)$

$$\Omega(x.y) = Q(x)y^p + x^p.\Omega(y) \in K_{\alpha+\beta}(X;Z/_p).$$

(b) If $\alpha \equiv \beta (\mathrm{mod}\ 2)$

$$\Omega(x.y) = \begin{cases} Q(x)y^p + x^p Q(y) & \text{if } \alpha \equiv 0 (\mathrm{mod}\ 2) \\ \lambda Q(x), \Omega(y) & \text{if } \alpha \equiv 1 (\mathrm{mod}\ 2). \end{cases}$$

in $K_o(X;Z/_p)/_{[\mathrm{Ind}_p(X)]^2}$, $(0 \neq \lambda \in Z/_p)$.

(c) If $z \in \ker \beta_2 \subset K_1(X;Z/_2)$ then $z^2 = 0$

and if $z \in K_1(X;Z/_p)$ then $z^p = 0$.

(iv) If $(\psi^k)_*: K_*(X;Z/_p) \to K_*(X;Z/_p)$

is the dual of the Adams operation, ψ^k, for k prime to p then

$$Q\ \psi^k_*(x) = \psi^k_* Q(x) \in K_*(X;Z/_p)/_{\mathrm{Ind}_p(X)}.$$

(v) (a) If $p \neq 2$ there exists $\gamma_i \in Z/_p$ independent of X, such that

$$\beta_p(\Omega(x)) = \begin{cases} \gamma_1 Q(\beta_p(x)) - x^{p-1}\beta_p x, & \text{if } \deg x \equiv 0 (\mathrm{mod}\ 2) \\ \gamma_o(\beta_p(x))^p, & \text{if } \deg x \equiv 1 (\mathrm{mod}\ 2). \end{cases}$$

(b) Let B_2 be the second mod 2 Bockstein.

For $x \in \ker \beta_2 \subset K_\alpha(X;Z/_2)$

$$\beta_2\Omega(x) = Q(B_2 x) \in K_1(X;Z/_2) \quad \text{if } \alpha \equiv 0 (\mathrm{mod}\ 2)$$

and

$$\beta_2\Omega(x) = B_2(x)^2 \qquad \text{if } \alpha \equiv 1 (\mathrm{mod}\ 2).$$

[Here $B_2(x)^2$ means z^2 for any $z \in B_2(x)$.]

(vi) Let $\sigma : K_\alpha(\Omega X; Z/_p) \to K_{\alpha-1}(X; Z/_p)$ be the suspension homomorphism. Let $x \in K_\alpha(\Omega X; Z/_p)$.

(a) If $p = 2$, $\alpha \equiv 0 \pmod 2$

$$\sigma Q(x) = Q\sigma(x) \in K_1(X; Z/_2).$$

(b) If $p = 2$, $\alpha \equiv 1 \pmod 2$

$$\sigma Q(x) = \sigma(x)^2 \in K_0(X; Z/_2).$$

(c) If $p \neq 2$ there exist $0 \neq \lambda, n_o \in Z/_p$ independent of X (c.f. §A3.5) such that

$$\sigma Q(x) = \begin{cases} (-\lambda/n_0) Q(\sigma(x)), & \text{if } \alpha \equiv 0 \pmod 2 \\ n_0 \sigma(x)^p, & \text{if } \alpha \equiv 1 \pmod 2 \end{cases}$$

Before proving Theorem 5.1 we state the corollaries of §4 which concern the case $p = 2$.

<u>Proposition 5.2:</u>

Let $x \in K_\alpha(X; Z/_2)$.

(i) There exist elements $\{Q(u_{\beta,q}) \in K_\beta(X; Z/_2)/_{\mathrm{Ind}_2(X)}$ and $\{v_r \in K_*(X; Z/_2)\}$ such that

$$Q(x^2) = \sum Q(u_{1,q})Q(u_{1,r}) + \sum Q(u_{o,q})v_r v_s\,, \quad \text{modulo}$$

$\mathrm{im}\{\tilde{K}_*(X; Z/_2)^{\otimes 4} \to \tilde{K}_o(X; Z/_2)\}$.

(ii) There exist elements $\{Q(w_r) \in K_1(X; Z/_2)\}$ and $\{v_r \in K_*(X; Z/_2)\}$ such that $Q(x.\beta_2 x) = \sum Q(w_r)v_s v_t$ modulo $\mathrm{im}\{\tilde{K}_*(X; Z/_2)^{\otimes 4} \to \tilde{K}_1(X; Z/_2)\}$.

Definition 5.3:

Let Ω be the operation given by the compositions

$$K_\alpha(X;Z/_p) \xrightarrow[q]{} K_\alpha^{\Sigma_p}(X^p;Z/_p)/_{\mathrm{Ind}_p(X)} \xrightarrow[(\mu_p)_*]{} K_\alpha(X;Z/_p)/_{\mathrm{Ind}_p(X)}$$

if $p \neq 2$

and

$$\begin{array}{ccccc} \ker \beta_2 & \xrightarrow{q} & K_\alpha^{\Sigma_2}(X^2;Z/_2)/_{\mathrm{Ind}_2(X)} & \xrightarrow{(\mu_2)_*} & K_\alpha(X;Z/_2)/_{\mathrm{Ind}_2(X)} \\ \cap & & & & \\ K_\alpha(X;Z/_2) & & & & \end{array}$$

where q is the operation of §4.

Proof of Theorem 5.1 and Proposition 5.2:

For the operation Q of Definition 5.3 the properties (§5.1(i)-(v)) and (§5.2) are immediate corollaries of §4, the fact that $(\mu_p \circ i)_*$ is the p-fold product and the commutative diagrams of [Ma,§1.4].

We now deduce (§5.1(vi) from the results of Appendix III. Let Y be a space. From [Ma,§6.10] there is a space,

$E_n(p,P\Omega^{n-1}Y,\Omega^nY)$, and a commutative diagram in which the right-hand vertical is the path fibration.

$$\begin{array}{ccccc} & \hookrightarrow & (\Omega^nY)^p\times_{\Sigma_p}\mathcal{C}_n(p) & \xrightarrow{\theta_{n,p}} & \Omega^nY \\ (\Omega^nY)^p\times_{\Sigma_p}C_{n-1}(p) & & \downarrow & & \downarrow \\ & \searrow \cap & E_n(p,P\Omega^{n-1}Y,\Omega^nY) & \xrightarrow{\tilde{\theta}_{n,p}} & P\Omega^{n-1}Y \\ & & \uparrow & P\theta_{n-1,p} & \\ & & (P\Omega^{n-1}Y)^p\times_{\Sigma_p}C_{n-1}(p) & & \downarrow s \\ & s^p\times_{\Sigma_p}1 & \downarrow & & \\ & & (\Omega^{n-1}Y)^p\times_{\Sigma_p}C_{n-1}(p) & \xrightarrow[\theta_{n-1,p}]{} & \Omega^{n-1}Y \end{array}$$

Hence we obtain a commutative diagram

$$\begin{array}{ccc}
\tilde{K}^*(\Omega^{n-1}Y;Z/_p) & \xrightarrow{\theta^*_{n-1,p}} & K^*((\Omega^{n-1}Y)^p\times_{\Sigma_p}C_{n-1}(p);Z/_p) \\
 & & \downarrow s' \\
\downarrow s^* & & K^*(\Gamma(P\Omega^{n-1}Y)^p,(\Omega^nY)^p\rceil\times_{\Sigma_p}C_{n-1}(p);Z/_p) \\
 & & \uparrow \\
K^*(P\Omega^{n-1}Y,\Omega^nY;Z/_p) & \longrightarrow & K^*(E_n(p,P\Omega^{n-1}Y,\Omega^nY),(\Omega^nY)^p\times_{\Sigma_p}C_n(p);Z/_p) \\
\delta_1\uparrow\cong & & \\
\tilde{K}^*(\Omega^nY;Z/_p) & \longrightarrow & K^*(\Gamma(\Omega^nY)^p,*\rceil\times_{\Sigma_p}C_n(p);Z/_p)
\end{array}$$

$$\delta_2: K^*(\Gamma(\Omega^nY)^p,*\rceil\times_{\Sigma_p}C_n(p);Z/_p)\to K^*(E_n(p,P\Omega^{n-1}Y,\Omega^nY),(\Omega^nY)^p\times_{\Sigma_p}C_n(p);Z/_p)$$

$$\alpha_n: K^*(\Gamma(\Omega^nY)^p,*\rceil\times_{\Sigma_p}C_n(p);Z/_p)\to K^*(\Gamma(\Omega^nY)^p,*\rceil\times_{\Sigma_p}C_{n-1}(p);Z/_p)$$

$$\delta_3\ (\cong): K^*(\Gamma(\Omega^nY)^p,*\rceil\times_{\Sigma_p}C_{n-1}(p);Z/_p)\to K^*(\Gamma(P\Omega^{n-1}Y)^p,(\Omega^nY)^p\rceil\times_{\Sigma_p}C_{n-1}(p);Z/_p)$$

If X is an infinite loopspace we may replace $\Omega^{n-1}Y$ by X in the above diagram and let $n\to\infty$ in which case $\alpha_\infty = 1$ and $\theta_{\infty,p} = \mu_p$.

Take a map $f\colon C\Omega X\to PX$ extending the inclusion $\Omega X\subset PX$. Hence there is a commutative diagram, induced by f,

$$\begin{array}{ccc}
K^*_{\Sigma_p}((PX,\Omega X)^p;Z/_p) & \xrightarrow{\cong} & K^*_{\Sigma_p}((C\Omega X,\Omega X)^p;Z/_p) \\
\downarrow & & \downarrow \\
K^*_{\Sigma_p}((PX)^p,(\Omega X)^p;Z/_p) & \xrightarrow{\cong} & K^*_{\Sigma_p}((C\Omega X)^p,(\Omega X)^p;Z/_p) \\
\delta_3\uparrow\cong & & \\
K^*_{\Sigma_p}((\Omega X)^p,*;Z/_p) & &
\end{array}$$

$$\delta: K^*_{\Sigma_p}((\Omega X)^p,*;Z/_p)\to K^*_{\Sigma_p}((C\Omega X)^p,(\Omega X)^p;Z/_p)$$

and we may use the analysis of δ from Appendix III to determine δ_3. Also the commutative diagram

$$\begin{array}{ccc}
K^*_{\Sigma_p}(X^p;Z/_p) & \xleftarrow{i_!} & K^*(X^p;Z/_p) \\
\downarrow s' & & \downarrow s'' \\
K^*_{\Sigma_p}((PX)^p,(\Omega X)^p;Z/_p) & \xleftarrow{i_!} & K^*(PX)^p,(\Omega X)^p;Z/_p)=K^*(P(X^p),\Omega(X^p);Z/_p) \\
\delta_3 \uparrow \cong & & \cong \uparrow \delta_3' \\
K^*_{\Sigma_p}((\Omega X)^p,*;Z/_p) & \xleftarrow{i_!} & \tilde{K}^*((\Omega X)^p;Z/_p) = \tilde{K}^*(\Omega(X^p);Z/_p)
\end{array}$$

shows that

$$(\delta_3^{-1})s'(i_!(x_1\otimes\ldots\otimes x_p)) = i_!((\delta_3'^{-1})s''(x_1\otimes\ldots\otimes x_p))$$

is zero unless all but one of the $\{x_i\}$ are equal to an element $Z/_p$, since the right-hand vertical is the transgression for X^p and hence kills decomposables.

For $x \in \tilde{K}_\alpha(\Omega X;Z/_p)$ we may now evaluate $< q(x),(\delta_3^{-1})s'(z) >$ for $z \in K^{\alpha+1}_{\Sigma_p}(X^p;Z/_p)$.

Case (a): $p = 2$, $x \in \ker \beta_2 \subset \tilde{K}_o(\Omega X;Z/_2)$.

$$<q(x),(\delta_3^{-1})s'(i_!(u\otimes v) + z_1^{\otimes 2}\otimes e_1)>$$

$$\doteq \quad <q(x),i_!(\delta_3')^{-1}s''(u\otimes v) + i_![(\delta_1)^{-1}s^*(z_1)^{\otimes 2}>$$

by Proposition A.3.2(ii)

$$= \quad < x^{\otimes 2},(\delta_3')^{-1}s''(u\otimes v) + [(\delta_1)^{-1}s^*(z_1)]^{\otimes 2}>$$

$$= \quad < x^{\otimes 2}, [(\delta_1)^{-1}s^*(z_1)]^{\otimes 2}>$$

$$= \quad < x, (\delta_1)^{-1}s^*(z_1)>$$

$$= \quad < \sigma x,z_1 >$$

$$= \quad < q\sigma(x), i_!(u\otimes v) + z_1^{\otimes 2}\otimes e_1 > .$$

Case (b): $p = 2$, $x \in \ker \beta_2 \subset \tilde{K}_1(\Omega X; Z/_2)$.

$$< q(x), (\delta_3^{-1})s'(i_!(u \otimes v) + z_o^{\otimes 2} \otimes e_o) >$$

$$= \quad < q(x),\ i_!((\delta_3')^{-1}s''(u \otimes v)) + [(\delta_1^{-1})s^*(z_o)]^{\otimes 2} \otimes e_o + i_!(z) >$$

by Proposition A.3.2(i)

$$= \quad < x,\ (\delta_1^{-1})s^*(z_o) >$$

$$= \quad < \sigma(x), z_o >$$

$$= \quad < i_*(\sigma(x)^{\otimes 2}), i_!(u \otimes v) + z_o^{\otimes 2} \otimes e_o > .$$

Case (c)(i) $p \neq 2$ $x \in \tilde{K}_o(\Omega X; Z/_p)$

$$< q(x),\ \delta_3^{-1}\ s'(i_!(u_1 \otimes \ldots \otimes u_p) + z^{\otimes P} \otimes e_{p-1}) >$$

$$= < q(x), i_!((\delta_3')^{-1}s''(u_1 \otimes \ldots \otimes u_p)) + (-{}^1/n_o)[\delta_1^{-1}s^*(z)]^{\otimes P} \otimes e_{2p-2} >$$

by Proposition A3.5

$$= \quad (-{}^{\lambda}/n_o) < \sigma(x), z > \quad (\text{where } (\lambda i_!(1) = e_{2p-2}))$$

$$= < (-{}^{\lambda}/n_o) q(\sigma(x)),\ i_!(u_1 \otimes \ldots \otimes u_p) + z^{\otimes P} \otimes e_{p-1} > .$$

Case (c)(ii) $p \neq 2$, $x \in \tilde{K}_1(\Omega X; Z/_p)$

Similarly $<q(x), \delta^{-1}s'((x))> = <n_o i_*(\sigma(x)^{\otimes P}), w>$ $(w \in K^*_{\Sigma_p}(X^P; Z/_p))$.

Hence for $y \in K^*(X; Z/_p)$

$$< \sigma Q(x), y >$$

$$= \quad < Q(x),\ \delta_1^{-1}s^*(y) >$$

$$= \quad < (\mu_p)_* q(x),\ \delta_1^{-1}\ s^*(y) >$$

$$= \quad < q(x),\ \mu_p^*\ \delta_1^{-1}\ s^*(y) >$$

$$= \quad < q(x),\ \delta_3^{-1}\ s'\ \mu_p^*(y) >$$

$$= \begin{cases} < q\sigma(x),\ \mu_2^*(y) > = < Q\sigma(x), y > & \text{in case (a),} \\ < i_*\sigma(x)^{\otimes 2},\ \mu_2^*(y) > = < \sigma(x)^2, y > & \text{in case (b),} \\ < (-\lambda/n_o)q(\sigma(x)), \mu_p^*(y) > = < (-\lambda/n_o)Q\sigma(x), y > & \text{in case (c)(i),} \\ < n_o i_*\sigma(x)^{\otimes P},\ \mu_p^* y > = < n_o\sigma(x)^P, y > & \text{in case (c)(ii).} \end{cases}$$

§6: Q on BU and Z × BU

§6.1: Let BU and Z × BU be the representing spaces for the groups, $\tilde{K}(-)$ and $K(-)$, respectively. Thus $BU = \varinjlim_n BU(n)$ and $K^*(BU(n)) = Z[[\gamma^1,\ldots,\gamma^n]]$ where γ^i is the i-th γ-operation [At 2] and $\deg \gamma^i \equiv 0 \pmod 2$. $K^*(BU(n))$ embeds, via the restriction homomorphism, as the invariants under the action of the Weyl group, Σ_n, in $K^*(BT(n))$ where $K^*(BT(n)) = Z[[c_1,\ldots,c_n]]$ and γ^i corresponds to the i-th elementary symmetric function in the $\{c_j\}$. Dually we have an epimorphism, $K_*(BT(n)) \to K_*(BU(n))$. Care is required here since the universal coefficient theorems of [An] behave well only for finite complexes, but in this case the skeletal filtration of $K^*(BU(n))$ and $K^*(BT(n))$ behave nicely and the dual of a monomorphism is indeed an epimorphism. Also $K_*(BT(n)) \to K_*(BU(n)$ factors through an isomorphism between the Σ_n-coinvariants of $K_*(BT(n))$ and the group $K_*(BU(n))$. Now form the disjoint union $BT_* = \coprod_{o \leq n} BT(n)$ and $BU_* = \coprod_{o \leq n} BU(n)$, which are monoids under the Whitney sum operation, $U(m) \times U(n) \to U(m+n)$. Now $K^*(BT(1)) = Z[[c]]$ with comultiplication $m^*(c) = c \otimes 1 + 1 \otimes c + c \otimes c$ in $K^*(BT(1) \times BT(1))$. Hence the dual $K_*(BT(1))$ has a basis $\{u_i\}$ $(i \geq 1)$ satisfying

$$\langle u_i, c^j \rangle = \delta_{i,j}$$

and $(i,j > 0)$

$$m_*(u_i \otimes u_j) = \sum_q \binom{q}{2q-i-j}\binom{2q-i-j}{q-j} u_q \qquad (6.1.1)$$

where the sum is taken over $(1 \leq q \leq i+j \leq 2q)$.

By the Kunneth formula $K_*(BT_*) \cong T(K_*(BT(1))$, the tensor algebra on $K_*(BT(1))$ and the epimorphism $T(K_o(BT(1))) \to K_*(BT_*)$

factors through the coinvariants under the action of $\coprod_{0\leq n} \Sigma_n$, the symmetric algebra $S(K_o(BT(1)))$, to give an isomorphism of algebras

$$\phi_o: S(K_o(BT(1))) \xrightarrow{\cong} K_*(BU_*).$$

If $u_o = 1 \in K_*(BT(1))$ then

$$S(K_o(BT(1))) = Z[u_o, u_1, u_2, \ldots, u_n, \ldots],$$

a polynomial algebra. Finally

$$BT(n) \to \{n\} \times BU(n) \to \{n\} \times BU \subset Z \times BU$$

induces $S(K_o(BT(1))) \to K_*(Z \times BU)$ where $u_o \in K_o(\{1\} \times BU)$ acts like a translation. Hence we have isomorphism

$$\phi_1: Z[u_o, u_o^{-1}, u_1, u_2, \ldots, u_n, \ldots] \xrightarrow{\cong} K_*(Z \times BU)$$

and

$$\phi_2: Z[v_1, v_2, \ldots, v_n, \ldots] \xrightarrow{\cong} K_*(BU)$$

related by $\quad \phi_1(u_j) = \phi_o(u_j) \qquad (j \geq 0)$

and $\quad \phi_2(v_j) = \phi_1(u_j/u_o)$.

§6.2:

Now consider the H^∞-action map, μ_p. In this section we will describe explicitly the action maps $\Theta_{2n,p}[M_a]$. This discussion is included for completeness but the description of $(\mu_p)^*$ and $(\mu_p)_*$ which can be obtained from this discussion will not be used in the computation of Q on $Z \times BU$.

We will use the method of [P] to obtain the Q-action on $Z \times BU$, although it is possible to obtain these results from μ_p^* and the results of [K].

Let $b \in \tilde{K}(S^2) \cong K(D^2, S^1)$ be the Thom class and let $b_n: S^{2n} \to BU$ correspond to $b^{2n} \in \tilde{K}(S^{2n})$. Let $w: BU \times BU \to BU$

be the Whitney sum, with respect to which H^{∞}-structure we are working, and let $m: BU \times BU \to BU$ be the multiplication map corresponding to $\gamma^1 \otimes \gamma^1$. The map

$$B_n = m_o(b_n \wedge 1): S^{2n} \wedge BU \to BU$$

induces a homotopy equivalence

$$\mathrm{adj}(B_n): BU \to (\Omega^{2n}BU)_o$$

where $(\ )_o$ is the component of the base point. An element $c \in C_{2n}(p)$ $[M_a]$ may be considered as a map $c: S^{2n} \to \bigvee_1^p S^{2n}$. In terms of this representation of $C_{2n}(p)$ the adjoint of $\Theta_{2n,p}$

$$\mathrm{adj}(\Theta_{2n,p}): S^{2n} \wedge [(BU)^p \times_{\Sigma_p} C_{2n}(p)] \to BU$$

is given by

$$\mathrm{adj}\ \Theta_{2n,p}(z \wedge [x_1,\dots,x_p,c]) = [\bigvee_{i=1}^{p} B_n(-,x_i)]\circ(c \wedge 1)(z),$$

$$(z \in S^{2n},\ x_i \in BU,\ c \in C_{2n}(p)).$$

As μ_p is $\lim\limits_{\overrightarrow{n}} \Theta_{2n,p}$ the homomorphism

$$\mu_p^*: K^*(BU) \to K^*((BU)^p \times_{\Sigma_p} E\Sigma_p)$$

is $\lim\limits_{\overleftarrow{n}} (\Theta_{2n,p})^*$.

Similarly $B_n': Z \times BU \xrightarrow{\simeq} \Omega^{2n}BU$ is defined by $B_n'|\{n\} \times BU = \mathrm{adj}\ B_n + n.\alpha$ where $\alpha: BU \to \Omega^{2n}BU$ is a constant map into $(\Omega^{2n}BU)_1$, the component of maps of degree one.

From this explicit description of $\Theta_{2n,p}$ it is simple to show that the map

$$BT(1)^p \times_{\pi_p} E\pi_p \to BT(1)^p \times_{\Sigma_p} E\Sigma_p \to (\{0\}\times BU)^p \times_{\Sigma_p} E\Sigma_p \xrightarrow{\mu_p} BU$$

represents $(i_1)_!(c \otimes 1^{\otimes p-1}) \in K^0(BT(1)^p \times_{\pi_p} E\pi_p)$, which, in principal determines Q on BU, since it determines $(\mu_p)^*$ and hence $(\mu_p)_*$.

§6.3:

Let $U(n)^p \subset U(np)$ be the canonical Whitney sum homomorphism. This extends to an inclusion $j\colon \Sigma_p \int U(n) \to U(np)$ inducing $E\Sigma_p \times_{\Sigma_p} (BU(n))^p = B(\Sigma_p \int U(n) \xrightarrow[Bj]{} BU(np)$. The following homotopy commutative diagram of Boardman [K-P, §2.4; P, §1] $(n, p \geq 1)$

$$(6.3.1) \qquad \begin{array}{ccc} E\Sigma_p \times_{\Sigma_p} BU(n)^p & \xrightarrow{Bj} & BU(np) \\ \downarrow & & \downarrow 1 \\ E\Sigma_p \times_{\Sigma_p} (\{n\}\times BU)^p & \xrightarrow[\mu_p]{} & \{np\}\times BU(np) \subset Z\times BU \end{array}$$

turns the determination of Q from a homotopy theoretic problem into a simple group-theoretic problem.

By the Cartan formula it suffices to determine $Q(u_i)$ $(i \geq 0)$ and $Q(u_o^{-1})$. Before doing this we prove some preliminary results.

Let $e_j \in K_o(B\pi_p; Z/_p)$, $(0 \leq j \leq p-1)$, satisfy $< e_j, \sigma^k > = \delta_{j,k}$, $(0 \leq k \leq p-1)$. Hence the image of e_j in $K_o(BT(1); Z/_p)$ is u_j, $(0 \leq j \leq p-1)$. If $\underline{t} = (t_1, \ldots, t_p)$ is an ordered set of positive integers let $G(\underline{t})$ denote the subgroup of Σ_p which stabilises $\underline{t}$. Let $\Delta\colon \pi_p \times S^1 \to \Sigma_p \int S^1$ be given by $\pi_p \subset \Sigma_p$ on π_p and the diagonal $S^1 \to (S^1)^p$ on S^1.

Lemma 6.3.2:

If $0 \neq \mu \in Z/_p$ satisfies

$$0 = (i_1)_!(1) + \mu\, \sigma^{p-1} \in R(\pi_p) \otimes Z/_p$$

then

$$x_p = 1/_\mu\, \Delta_*(i_2)_*(e_{p-1} \otimes u_{kp}) - \sum_{\underline{t}} \frac{1}{|G(\underline{t})|}\, i_*(u_{t_1} \otimes \ldots \otimes u_{t_p})$$

is an element of $q(u_k) \subset K_*^{\Sigma_p}((BS^1)^p; Z/_p)$, (where $\underline{t}$ runs over partitions of kp not equal to $(k,k,\ldots,k)$).

Proof:

$$\begin{aligned}
&< x_p, i_!(c^{s_1} \otimes \ldots \otimes c^{s_p})> \\
= \quad & (1/_\mu) < e_{p-1} \otimes u_{kp}\, ,\, i_2^*(i_2)_!(i_1)_!\Delta^*(c^{s_1} \otimes \ldots \otimes c^{s_p})> \\
& - \sum_{\underline{t}} \frac{1}{|G(\underline{t})|} < u_{t_1} \otimes \ldots \otimes u_{t_p}\, ,\, i^* i_!(c^{s_1} \otimes \ldots \otimes c^{s_p})> \\
= \quad & 1/_\mu < (i_2)^!(i_2)_* e_{p-1} \otimes u_{kp}, (-\mu)\sigma^{p-1} \otimes c^s> \\
& - \sum_{\underline{t}} \frac{1}{|G(\underline{t})|} < u_{t_1} \otimes \ldots \otimes u_{t_p}\, ,\, \sum_{g \in \Sigma_p} g^*(c^{s_1} \otimes \ldots \otimes c^{s_p})> \\
& \qquad \text{(where } s = \sum_k s_k\text{)} \\
= \quad & - <(p-1)!\, e_{p-1} \otimes u_{kp}, \sigma^{p-1} \otimes c^s > \\
& - \sum_{\underline{t}} 1/_{|G(\underline{t})|} < u_{t_1} \otimes \ldots \otimes u_{t_p}\, ,\, \sum_{g \in \Sigma_p} g^*(c^{s_1} \otimes \ldots \otimes c^{s_p})> \\
= \quad & < u_k^{\otimes p}\, ,\, c^{s_1} \otimes \ldots \otimes c^{s_p}> \\
= \quad & < q(u_k)\, ,\, i_!(c^{s_1} \otimes \ldots \otimes c^{s_p})> .
\end{aligned}$$

There is a commutative diagram of homomorphisms

$$
\begin{array}{ccccc}
\pi_p \times S^1 & \xrightarrow{\Delta} & \Sigma_p \int S^1 & \xrightarrow{j} & U(p) \\
\alpha \downarrow & & & & \downarrow \beta \\
(S^1)^p & & \xrightarrow[j]{\hspace{3em}} & & U(p)
\end{array} \tag{6.3.3}
$$

where β is an inner automorphism and

$$\alpha(g,z) = (z, g.z, g^2 z, \ldots, g^{p-1}.z), \quad (\pi_p \subset S^1).$$

To prove this merely observe that S^1 goes in by $j \circ \Delta$ via either route and a generator of π_p goes to a matrix with eigenvalues $\{e^{2\pi ik/p} ; \ 0 \leq k \leq p-1\}$ via either route. Notices that $B(\beta)_*$ is the identity endomorphism of $K_*(BU(p);Z/_p)$.

<u>Lemma 6.3.4</u>:

$$(B\alpha)_* (e_{p-1} \otimes u_{kp})$$

$$= \sum_{\underline{s},\underline{r}} u_{r_1} \otimes m_*(\psi_*^1(u_{s_1}) \otimes u_{r_2}) \otimes \ldots \otimes m_*(\psi_*^{p-1}(u_{s_{p-1}}) \otimes u_{r_p})$$

in $K_*(BT(1)^p;Z/_p)$, where the summation is over partitions $\underline{s} = (s_1,\ldots,s_{p-1})$ of $(p-1)$ and $\underline{r} = (r_1,\ldots,r_p)$ of kp.

<u>Proof</u>:

On $K_*(BS^1;Z/_p)\psi_*^j$ $(1 \leq j \leq p-1)$ is induced by the j-th power map on S^1. However α is the composition

$$\pi_p \times S^1 \longrightarrow S^1 \times (S^1 \times S^1)^{p-1} \xrightarrow[1\times\pi_1^{p-1}\psi^j]{} S^1 \times (S^1 \times S^1)^{p-1} \xrightarrow[1\times m^{p-1}]{} (S^1)^p .$$

<u>Theorem 6.3.5</u>:

In the notation of Lemma 6.3.2 and Lemma 6.3.4

$$({}^{1}/_{\mu})\ [\sum_{\underline{s},\underline{r}} u_{r_1} \prod_{j=1}^{p-1} m_*(\psi_*^j(u_{s_j}) \otimes u_{r_{j+1}})]$$

$$- \sum_{\underline{t}} \frac{1}{|G(t)|} \prod_{j=1} u_{t_j}$$

is in $Q(u_k) \subset K_o(\{p\} \times BU; Z/_p)$.

Proof:

Immediate from §6.3.1, §6.3.3 and the formulae of Lemma 6.3.2 and Lemma 6.3.4.

Lemma:

Let p be an odd prime. There exists an integer, $j(p)$, $(0 \leq j(p) < p-1)$ such that

$$\sum_{1}^{p-1} (\psi^j)_*(u_{p-1}) = \sum_{t=1}^{j(p)} \lambda_t u_t \in K_o(BS^1; Z/_p)$$

where $0 \neq \lambda_{j(p)} \in Z/_p$ except when $j(p) = 0$, in which case the sums are zero.

For example $j(3) = 1$ and $\lambda_{j(3)} \equiv 2 \pmod 3$.

Proof:

If $c = y-1 \in K^o(BS^1)$ then $\psi^j(c) = jc + j\frac{(j-1)}{2}c^2 + c^3f(c)$.

Hence $\langle \sum(\psi^j)_*(u_{p-1}), c^m \rangle$

$$= \langle u_{p-1}, \sum[jc + j\frac{(j-1)}{2}c^2 + c^3f(c)]^m \rangle$$

$$= \begin{cases} 0 & , \text{ if } m \geq p, \\ \sum j^{p-1} \equiv \sum_j = \frac{(p-1)p}{2} & , \text{ if } m = p-1, \\ \sum (p-2) j^{p-2}\frac{(j-1)}{2} \equiv \frac{p(p-1)(2p-1)(2-p)}{12}, & \text{ if } m = p-2. \end{cases}$$

Corollary 6.3.6:

Let Decomp $\subset K_o(Z\times BU;Z/_p)$ denote the translates of the decomposables in the algebra $K_o(O\times BU;Z/_p)$. For p an odd prime let $0 \leq j(p) < p-1$ be defined as in the previous lemma and put $j(2) = 1$.

Then for $u_k \in K_o(1\times BU;Z/_p)$

$$u_o^{-p+1}\Omega(u_k) \equiv \sum_{1}^{kp+j(p)} \nu_t u_t \qquad \text{(mod Decomp)}$$

where $\nu_{kp+j(p)} \not\equiv 0 \pmod p$.

In fact for $p=2$ $u_o^{-1}\Omega(u_k) \equiv u_{2k}+u_{2k+1}$ (mod Decomp).

Proof:

The only indecomposable monomials in the formula of Theorem 6.3.5 are the ones arising from $\underline{t} = (kp,0,0,\ldots)$ and the $(p-1)$ terms which arise when $s_u = p-1$, $s_v = 0$ $(v\neq u)$ and $r_{u+1} = kp$, $r_w = 0$ $w \neq u+1$ for $1 \leq u \leq p-1$.

Thus $u_o^{-p+1}\Omega(u_K) \equiv \frac{1}{\mu}\sum_{j=1}^{p-1} m_*(\psi^j(u_{p-1})\otimes u_{kp}) - u_{kp}$ (mod Decomp).

However, if $1 \leq t \leq p-1$, then

$$< m_*(u_t\otimes u_{kp}),\ c^m >$$

$$= \ < u_t\otimes u_{kp},\ (c\otimes c + c\otimes 1 + 1\otimes c)^m >$$

$$= \begin{cases} 0 & \text{if } m > kp + t, \\ \binom{m}{t} \not\equiv 0 \pmod p & \text{if } m = kp + t. \end{cases}$$

Hence the general result follows. When $p=2$

$$< u_1\otimes u_{2k},\ (c\otimes c + c\otimes 1 + 1\otimes c)^m >$$

$$= \begin{cases} 0 & \text{if } m > 2k+1 \text{ or } m < 2k, \\ 2k \equiv 0 \pmod 2 & \text{if } m = 2k \\ 2k+1 \equiv 1 \pmod 2 & \text{if } m = 2k+1. \end{cases}$$

so that $m_*(u_1 \otimes u_{2k}) = u_{2k+1}$.

Lemma 6.3.7:

(i) $(1/_\mu) \sum_{\underline{s}} \prod_{j=1}^{p-1} \psi_*^j(u_{s_j}) \in \Omega(u_o) \subset K_o(\{p\} \times BU; Z/_p)$

where the sum is taken over $\underline{s} = (s_1, \ldots, s_{p-1})$ which partition $(p-1)$.

(ii) $\Omega(u_o^{-1}) = u_o^{-2p}\, \Omega(u_o) \in K_o(Z \times BU; Z/_p) /_{\text{Ind } p}(Z\times BU)$.

Proof:

(i) This follows from the fact that

$$1/_\mu\ \Delta_*(i_2)_*(e_{p-1}) \in q(u_o).$$

(ii) This follows from the Cartan formula since $u_o . u_o^{-1} = 1$

and $0 = (\mu_p)_*: \tilde{K}_*(B\Sigma_p; Z/_p) \longrightarrow K_*(\{0\} \times BU; Z/_p)$.

We conclude this section by obtaining new proofs of the results of [H2] on $K_*(\Omega S^o; Z/_p)$. Let $(QS^o)_q$ be the component consisting of maps of degree q.

Theorem 6.3.8:

Let $\alpha: QS^o \to Z \times BU$ be the infinite loop map derived from $\alpha': S^o \to Z \times BU$ given by $\alpha'(i) = (i, b_o)$ $(i = 0,1)$.

Let $1 = \theta_1 \in K_o((\Omega S^o)_1; Z/_p)$ and

$$\theta_{pj} \in q^j(\theta_1) \subset \text{im}\{K_o(B\Sigma_{pj}; Z/_p) \to K_o((QS^o)_{pj}; Z/_p)\}$$

then

(i) $\alpha_*(\Theta_1) = u_o$,

(ii) α_* is a monomorphism,

(iii) $K_*(QS^o;Z/_p) \cong Z/_p[\Theta_1,\Theta_1^{-1},\Theta_p,\ldots,\Theta_{p^k},\ldots]$.

Proof:

(i) This is obvious from the definition of α'.

We prove (ii) and (iii) together. The action map

$$\mu_p\colon C_\infty(S^o) = \coprod_n B\Sigma_n = B\Sigma_* \to QS^o$$

is easily shown to induce an epimorphism in

$K_*(-;Z/_p)[\Theta_1^{-1}]$, [H2]. However from the behaviour of the transfer (§2) $K_*(B\Sigma_*;Z/_p)$ is easily seen to be generated by the images of $K_*(BG;Z/_p)$ where G runs through iterated wreath products of Σ_p. Hence $K_*(B\Sigma_*;Z/_p)$ is generated by iterates of Θ_1 under the $q(-)$ operation.

Now $\psi^j_* u_{p-1} + u_{p-1} = \sum_{k=1}^{p-2} y_k u_u$, $(y_k \in Z/_p)$,

and by Lemma 6.3.7(i) and Corollary 6.3.6 the $\{\alpha_*(\Theta_{p^k} / \Theta_1^{p^k}); k \geq 0\}$ are polynomially independent in $K_*(Z \times BU;Z/_p)$. Thus the epimorphism

$Z/_p[\Theta_1,\Theta_1^{-1},\ldots,\Theta_{p^k},\ldots] \to K_*(QS^o;Z/_p)$ is a monomorphism and so is α_* .

§7: Mod p spherical characteristic classes in K-theory

Let G_n be the set of homotopy equivalences, $f:S^{n-1} \to S^{n-1}$, topologised via the pairing $\langle f,g\rangle = \sup_{x\in S^{n-1}} \langle f(x),S(x)\rangle$. The composition map $\circ: G_n \times G_n \to G_n$ makes G_n into an associative H-space with a unit. Let $F_{n-1} \subset G_n$ be the subspace of base-point preserving maps. Put $G = \varinjlim_n G_n$ and $F = \varinjlim_n F_n$, so that $F = \varinjlim_n (\Omega^n S^n)_{\pm 1} = (QS^0)_{-1} \cup (QS^0)_{+1}$ with the H-space multiplication given by composition. The quasi-fibration $F_i \xrightarrow{j} G_{i+1} \longrightarrow S^i$ shows that there is an isomorphism of Pontrjagin algebras $K_*(F;Z/_p) \xrightarrow[j_*]{} K_*(G;Z/_p)$.

The components of the identity map are called SF and SG. From the point of view of the induced structures on $K_*(F;Z/_p)$ the composition product and the smash product induce the same Pontrjagin algebra and the same Q operation from their H^∞-structures.

If Σ_m permutes objects $(e_1,\ldots,e_m)$ which are basis elements of some vector space V then $(\Sigma_m)^n$ permutes the basis $\{e_{i_1}\otimes\ldots\otimes e_{i_n}\}$ of $V^{\otimes n}$ by $(\sigma_i \in \Sigma_m)$

$$(\sigma_1\times\ldots\times\sigma_n)(e_{i_1}\otimes\ldots\otimes e_{i_n}) = e_{\sigma_1(i_1)}\otimes\ldots\otimes e_{\sigma_n(i_n)}$$

and Σ_n permutes this basis by $(\tau\in\Sigma_n)$

$$\tau(e_{i_1}\otimes\ldots\otimes e_{i_n}) = e_{i_{\tau(1)}}\otimes\ldots\otimes e_{i_{\tau(n)}}.$$

Let $\Sigma_m \int \Sigma_n \overset{i}{\subset} \Sigma_{m^n}$ denote the semi-direct product group generated by $(\Sigma_m)^n$ and Σ_n acting on the ordered set of m^n

basis elements $\{e_{i_1}\otimes\ldots\otimes e_{i_n}\}$ with the following ordering. If $n=1$ the order is $(e_1,e_2,\ldots,e_m)$ and if the basis of $V^{\otimes k}$ is ordered as $(y_1,\ldots,y_t)$ $(t = m^k)$ then the basis of $V^{\otimes k}\otimes V$ is ordered as

$$(y_1\otimes e_1, y_2\otimes e_1,\ldots,y_t\otimes e_1, y_1\otimes e_2, y_2\otimes e_2,\ldots,y_t\otimes e_1,\ldots,y_t\otimes e_m).$$

If we consider QS^0 as a monoid under the composition product, rather than considering only $F \subset QS^0$, then QS^0 admits an action χ, by the operad $\{C_\infty(j); j\geq 1\}$ [Ma] extending the H^∞-action on F. Furthermore if $\mu: C_\infty(S_0) = \coprod_n B\Sigma_n \to QS^0$ is the H^∞-action map for the loop-space multiplication on $\lim_{\substack{\to\\ n}} \Omega^n S^n = QS^0$ [c.f.§6.3.8] then there are homotopy commutative diagrams (identifying $C_\infty(j)/_{\Sigma_j} = B\Sigma_j$)

(7.1)
$$\begin{array}{ccc}
B\Sigma_n \times_{\Sigma_n} (B\Sigma_m)^n = B(\Sigma_m \int \Sigma_n) & \xrightarrow{Bi} & B\Sigma_{m^n} \\
\downarrow 1\times\mu^n & & \downarrow \mu \\
B\Sigma_n \times_{\Sigma_n} ((QS^0)_m)^n & \xrightarrow{\chi} & (QS^0)_{m^n}
\end{array},$$

where $(QS^0)_j$ = (maps of degree j). A similar diagram, for the composition

$$B\Sigma_p \times B\Sigma_p \longrightarrow (QS^0)_p \times (QS^0)_p \xrightarrow{\circ} (QS^0)_{p^2}, \quad \text{is}$$

proved in [Mi, §5.1]. The general proposition is more easily seen in the framework of the C_∞-operad actions. The diagram (7.1) has also been used by J.P. May and S.B. Priddy in their computations of Dyer-Lashof operations in $H_*(F; Z/_p)$.

We now proceed to calculate the algebra $K_*(SF; Z/_p)$, after the following programme. From (§6.3.8) we know the structure of

$K_*(\Omega S^o;Z/_p)$ additively and from (7.1) we determine some o-products explicitly. We then classify all primitives in $K_*((\Omega S^o)_1;Z/_p)$ and proceed by induction on a suitable partial ordering to compare the o-product with the loop product, by considering $\Delta_*(x) - x\otimes 1 - 1\otimes x$ $(x\in K_*(SF;Z/_2), \Delta\equiv\text{diagonal})$. This process leads to the determination of a set of algebra generators for $K_*(SF;Z/_p)$ in terms of translates from $K_*((\Omega S^o)_o;Z/_p)$.

For the remainder of this section we choose, once for all, an element $(j \geq 1)$

$$\Theta_{p^j}\in q^j(\Theta_1) \in K_o(B\Sigma_{p^j};Z/_p) \qquad [\text{c.f.}\S 6.3.8]$$

and also denote by Θ_{p^j} the image of this element under $(\mu)_*$. Hence $\Theta_{p^j} \in \tilde{K}_o((\Omega S^o)_{p^j};Z/_p)$ and put $\tilde{\Theta}_j = \Theta_{p^j}.(\Theta_1^{-1})^{p^j}\in\tilde{K}_o((\Omega S^o)_o;Z/$ $(j \geq 1)$. By Theorem 6.3.8 we have

$$\tilde{K}_*((\Omega S^o)_o;Z/_p) \cong Z/_p[\tilde{\Theta}_1,\tilde{\Theta}_2,\ldots,\tilde{\Theta}_j,\ldots].$$

Also $\tilde{\Theta}_j$ is well-defined up to addition of polynomials in $\{\tilde{\Theta}_1,\ldots,\tilde{\Theta}_{j-1}\}$ which are p-th powers in $\tilde{K}_o((\Omega S^o)_o;Z/_p)$. In particular $\tilde{\Theta}_1$ is an element. If Δ is the diagonal map then by Proposition 4.3, since $\Theta_{p^j}\in q(\Theta_{p^{j-1}})$ we have

$$(7.2) \qquad \Delta_*\tilde{\Theta}_j = \tilde{\Theta}_j\otimes 1 + 1\otimes\tilde{\Theta}_j - \sum_{i=1}^{p-1}\binom{p}{i}\Big/_p\, \tilde{\Theta}_{j-1}^{\,i}\otimes\tilde{\Theta}_{j-1}^{\,p-i} + z_j^p$$

$(j \geq 2)$ in $\tilde{K}_o((\Omega S^o)_o;Z/_p)^{\otimes 2}$,

where z_j is a polynomial in $\{\tilde{\Theta}_1,\ldots,\tilde{\Theta}_{j-1}\}$. The element $\tilde{\Theta}_1$ is primitive, by §4.3, and hence so are the $(\tilde{\Theta}_1)^{p^j}$.

Proposition 7.3.

A basis for the vector space of primitive elements of $K_o((\Omega S^o)_o;Z/_p)$ is $\{(\tilde{\Theta}_1)^{p^j}\}$.

Proof: This is a straightforward consequence of (7.2). The case p=2 adequately illustrates the method of proof and is notationally less tedious than the case p odd, so we will give only the proof of the case p=2.

First it is clear that the only primitives in the vector space spanned by $\{(\Theta_1)^n;\ n\geq 1\}$ are those in the statement of the proposition. Suppose $p(\tilde{\Theta}_1,\ldots,\tilde{\Theta}_k)$ is a primitive polynomial involving $\tilde{\Theta}_k$ (k>1) non-trivially. If $p(\tilde{\Theta}_1,\ldots,\tilde{\Theta}_k)= p_1(\tilde{\Theta}_1,\ldots,\tilde{\Theta}_k)^2$ then it is also clear that p_1 is primitive. Write

$$p(\tilde{\Theta}_1,\ldots,\tilde{\Theta}_k) = \sum_{\underline{\beta}} \tilde{\Theta}_u^{\beta_u}\cdots\tilde{\Theta}_1^{\beta_1} + p_2(\tilde{\Theta}_1,\ldots,\tilde{\Theta}_k)^2$$

where the sum is taken over monomials $\tilde{\Theta}^{\underline{\beta}}$ in $p(\tilde{\Theta}_1,\ldots,\tilde{\Theta}_k)$ which are non-squares. If p is primitive we have

$$(7.4)\qquad p\otimes 1 + 1\otimes p = \sum_{\underline{\beta}} \prod_{j=1}^{k} (\tilde{\Theta}_j\otimes 1 + 1\otimes\tilde{\Theta}_j + \tilde{\Theta}_{j-1}\otimes\tilde{\Theta}_{j-1} + z_j^2)^{\beta_j}$$

$$+ (\Delta_* p_2)^2,$$

by (7.2) and each z_j is a sum of monomials of the form $\tilde{\Theta}^{\underline{\gamma}}\otimes\tilde{\Theta}^{\underline{\delta}}$ where $\tilde{\Theta}^{\underline{\gamma}}$ and $\tilde{\Theta}^{\underline{\delta}}$ are both squares of monomials which are not one. If possible choose $\underline{\beta} = (\beta_1,\ldots,\beta_{j+1},\ldots,\beta_k)$ with $\beta_k \neq 0$ and β_j odd. In $\Delta_*\tilde{\Theta}^{\underline{\beta}}$ the monomial

$$\tilde{\Theta}_k^{\beta_k}\,\tilde{\Theta}_{k-1}^{\beta_{k-1}}\cdots\tilde{\Theta}_j^{\beta_j-1}\cdots\tilde{\Theta}_1^{\beta_1}\otimes\tilde{\Theta}_j$$

is a non-square occuring with coefficient $\binom{\beta_j}{1} \equiv 1 (\text{mod } 2)$.

If $\beta_j = 1$ the monomial cannot be cancelled by any other on the right of (7.4) and if $\beta_j > 1$ it can only be cancelled by the contribution from a monomial $\tilde{\Theta}_k^{\beta_k}\ldots\tilde{\Theta}_{j+1}^{\beta_{j+1}+1}\tilde{\Theta}_j^{\beta_j-2}\ldots\tilde{\Theta}_1^{\beta_1}$ where β_{j+1} must be even. However if β_{j+1} is even and $\beta_j > 1$ is odd then $\tilde{\Theta}_k^{\beta_k}\ldots\tilde{\Theta}_{j+1}^{\beta_{j+1}+1}\tilde{\Theta}_j^{\beta_j-2}\ldots\tilde{\Theta}_1^{\beta_1}$ contributes a monomial

$$\tilde{\Theta}_k^{\beta_k}\ldots\tilde{\Theta}_{j+1}^{\beta_{j+1}+1}\tilde{\Theta}_j^{\beta_j-3}\ldots\tilde{\Theta}_1^{\beta_1}\otimes\tilde{\Theta}_j$$

which will not cancel out since $\beta_{j+1}+1$ is odd.

Thus $p(\Theta_1,\ldots,\Theta_k) = p_3(\tilde{\Theta}_1,\ldots,\tilde{\Theta}_\ell) + p_4(\tilde{\Theta}_1,\ldots,\tilde{\Theta}_k)^2$ for some $\ell <$

Now repeat the preceding argument for non-square monomials $\tilde{\Theta}_\ell^{\beta_\ell}\ldots\tilde{\Theta}_1^{\beta_1}$ in p. By induction on ℓ this shows p is a square and by induction on the degree of p we see that $p = p(\tilde{\Theta}_1)$, whic completes the proof.

Let k denote the inclusion $k:(\Sigma_p)^p = \Sigma_p\int 1 \subset \Sigma_p\int\Sigma_p \subset \Sigma_{p2}$ and let j denote $j:\pi_p\times\pi_p \subset \Sigma_p\times\Sigma_p \subset \Sigma_p\int\int\Sigma_2 \subset \Sigma_{p2}$. Let $\lambda\in R(\Sigma_p)$ satisfy

$$\lambda(g) = \begin{cases} p & \text{if } g\in\Sigma_p \text{ is a p-cycle} \\ 0 & \text{otherwise.} \end{cases}$$

<u>Lemma 7.4.</u>

If $0 < i \leq p$ then

$$k_!(\lambda^{\otimes i}\otimes 1^{\otimes p-i}) = (p-1)!p^{i-1}[p(\lambda\otimes 1+1\otimes\lambda) - \lambda^{\otimes 2}] \text{ in } R(\pi_p)^{\otimes 2}.$$

<u>Proof:</u>

Let λ_i be the function on Σ_{p^2} defined by

$$\lambda_i(x) = \begin{cases} (\lambda^{\otimes i}\otimes 1^{\otimes p-i})(x), & \text{if } x\in(\Sigma_p)^p \\ 0 & \text{otherwise.} \end{cases}$$

Thus $k_!(\lambda^{\otimes i}\otimes 1^{\otimes p-i})(x) = \frac{1}{(p!)^p}\sum_{g\in\Sigma_{p^2}} \lambda_i(g^{-1}xg)$, $(x\in\Sigma_{p^2})$

If $1 \neq x\in\pi_p\times\pi_p$ then x is the product of p disjoint p-cycles. Let $C(x)$ be the centraliser of x in Σ_{p^2}. There are $((p-1)!)^p$ distinct conjugates of x in $(\Sigma_p)^p$ and, since $C(x) = \pi_p\int\Sigma_p$,

$$k_!(\lambda^{\otimes i}\otimes 1^{\otimes p-i})(x) = \begin{cases} (p-1)!p^{i+1} & \text{if } 1 \neq x\in\pi_p\times\pi_p \\ 0 & \text{if } x = 1, \end{cases}$$

which is the value of the element in the statement when evaluated on an element of $\pi_p\times\pi_p$.

Lemma 7.5.

$$(B_j)_*(\theta_p\otimes\theta_p) = \begin{cases} \theta_2.\theta_2 + \theta_2.(\theta_1)^2 & \text{if } p = 2, \\ v.\theta_p.(\theta_1)^{p(p-1)} & \text{if } p \neq 2 \end{cases}$$

in $K_o(B\Sigma_{p^2};Z/_p)$, where $0 \neq v\in Z/_p$.

Proof:

Since θ_1 and θ_p generate $K_o(B\Sigma_p;Z/_p)$ the generators of $K_o(B\Sigma_{p^2};Z/_p)$ are

$$\{(\theta_p)^j(\theta_1)^{p(p-j)}, 0 \leq j \leq p\} \text{ and } q(\theta_p).$$

The only other possible generator is $q(\theta_1^p)$ which is the image of

$\Theta_p \epsilon \tilde{K}_o(B\Sigma_p;Z/_p)$ under $\Sigma_p = 1\int\Sigma_p \to \Sigma_p\int\Sigma_p \to \Sigma_{p^2}$, which is zero because Θ_p is primitive or by the Cartan formula.

Case (i) $p\neq 2$

$\Theta_p \epsilon K_o(B\Sigma_p;Z/_p)$ is carried by the $2(p-1)$-skeleton so $(B_j)_*(\Theta_p \otimes \Theta_p)$ is carried by the $4(p-1)$-skeleton. However $(\Theta_p)^j(\Theta_1)^{p(p-j)}$ is carried by the $2j(p-1)$-skeleton and $q(\Theta_p)$ is carried by the $2(p^2-1)$-skeleton. Since these generators are not carried by any lower skeleton than those specified above the only possibility for $(B_j)_*(\Theta_p \otimes \Theta_p)$ is a multiple of $\Theta_p(\Theta_1)^{p(p-1)}$. This multiple is non-zero because

$$< (B_j)_*(\Theta_p \otimes \Theta_p), k_!(\lambda \otimes 1^{\otimes p-1})>$$

$$\equiv < \Theta_p, \lambda >^2 \pmod p), \text{ which is non-zero.}$$

Case (ii): $p=2$.

Suppose $(B_j)_*(\Theta_2 \otimes \Theta_2) = aq(\Theta_2) + b\Theta_2^2 + c\Theta_2(\Theta_1)^2$

then, by considering the lowest skeletons carrying these elements we see $a \equiv 0 \pmod 2$. Let det be the representation of Σ_4 which assigns to each permutation its sign. Hence det is trivial when restricted to $\Pi_2 \times \Pi_2$ via j. However, in $R(\Sigma_2)\otimes R(\Sigma_2)$

$$k^*(\det - 1) = y \otimes y - 1 \otimes 1$$

$$\equiv \sigma\otimes\sigma + 1\otimes\sigma + \sigma\otimes 1 \pmod 2)$$

and

$$0 = < \Theta_2 \otimes \Theta_2, (B_j)^*(\det - 1)>$$

$$= < c\Theta_2 \otimes (\Theta_1)^2 + b\Theta_2 \otimes \Theta_2, \sigma\otimes\sigma + \sigma\otimes 1 + 1\otimes\sigma >$$

$$\equiv b + c \pmod 2.$$

Hence $b \equiv c \equiv 1 \pmod 2$ since

$$0 \neq \langle (B_j)_*(\Theta_2 \otimes \Theta_2),\ k_!(\lambda \otimes 1)\rangle \in Z/_2.$$

Corollary 7.6: If $k \geq 1$ then

$$(\Theta_p^k)\circ(\Theta_p) = \begin{cases} k(\Theta_2^{k+1}\Theta_1^{2(k-1)} + \Theta_2^k\Theta_1^{2k}), & \text{if } p = 2, \\ vk(\Theta_p^k(\Theta_1)^{kp(p-1)}) & , \text{if } p \neq 2, \end{cases}$$

where $0 \neq v \in Z/_p$.

Proof: From the relation (c.f.(7.1)) between o-product and the $\otimes$-product of symmetric groups it suffices to compute the image of $\Theta_p \otimes \Theta_p^k \in K_o(B\Sigma_p \times B(\Sigma_p)^k; Z/_p)$ under the homomorphism induced by

$$\Sigma_p \times (\Sigma_p)^k \xrightarrow[\Delta\times 1]{} (\Sigma_p)^k \times (\Sigma_p)^k \xrightarrow[\text{shuffle}]{} (\Sigma_p \times \Sigma_p)^k \xrightarrow[j^k]{} (\Sigma_{p^2})^k \subset \Sigma_{p^2k}.$$

This image is

$$(j_*)^{\otimes k}(\Delta_*(\Theta_p)\otimes \Theta_p^{\otimes k})$$

$$= (j_*)^{\otimes k}\Big(\sum_{i=o}^{k-1} (\Theta_p \otimes 1)^{\otimes i} \otimes (\Theta_p \otimes \Theta_p) \otimes (\Theta_p \otimes 1)^{\otimes p-i-1}\Big)$$

$$= \begin{cases} k(\Theta_2^{k+1}\Theta_1^{2(k-2)} + \Theta_2^k\Theta_1^{2k}), & \text{if } p = 2, \\ kv\Theta_p^k(\Theta_1)^{kp(p-1)}, & \text{if } p \neq 2. \end{cases}$$

Let $a_k = \tilde{\Theta}_1^k\, \Theta_1 \in K_o((\Omega S^o)_1; Z/_p$.

Lemma 7.7: The following formulae hold in $K_o((\Omega S^o); Z/_p)$:

(i) $(\Theta_p)\circ(\Theta_p^k) = [\tilde{\Theta}_1 \circ (\tilde{\Theta}_1^k)]\Theta_1^{kp^2}$

(ii) $a_1 \circ a_k = a_{k+1} + [\tilde{\Theta}_1 \circ (\tilde{\Theta}_1)^k]\Theta_1$

(iii) $a_1 \circ a_k = \begin{cases} (k+1)a_{k+1} + ka_k, & \text{if } p = 2 \\ a_{k+1} + kva_k, & \text{if } p \neq 2, \end{cases}$

where $0 \neq v \in Z/_p$.

<u>Proof</u>: (i),(ii) and Lemma 7.6 imply (iii). Parts (i) and (ii) follow from the formula [Mi, §2.2(ii)] for $x \circ (y.z)$.

(ii) We have $\Delta_*(a_1) = a_1 \otimes \Theta_1 + \Theta_1 \otimes a_1$ and

$$a_1 \circ (\tilde{\Theta}_1^k . \Theta_1)$$

$$= (a_1 \circ \tilde{\Theta}_1^k)(\Theta_1 \circ \Theta_1) + (\Theta_1 \circ \tilde{\Theta}_1^k)(a_1 \circ \Theta_1)$$

$$= (\tilde{\Theta}_1 . \Theta_1) \circ (\tilde{\Theta}_1^k) + a_{k+1}$$

$$= a_{k+1} + \Sigma \binom{k}{j} (\tilde{\Theta}_1^j \circ \tilde{\Theta}_1)\ (\tilde{\Theta}_1^{k-j} \circ \Theta_1)\Theta_1$$

$= a_{k+1} + [\tilde{\Theta}_1^k \circ \tilde{\Theta}_1]\Theta_1$, since $x \circ \Theta_1 = \varepsilon(x)$ where ε is the augmentation.

(i) Also, in $K_0(QS^0; Z/_p)$, $\tilde{\Theta}_1^k \circ \Theta_1^p$ is zero since $(- \circ \Theta_1^p)$ is the same homomorphism as that induced by

$$(QS^0) \xrightarrow[\Delta]{} (QS^0)^p \xrightarrow[\text{multiplication}]{} (QS^0).$$

Hence

$$\Theta_p \circ (\tilde{\Theta}_1^k . \Theta_1^{kp})$$

$$= (\Theta_p \circ \tilde{\Theta}_1^k)(\Theta_1^p \circ \Theta_1^{kp}) + (\Theta_1^p \circ \tilde{\Theta}_1^k)(\Theta_p \circ \Theta_1^{kp})$$

$$= \tilde{\Theta}_1^k \circ (\tilde{\Theta}_1 . \Theta_1^p)\Theta_1^{kp^2}$$

$$= [\Sigma \binom{k}{j} (\tilde{\Theta}_1^j \circ \tilde{\Theta}_1)(\tilde{\Theta}_1^{k-j} \circ \Theta_1^p)]\Theta_1^{kp^2}$$

$$= [\tilde{\Theta}_1^k \circ \tilde{\Theta}_1]\Theta_1^{kp^2}.$$

Lemma 7.8: In $K_o(QS^o;Z/_2)$

$$a_k \circ a_{2^n} = a_{2^n+k} \quad , \text{ if } n > 0.$$

Proof: Let β_1 be the composition

$$(\Sigma_2)^{2^n} \times (\Sigma_2)^k \longrightarrow [(\Sigma_2)^{2^n}]^k \times (\Sigma_2)^k \xrightarrow[\beta^k]{} (\Sigma_{2^{n+2}})^k \longrightarrow \Sigma_{k2^{n+2}}$$

where β is the homomorphism used in the proof of Corollary 7.6 to compute $(\Theta_2^k) \circ (\Theta_2)$. Thus

$(\beta_1)_*(\Theta_2^{\otimes 2^n} \otimes \Theta_2^{\otimes k}) \equiv 0 \pmod 2$ and $\Theta_2^{2^n} \circ \Theta_2^k = 0$. Using again the formula of [Mi, §2.2(ii)] we obtain

$$a_k \circ a_{2^n} = (\tilde{\Theta}_1^k \circ \tilde{\Theta}_1^{2^n})\Theta_1^{k2^{n+2}} + a_{2^n+k}$$

and

$$(\Theta_2^{2^n}) \circ (\Theta_2^k) = (\tilde{\Theta}_1^k \circ \tilde{\Theta}_1^{2^n})\Theta_1^{k2^{n+2}}$$

from which the result follows.

Let $\psi_j = \tilde{\Theta}_j \cdot \Theta_1 \in K_o(SF;Z/_p)$ and let $\nu = \tilde{\Theta}_1^2 \Theta_1 \in K_o(SF;Z/_2)$.

By §§7.7, 7.8 the o-product subalgebras generated by the elements $\{\tilde{\Theta}_1^j \cdot \Theta_1; j \geq 0\}$ are $Z/_p[\psi_1]$ if $p \neq 2$ and $Z/_2[\psi_1,\nu] \big/ (\psi_1^2+\psi_1)$ if $p = 2$.

Consider the map

$$\Delta_1 : K_o((QS^o)_o\ ;Z/_p) \longrightarrow K_o((QS^o)_o;Z/_p)^{\otimes 2}$$

defined by

$$\Delta_1(x) = \Delta_*(x) - x\otimes 1 - 1\otimes x .$$

This map has the following properties:

(a) the kernel of Δ_1 is generated by $\{\tilde{\Theta}_1^{p^m}; m \geq 0\}$

(b) Δ_1 sends a monomial

$\tilde{\Theta}_1^{\beta_1}\ldots\tilde{\Theta}_k^{\beta_k}$ of weight $|\beta| = \Sigma\beta_i$ to an element which is the sum of elements of the form

(i) terms of weight $|\beta|$ involving $\tilde{\Theta}_1,\ldots,\tilde{\Theta}_k$

(ii) terms involving only $\tilde{\Theta}_1,\ldots,\tilde{\Theta}_{k-1}$.

(c) If $\Delta_1(x) = p(\tilde{\Theta}_1,\ldots,\tilde{\Theta}_k)$ then x involves only $\tilde{\Theta}_1,\ldots,\tilde{\Theta}_k$.

The properties (a),(b) and (c) are consequences of (7.2) and Proposition 7.3. Hence, using the fact that Δ_* is an algebra map for both loop and composition product, we have by a suitable induction on the maximum suffix k such that $\tilde{\Theta}_k$ appears in $\tilde{\Theta}^{\underline{\beta}}$ we have:

<u>Lemma 7.9</u>: If $\max(i_j) = k > 1$ and at most one i_j is equal to one then

$$\begin{aligned} x &= (\tilde{\Theta}_{i_1}^{\beta_1}.\Theta_1)\circ(\tilde{\Theta}_{i_2}^{\beta_2}.\Theta_1)\circ\ldots\circ(\tilde{\Theta}_{i_t}^{\beta_t}.\Theta_1) \\ &= (\tilde{\Theta}_{i_1}^{\beta_1}\tilde{\Theta}_{i_2}^{\beta_2}\ldots\tilde{\Theta}_{i_t}^{\beta_t})\Theta_1 + p(\tilde{\Theta}_1,\ldots,\tilde{\Theta}_{k-1})\Theta_1 \\ &\qquad + (\text{a sum of monomials involving fewer } \tilde{\Theta}_k) \end{aligned}$$

in $K_o(SF;Z/_p)$.

<u>Proof</u>: The terms involving $\tilde{\Theta}_k$, in which $\tilde{\Theta}_k$ occurs the maximum number of times in the expressions $\Delta_1(x)$ and $\Delta_1(y)$

$(y = \tilde{\Theta}_{i_1}^{\beta_1} \ldots \tilde{\Theta}_{i_t}^{\beta_t}\Theta_1)$ in $K_o(SF;Z/_p)^{\otimes 2}$ are the same except that o-products in the first become loop-products in the second. Hence $\Delta_1(x-y)$ involves only monomials in $\tilde{\Theta}_1,\ldots,\tilde{\Theta}_k$ in which $\tilde{\Theta}_k$ occurs strictly less often than in x, if the proposition is true for monomials in $\tilde{\Theta}_1,\ldots,\tilde{\Theta}_k$ which involve strictly fewer occurrences of $\tilde{\Theta}_k$. By Properties (a),(b) and (c) of Δ_1 this means $x-y = p'(\tilde{\Theta}_1,\ldots,\tilde{\Theta}_k).\Theta_1$ involving monomials of strictly smaller weight in $\tilde{\Theta}_k$ and the induction step is complete.

To commence the induction we have to consider $(\tilde{\Theta}_2.\Theta_1)\circ(\tilde{\Theta}_1^s.\Theta_1)$ and since $\Delta_*(\tilde{\Theta}_1^s) = \Sigma\binom{s}{j}\tilde{\Theta}_1^j\otimes\tilde{\Theta}_1^{s-j}$ it suffices, by induction on s, to consider $(\tilde{\Theta}_2.\Theta_1)\circ(\tilde{\Theta}_1.\Theta_1) = x$. However in this case

$$\begin{aligned}\Delta_1(x) &= \tilde{\Theta}_2.\Theta_1\otimes\tilde{\Theta}_1.\Theta_1 + \tilde{\Theta}_1.\Theta_1\otimes\tilde{\Theta}_2.\Theta_1 \\ &\quad + [(\tilde{\Theta}_1.\Theta_1)\circ(\tilde{\Theta}_1.\Theta_1)]\otimes(\tilde{\Theta}_1.\Theta_1) \\ &\quad + (\tilde{\Theta}_1.\Theta_1)\otimes[(\tilde{\Theta}_1.\Theta_1)\circ(\tilde{\Theta}_1.\Theta_1)] + \lambda(\tilde{\Theta}_1.\Theta_1)^{\otimes 2} \\ &= \Delta_1(y) + \Delta_1(p(\tilde{\Theta}_1)\Theta_1).\end{aligned}$$

<u>Theorem 7.10</u>: $K_1(SG;Z/_p) = 0$,

$$K_o(SG;Z/_2) \cong \frac{Z/_2[\psi_1,\nu,\psi_2,\psi_3,\ldots]}{(\psi_1\circ\psi_1+\psi_1)}$$

and if $p \neq 2$

$$K_o(SG;Z/_p) \cong Z/_p[\psi_1,\psi_2,\ldots,\psi_j,\ldots].$$

<u>Proof</u>: By §§7.7, 7.8 and 7.9 the $\{\psi_j,\nu\}$ on a set of algebra

generators. By considering the monomials in a polynomial in $\psi_1, \psi_2, \ldots, \psi_k$ (and possibly ν) which involve the maximum number of occurrences of ψ_k and using Lemma 7.9 it is clear that there are no further relations than the one stated.

<u>Theorem</u>: If $\sigma: K_o(SG;Z/_p) \to K_1(BSG;Z/_p)$ is the transgression then

$$K_*(BSG;Z/_p) \cong \begin{cases} E(\sigma(\nu), \sigma(\psi_2), \sigma(\psi_3), \ldots \sigma(\psi_j) \ldots) & \text{if } p = 2 \\ E(\sigma(\psi_1), \sigma(\psi_2), \ldots) & \text{if } p \neq 2. \end{cases}$$

<u>Proof</u>: The spectral sequences

$$E^2_{p,\alpha} = \mathrm{Tor}^{K_*(SG;Z/_p)}_{p,\alpha}(Z/_p, Z/_p) \Longrightarrow K_*(BSG;Z/_p)$$

collapse since the E^2-term is an exterior on generators in $E^2_{1,o}$ which are therefore permanent cycles. Hence the algebra structure of the spectral sequence shows that it collapses. The map

$$\frac{K_o(SG;Z/_p)}{K_o(SG;Z/_p)^2} \cong \mathrm{Tor}_{1,o} \to \tilde{K}_1(BSG;Z/_p)$$

is the transgression, σ, [c.f. R-S].

§8: Q on $K_*(-;Z/_2)$ for the spaces BSO, BO, BSpin, SO and Spin.

Lemma 8.1: Let $\tau : BS^1 \to BS^1$ be the map induced by complex conjugation on S^1. In the notation of §6.1

$$\tau_* u_k = \sum_{j=1}^{k} \binom{k-1}{j-1} u_j \in K_o(BS^1;Z/_2), \quad (k > 0).$$

Proof: For $c = y-1 \in K^o(BS^1) = Z[[c]]$ we have

$$\tau^* c = \tau^* y - 1 = (y^{-1}) - 1 = (-c)/_{(1+c)} \ .$$

Hence $\tau^*(c^j) \equiv c^j(1+c)^{-j} \pmod 2$ and

$$< \tau_* u_k, c^j > = < u_k, c^j(1+c)^{-j} >$$

$$= < u_k, c^j(1 + \sum_{t \geq 1} c^t \binom{j+t-1}{t}) >$$

$$= \begin{cases} \binom{k-1}{j-1} & \text{if } 1 \leq j \leq k , \\ 0 & \text{otherwise} . \end{cases}$$

Lemma 8.2: Let $a_{k,t} \in Z/_2$ $(k,t \geq 1)$ be defined by

$$a_{k,t} = \Sigma \ t!/_{(p!q!(t-p-q)!)}$$

where the sum is taken over $(p,q \geq 0)$ satisfying

$$p + q \leq t = k-q-2p.$$

If $(\psi^3)_* : K_o(BS^1;Z/_2) \to K_o(BS^1;Z/_2)$ is the dual of the Adams operation, ψ^3, then

$$(\psi^3)_*(u_k) = \sum_{1 \leq t} a_{k,t} u_t .$$

<u>Proof</u>: Since $\psi^3(c) = v^3-1 \equiv (c + c^2 + c^3) \pmod 2$

$$< (\psi^3)_* u_k, c^t > = < u_k, (c + c^2 + c^3)^t >$$

$$= < u_k, \Sigma \frac{t!}{p!q!(t-p-q)!} c^{t+q+2p} >$$

$$= a_{k,t}.$$

<u>Lemma 8.3</u>: Let $(\psi^3-1):BU \to BU$ be the natural transformation on $\widetilde{K}U$ which sends $x \in \widetilde{K}(X)$ to $\psi^3(x)-x$. In the notation of §6.1 and §8.2 then

$$(\psi^3-1)_*(u_k) = \Sigma a_{k,t} u_t + \sum_{p=1}^{k-1} (\psi^3-1)_*(u_p).u_{k-p} \quad \text{and} \quad (\psi^3-1)_*(u_1) =$$

<u>Proof</u>: In $K_o(BU;Z/_2)$ we have $\Delta_* u_k = \sum_{p+q=k} u_p \otimes u_q$. Hence the result follows from the homotopy commutative diagram

$$\begin{array}{ccc} BU & \xrightarrow{\Delta} & BU \times BU \\ \psi^3 \downarrow & & \downarrow (\psi^3-1)\times 1 \\ BU & \xleftarrow[m]{} & BU \times BU \end{array} .$$

<u>Lemma 8.4</u>: The inclusion $S^1 = SO(2) \subset SO(3)$ induces an epimorphis $K_o(BS^1;Z/_2) \to K_o(BSO(3);Z/_2)$ and the images of the elements $\{u_{2k}; k \geq 0\}$ form a vector space basis of $K_o(BSO(3);Z/_2)$.

<u>Proof</u>: Since $SO(2) \subset SO(3)$ is a maximal torus this inclusion induces a monomorphism between representation rings and a monomorph

$K^o(BSO(3);Z/_2) \to K^o(BS^1;Z/_2)$. Dually, this map induces an epimorphism in $K_o(-;Z/_2)$. As in §6.1 care is required in dualisin monomorphisms into epimorphisms for infinite dimensional complexes.

The Weyl group of $SO(3)$ is $Z/_2$ acting on S^1 by complex conjugation. Hence, by Lemma 8.1, in $K_o(BSO(3);Z/_2)$

$$0 = \tau_*(u_{2k+2}) + u_{2k+2}$$

$$= u_{2k+1} + \sum_{j=1}^{2k} \binom{2k+1}{j-1} u_j$$

so $\{u_{2k}; k \geq 0\}$ generate $K_o(BSO(3);Z/_2)$.
Also $K^o(BSO(3);Z/_2) = Z/_2[[w]]$ when w restricts to $(y-1+\frac{1}{y}-1) \equiv \sum_{j\geq 2} c^j \pmod 2$ and from the pairing $K_o(BSO(3);Z/_2) \otimes K^o(BSO(3);Z/_2) \to Z/_2$ it is clear that the $\{u_{2k}; k \geq 0\}$ are linearly independent.

<u>Proposition 8.5</u>: Let $w_k \in K_o(BSO;Z/_2)$ be the image of $u_{2k} \in K_o(BSO(2);Z/_2)$ under the map induced by

$$SO(2) \subset SO.$$

The algebra $K_*(BSO;Z/_2)$, with multiplication induced by Whitney sum, is

$$Z/_2[w_1, w_2, \ldots, w_n, \ldots].$$

<u>Proof</u>: As for BU [c.f.§6.1] there is an algebra epimorphism $S(K_o(BSO(2);Z/_2)) \to K_*(BSO;Z/_2)$. Since $SO(2) \subset SO(3) \subset SO$ the epimorphism factors through an epimorphism

$$Z/_2[w_1, \ldots, w_n, \ldots] = S(K_o(BSO(3);Z/_2)) \xrightarrow{\alpha} K_*(BSO;Z/_2).$$

However under complexification $K_o(BSO;Z/_2) \to K_o(BU;Z/_2)$ the image of u_{2k} is

$(u_o^{-2})(\sum_{p+q=2k} u_p.\tau_*(u_q))$

in $K_o(BU{\times}Z;Z/_2) = Z/_2[u_o,u_o^{-1},u_1,\ldots,u_n,\ldots]$

so α is also a monomorphism.

We now calculate the algebra $K_*(BO{\times}Z;Z/_2)$ where $BO{\times}Z$ is the infinite loop-space representing $KO(-)$. Firstly we consider the zero-component $BO = \lim BO(2n+1)$. Since $BO(2n+1) = B\Sigma_2{\times}BSO(2n+1)$ where $\Sigma_2 = O(1)$ we have $K_*(BO;Z/_2) = \lim[K_o(B\Sigma_2;Z/_2)\otimes K_o(BSO(2n+1)$ The element $o \neq e_1 \epsilon \tilde{K}_o(B\Sigma_2;Z/_2)$ is mapped under $\Sigma_2{\times}SO(2n-1) \to \Sigma_2{\times}SO(2$ to itself. For this map sends a generator, τ, of Σ_2 given by $(-I_{2n-1})$ to $(-I_{2n-1})\otimes(I_2)$ which is $(-I_{2n+1})(I_{2n-1}\otimes(-I_2))$ in $\Sigma_2{\times}SO(2n+1)$. Hence $\Sigma_2 \to \Sigma_2{\times}SO(2n+1)$ is given by

$$\Sigma_2 \xrightarrow{\Delta} \Sigma_2{\times}\Sigma_2 \xrightarrow{1\times\beta} \Sigma_2{\times}SO(2) \longrightarrow \Sigma_2{\times}SO(2n+1)$$

where $\beta(\tau) = (-I_2)$ and the image of e_1 is

$$\begin{aligned}(1{\times}\beta)_*\Delta_*(e_1) &= (1{\times}\beta)_*(e_1\otimes 1 + 1\otimes e_1)\\ &= e_1\otimes 1 + 1\otimes\beta_*e_1\\ &= e_1\otimes 1\,.\end{aligned}$$

Hence additively $K_*(BO;Z/_2) \cong K_o(B\Sigma_2;Z/_2)\otimes K_o(BSO;Z/_2)$, via the maps defined above.

<u>Proposition 8.6</u>: Let $1 = u_o \epsilon K_o(BO{\times}1;Z/_2)$, let x be the image of the generator of $\tilde{K}_o(BO(1);Z/_2)$ in $K_o(BO{\times}1;Z/_2)$ and let $y_k = w_k{\cdot}u_o^2 \epsilon K_o(BO{\times}2;Z/_2)$, $(k \geq 1)$.

In $K_o(BO{\times}Z;Z/_2)$

$0 = x^2 + x.u_o + y_1$.

Proof: Consider the homomorphism, γ, given by Whitney sum

$$Z/_2 \times Z/_2 = O(1)\times O(1) \subset O(2) \subset O(3) = \Sigma_2 \times SO(3).$$

Thus, for $a,b\in Z/_2$,

$$\gamma(a,b) = \left(\tau^{a+b}, \begin{pmatrix} (-1)^b & & \\ & (-1)^a & \\ & & (-1)^{a+b} \end{pmatrix}\right).$$

If $\gamma_1 = \pi_2 \cdot \gamma$ where π_2 is projection into $SO(3)$ then γ is the composition

$$(\Sigma_2\times\Sigma_2) \xrightarrow{\Delta} (\Sigma_2\times\Sigma_2)^2 \xrightarrow[\text{shuffle}]{} (\Sigma_2\times\Sigma_2)\times(\Sigma_2\times\Sigma_2) \xrightarrow[m\times\gamma_1]{} \Sigma_2\times SO(3)$$

Where m is multiplication.

Hence $\gamma_*(e_1\otimes e_1)$

$$= m_*(e_1\otimes 1)\otimes(\gamma_1)_*(1\otimes e_1) + m_*(e_1^{\otimes 2})\otimes 1$$

$$+ m_*(1\otimes e_1)\otimes(\gamma_1)_*(e_1\otimes 1) + 1\otimes(\gamma_1)_*(e_1^{\otimes 2})$$

$$= e_1\otimes 1 + 1\otimes(\gamma_1)_*(e_1^{\otimes 2}),$$

since $m_*(e_1^{\otimes 2}) = e_1$ by §6.1.

If $x = (\mathbb{C}^3-3)\in R(SO(3))$ then

$$(\gamma_1)^*(x) = (y\otimes 1 + 1\otimes y + y\otimes y - 3) = \sigma\otimes\sigma\in R(\Sigma_2)^{\otimes 2}\otimes Z/_2.$$

Hence $\langle (\gamma_1)_*(e_1^{\otimes 2}), x^j \rangle$

$$= \langle e_1^{\otimes 2}, \sigma^j\otimes\sigma^j \rangle$$

$$= \begin{cases} 1 & \text{if } j = 1, \\ 0 & \text{otherwise,} \end{cases}$$

and so $(\gamma_1)_*(e_1^{\otimes 2})$ is the image of u_2 under

$$K_o(BSO(2);Z/_2) \to K_o(BSO(3);Z/_2).$$

<u>Proposition 8.7</u>: In the notation of §8.6 the algebra $K_*(BO\times Z;Z/_2)$ is isomorphic to

$$Z/_2[u_o,u_o^{-1}, x, y_1,y_2,\ldots,y_n,\ldots] \Big/ (x^2 + x.u_o + y_1).$$

<u>Proof</u>: The elements $\{u_o,u_o^{-1},x, y_j(j \geq 1)\}$ are clearly a set of algebra generators since we know $K(BO;Z/_2)$ additively and for the same reason there can be no further relations between these generato[rs] the the one given.

<u>§8.8. Q on $K_o(BO\times Z;Z/_2)$</u>

As for $BU\times Z$ (§6.3) we have homotopy commutative diagrams

$$(8.8.1)\qquad \begin{array}{ccc}
B\Sigma_p\times_{\Sigma_p} BO(n)^p = B(O(n)\int\Sigma_p) & \longrightarrow & BO(np) \\
\downarrow & & \downarrow \\
 & & BO\times(np) \\
 & & \cap \\
B\Sigma_p\times_{\Sigma_p}(BO\times(n))^p & \xrightarrow[\mu_p]{} & BO\times Z
\end{array}$$

by means of which we may compute Q on the generators of $K_o(BO\times Z;Z/_2)$. This will determine Q for the usual infinite loop-space structure on BSO for which $BSO \to BO\times(0)$ is a map of infinite loopspaces. However the generators Y_j, $(j \geq 1)$, origina[te] in $K_o(BSO(2)\times 2;Z/_2)$ and the commutative diagram of homomorphisms

$$\begin{array}{ccc} S^1 \wr \Sigma_2 = \mathbb{U}(1) \wr \Sigma_2 & \longrightarrow & \mathbb{U}(2) \\ \downarrow 1 & & \downarrow \\ SO(2) \wr \Sigma_2 & \longrightarrow & O(4) \end{array}$$

shows that $Q(Y_k)$ may be read off from the formula for $Q(u_{2k})$ of Theorem 6.3.5, translated into the correct component. Hence we find, since $m_*(u_1 \otimes u_p) = (p+1)u_{p+1} + pu_p$ in $K_o(BS^1; Z/_2)$,

(8.8.2) $$\sum_{o \leq t < 4k-t} u_t \cdot u_{4k-t} + \sum_{p=o}^{4k} ((p+1)u_{p+1} + pu_p)u_{4k-p} \in \Omega(u_{2k})$$

in $K_o(BU \times (2); Z/_2)$. To obtain the formula for $\Omega(y_k)$ from (8.8.2) in the left side replace u_{2s} by y_s, express u_{2s+1} in terms of the $\{u_{2s}\}$ by means of the formula

$$0 = \tau^*(u_{2s+2}) + u_{2s+2} \in K_o(BSO(2) \times (2); Z/_2)$$

where τ^* is defined in Lemma 8.1 and finally replace u_o by u_o^2.

<u>Lemma 8.8.3</u>:

$$x.u_o \in Q(u_o) \in K_o(BO \times (2); Z/_2)$$

<u>Proof</u>: By (8.8.1) an element of $\Omega(u_o)$ is given by the image of $0 \neq e_1 \in \tilde{K}(B\Sigma_2; Z/_2)$ under the homomorphism induced by

$$\Sigma_2 = 1 \wr \Sigma_2 \longrightarrow O(1) \wr \Sigma_2 \longrightarrow O(2).$$

However this homomorphism is conjugate to the inclusion $\Phi(1) \subset O(2)$ which, by definition, sends e_1 to x.

<u>Remark 8.8.4</u>: The computation of $\Omega(y_k)$ from (8.8.2) and Lemma

8.8.3 determine the Ω-operation on $K_o(BO\times Z;Z/_2)$ completely since Lemma 8.8.3 implies $x, u_o^{-3} \in Q(u_o^{-1})$ and $\Omega(x)$ is determined by $\Omega(x.u_o) = Q(x^2 + y_1)$.

<u>8.9 Remark:</u> <u>$(\psi^3-1)_*$ on $K_o(BSO;Z/_2)$ and $K_o(BO;Z/_2)$</u>

The H-map $BSO \to BU$ commutes with ψ^3 and hence with $(\psi^3-1)_*$ on $K_o(-;Z/_2)$. However this map induces a monomorphism $K_o(BSO;Z/_2) \to K_o(BU;Z/_2)$, (Proposition 8.5 (proof)) so $(\psi^3-1)_*$ on $K_o(BSO;Z/_2)$ may be determined from Lemma 8.3. Since ψ^3 is the identity on $K_o(B\Sigma_2)$ then $(\psi^3-1)_*(x) = 0$ which completes the implicit description of $(\psi^3-1)_*$ on $K_o(BO;Z/_2)$.

<u>§8.10: SO and Spin.</u>

From the calculation of $K^*(SO(m))$ in [Sn 1,§5] we have an isomorphism $K_*(Spin;Z/_2)\otimes K_o(B\Sigma_2;Z/_2) \cong K_*(SO;Z/_2)$ where the inclusion of $K_*(Spin;Z/_2)$ is induced by the map in the multiplicative fibration $Spin \xrightarrow{\pi} SO \longrightarrow B\Sigma_2$ and $0 \neq 1\otimes e_1 \in 1\otimes\tilde{K}_o(B\Sigma_2;Z/_2)$ maps to the generator of $K_o(B\Sigma_2;Z/_2)$. Hence the algebra structure of $K_*(Spin;Z/_2)$ is an exterior algebra on odd dimensional primitive generators, $E(a_1,a_2,\ldots,a_n,\ldots)$ since $K^*(Spin)$ is an exterior algebra on odd dimensional generators. To be explicit we have from [Sn 1;§5]

$$K^*(SO) \cong K^*(Spin)\otimes K(B\Sigma_2)$$

$$= E(v_1,\ldots,v_n,\ldots)\otimes \frac{Z[[e]]}{(e^2 + 2e)}$$ and

$$K^*(SO;Z/_2) = E(v_1,\ldots,v_n,\ldots)\otimes Z/_2[e] \Big/ (e^2).$$

Put a_j as the dual of v_j and b as the dual of e with respect to the basis given by monomials in the $\{v_i\}$ and e.

Since $\qquad m^*(e) = e\otimes 1 + 1\otimes e + e\otimes e \in K^0(SO)$

then $\qquad K_0(SO;Z/_2) \cong E(a_1,\ldots,a_n,\ldots)\otimes Z/_2[b] \Big/ (b^2 + b)$

as an algebra.

The maximal torus of $Spin(n)$ maps onto the maximal torus of $SO(n)$ by the squaring map. Hence the following diagram commutes

$$\begin{array}{ccc} BS^1 & \longrightarrow & BSpin \\ {\scriptstyle Bf}\downarrow & & \downarrow{\scriptstyle B\pi} \\ BS^1 & \longrightarrow & BSO \end{array}$$

when $f(z) = z^2 \in S^1$ and $(Bf)_*: K_0(BS^1;Z/_2) \to K_0(BS^1;Z/_2)$ is given by

$$(Bf)_*(u_k) = \begin{cases} 0 & \text{if } k \text{ odd}, \\ u_s & \text{if } k = 2s. \end{cases}$$

<u>Proposition 8.11</u>:

(i) As a map of algebras π_* is given by

$$K_*(Spin;Z/_2) \cong E(a_1,\ldots) \subset E(a_1,\ldots)\otimes Z/_2[b] \Big/ (b^2 + b) \cong K_*(SO;Z/_2).$$

(ii) $(B\pi)_*: K_*(BSpin;Z/_2) \longrightarrow K_0(BSO;Z/_2) \cong Z/_2[u_2,u_4,\ldots]$ is an algebra isomorphism. In terms of elements $u_j \in K_0(BS^1;Z/_2)$

where S^1 is the maximal torus of Spin(2)

$$K_*(BSpin;Z/_2) = Z/_2[u_4,u_8,\dots,u_{4k},\dots].$$

Proof:

(ii) There is a map between the Rothenberg-Steenrod spectral sequences

$$E_2 = Tor_{K_*(G;Z/_2)}(Z/_2,Z/_2) \Longrightarrow K_*(BG;Z/_2)$$

for G = Spin, SO induced by π. However

$$Tor_{K_*(SO;Z/_2}(Z/_2,Z/_2) \xleftarrow[\pi_*]{\cong} Tor_{K_*(Spin;Z/_2)}(Z/_2,Z/_2)$$

since
$$Tor^j_{Z/_2[b]/_{(b^2+b)}}(Z/_2,Z/_2) = \begin{cases} 0 & j > 0 \\ Z/_2 & j = 0 . \end{cases}$$

Hence π_* is an isomorphism of spectral sequences. The description of the generators in terms of $K_o(BS^1;Z/_2)$ follows from the previous discussion, as does part(i).

Remark 8.12: The Q-operation on $K_o(BSpin;Z/_2)$ follows from that on $K_o(BSO;Z/_2)$ for the usual infinite loopspace structure on BSpin since B_π is an H^∞-map for this structure. The Q-operation in $K_*(Spin;Z/_2)$ can be examined in terms of the suspension

$$K_*(Spin;Z/_2) \xrightarrow{\sigma} K_*(BSpin;Z/_2)$$

and in $K_*(SO;Z/_2)$ $Q(b)$ is not very interesting since $Q(b) = Q(b^2) \equiv 0 \ (Ind_2(SO))$.

§9: The J-homomorphism

Let p be a prime and let X_p denote the p-localisation of the space X, [Su]. Since Sullivan's solution of the conjecture of Adams [A] a certain amount of interest has centred around the infinite loopspaces that can be obtained as the fibre of maps of the form

$$(\psi^k - 1) : (BG)_p \longrightarrow (BH)_p$$

where H and G are appropriate choices from the set $\{O, SO, Spin, U, SU\}$ and in general k is a primitive root of unity mod p^2.

Let p be an odd prime then if $k^{p-1} \not\equiv 1 \pmod{p^2}$ let J_p be the fibre of $(\psi^k - 1) : BU_p \to BU_p$. The homotopy groups of J_p are related to the p-primary part of the image of the J-homomorphism [A;C-S;Q;Su]. Also J_p is independent of the choice of k and [Su] the J-homomorphism at p factors through a split monomorphism of loop-spaces

$$j : J_p \longrightarrow (SG)_p .$$

That is, there exists a loopspace Coker J_p and an equivalence of loopspaces

$$(SG)_p \simeq J_p \times \text{Coker } J_p . \qquad (9.1)$$

The following lemma is a consequence of the universal coefficient theorems of [An].

<u>Lemma 9.2:</u> If X is a connected space with $\tilde{K}_*(X;\mathbb{Q}) = \tilde{K}_*(X;Z/_p) = 0$ for all primes p then $\tilde{K}_*(X) = \tilde{K}^*(X) = 0$.

<u>Theorem 9.3</u>: Let p be an odd prime.

(i) $j_*: K_*(J_p; Z/_p) \longrightarrow K_*(SG_p; Z/_p)$ is an isomorphism of algebr

(ii) $(Bj)_*: K_*(BJ_p; Z/_p) \longrightarrow K_*(BSG_p; Z/_p)$ is an isomorphism

(iii) $\tilde{K}_*(\text{Coker } J_p) = 0 = \tilde{K}^*(\text{Coker } J_p)$

(iv) $\tilde{K}_*(B \text{ Coker } J_p) = 0 = \tilde{K}^*(B \text{ Coker } J_p)$.

<u>Proof</u>:

<u>(ii)</u>: This follows from (i) by comparing Rothenberg-Steenrod spectral sequences.

<u>(ii) and (iv)</u>: These follow from (i) and (ii) by Lemma 9.2 and th Kunneth formula applied to (9.1).

(i) The operation $\psi^k - 1: (KU)_p \to (\tilde{KU})_p$ produces a fibration of infinite loopspaces

$$Z \times J_p \longrightarrow Z \times BU_p \longrightarrow BU_p \ ,$$

where $o \times J_p$ and $O \times BU_p$ have the usual loopspace structure. Hence the map $S^o \to Z \times J_p$ which sends the basepoint to $O \times x_o$ and the othe point to $1 \times x_o$ induces a map commutative diagram of infinite loop-m

$$\begin{array}{ccc} (QS^o)_p & \xrightarrow{\alpha'} & Z \times J_p \\ & \alpha \searrow \quad \swarrow & \\ & Z \times BU_p & \end{array}$$

where α is the map of Theorem 6.3.8. Hence $(\alpha')_*$ is a monomorph in $K_*(-; Z/_p)$. However [Su] in the splitting (9.1) the projectio $(SG)_p \to J_p$ coincides with α' $[(QS^o)_o]_p$ and j_* is an isomorphis

<u>Remark 9.4</u>: Of course, from the point of view of the Q-operation

the isomorphism

$$(\alpha')_*: K_*((QS^0)_0; Z/p) \xrightarrow{\cong} K_*(J_p; Z/p)$$

is more important than j_* since α' is a map of infinite loopspaces.

We turn our attention, for the remainder of this section, to the J-homomorphism at the prime $p = 2$. <u>For the rest of this section spaces and maps will be 2-localised</u>. At $p = 2$ there is some uncertainty about the correct definition of the infinite loopspace (or even H-space!) J. The space J should have homotopy groups which approximate very closely to the 2-primary part of the image of the J-homomorphism, the J-homomorphism should factor through a map $J \to (SG)$ which should preferably be split, as in (9.1). It is for these reasons that the two serious candidates for J considered here (the same space with two H-space structures) are chosen. In particular it is claimed in [C-S] that these candidates are part of a splitting like that of (9.1).

<u>Definition 9.5</u>: Let the 2-localised spaces J_i, $1 \leq i \leq 5$, be defined as by the following 2-localised fibrations of infinite loopspaces:

(i) $J_1 \longrightarrow BO \xrightarrow{(\psi^3-1)} BSpin$

(ii) $J_2 \longrightarrow BSO \xrightarrow{(\psi^3-1)} BSO$

(iii) $J_3 \longrightarrow BSO \xrightarrow{(\psi^3-1)} BSpin$

(iv) $J_4 \longrightarrow BO \xrightarrow{(\psi^3-1)} BSO$

(v) $J_5 \times Z \longrightarrow BO \times Z \xrightarrow{(\psi^3-1)} BO.$

In (v) (ψ^3-1) represents $(x \to \psi^3(x)-x)$ as an operation $KO \to \widetilde{KO}$ and in (i)-(iv) (ψ^3-1) is given by the appropriate liftings of this map on the zero-component.

Lemma 9.6:

(a) J_5 has two components each homotopy equivalent to J_4

(b) There is a homotopy commutative diagram

$$\begin{array}{ccccc}
B\Sigma_2\times(B\Sigma_2\times J_3) & \xrightarrow{\simeq} & J_4 & \longrightarrow & BO \\
{\scriptstyle \text{incl.}}\uparrow & & \uparrow & & \uparrow{\scriptstyle 1} \\
(x_0)\times(B\Sigma_2\times J_3) & \xrightarrow[\simeq]{} & J_1 & \longrightarrow & BO
\end{array}$$

(c) There is a homotopy equivalence $B\Sigma_2\times J_3 \xrightarrow{\simeq} J_2$.

Proof:

(a) From the map of the homotopy sequence of fibration (iv) into that of the zero-component part of (v).

(b) From the homotopy sequence of (iii) mapping into that of (i) and (iv) we see that the maps induce isomorphisms

$$\pi_i(J_3) \longrightarrow \pi_i(J_4)$$

and $$\pi_i(J_3) \longrightarrow \pi_i(J_1) \qquad \text{if } i > 1.$$

Also if $B\Sigma_2 \to BO$ represents the canonical line bundle then there exists a lifting

$$\begin{array}{ccccc}
 & & & & SO \\
 & & & & \downarrow \\
 & & J_1 & \longrightarrow & J_4 \\
 & \nearrow & \downarrow & & \downarrow \\
B\Sigma_2 & \longrightarrow & BO & \xrightarrow{1} & BO \\
 & & \downarrow{\scriptstyle (\psi^3-1)} & & \downarrow \\
 & & BSpin & \longrightarrow & BSO
\end{array}$$

which induces $\pi_1(B\Sigma_2) \xrightarrow{\simeq} \pi_1(J_1)$. Also there exists $B\Sigma_2 \to SO \to J_4$ which is non-trivial on π_1 so

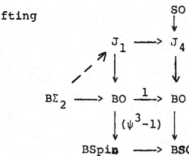

$$B\Sigma_2 \times B\Sigma_2 \longrightarrow SO \times J_1 \longrightarrow J_4 \times J_4 \xrightarrow{\text{mult}} J_4$$

gives an isomorphism $\pi_1(B\Sigma_2 \times B\Sigma_2) \longrightarrow \pi_1(J_4) = Z/_2 \otimes Z/_2$.

Hence $$B\Sigma_2 \times B\Sigma_2 \times J_3 \longrightarrow J_4 \times J_4 \xrightarrow{\text{mult}} J_4$$

and $$B\Sigma_2 \times J_3 \longrightarrow J_1 \times J_1 \xrightarrow{\text{mult}} J_1$$

are homotopy equivalences which fit into the required homotopy commutative diagram, since the canonical maps are maps of multiplicative fibrations.

(c) As for J_1 using $B\Sigma_2 \longrightarrow SO \longrightarrow J_2$.

Following the practice of [C-S] the H-spaces J_1 and J_2 are the two serious candidates for "J at the prime p = 2".

<u>Lemma 9.7</u>: In $K_o(J_4;Z/_2)$ there exist non-zero elements $\tilde{\Theta}_j$ ($1 \leq j$) satisfying

(a) $Z/_2[\tilde{\Theta}_1, \tilde{\Theta}_2, \ldots] \subset K_o(J_4;Z/_2)$

(b) The image of $\tilde{\Theta}_j$ under $J_4 \to BO = BO \times (0)$ is an element of $(u_o^{-2^j})Q^j(u_o)$.

(c) If Δ_* is the comultiplication then $\tilde{\Theta}_1$ is primitive and

$$\Delta_*(\tilde{\Theta}_j) = \tilde{\Theta}_j \otimes 1 + 1 \otimes \tilde{\Theta}_j + \tilde{\Theta}_{j-1} \otimes \tilde{\Theta}_{j-1} + z_j^{\otimes 2}$$

$(1 < j)$ where z_j is a polynomial in $\{\tilde{\Theta}_1, \ldots, \tilde{\Theta}_{j-1}\}$.

<u>Proof</u>: The map $S^o \to BO \times Z$ which maps the base-point into the zero-component and the other point into the one-component produces a diagram of maps of infinite loopspaces

(9.7.1)

$$\begin{array}{ccc} & & J_5\times Z \\ & \overset{\alpha_1}{\nearrow} & \downarrow \\ \Omega S^0 & \xrightarrow{\alpha_2} & BO\times Z \\ & \underset{\alpha}{\searrow} & \downarrow \\ & & BU\times Z \end{array}$$

where α is the map of §6.3.8. Since α_* is monic so is $(\alpha_1)_*$ and the elements $(\alpha_1)_*(\tilde{\theta}_j) = \tilde{\theta}_j$ satisfy (a),(b) and (c) and lie in $K_*(J_4;Z/_2)$ since $(\Omega S^0)_o$ maps into $J_4 \subset J_5\times(0)$.

<u>Corollary 9.8</u>: In $K_o(J_1;Z/_2)$ there exist non-zero elements ν_j $(1 \leq j)$ such that

(a) The image of ν_j under $J_1 \to BO$ is an element of $(u_o^{-2^j})\Omega^j(u_o)$.

(b) The element ν_1 is primitive and $(1 < j)$

$$\Delta_*\nu_j = \nu_j\otimes 1 + 1\otimes\nu_j + \nu_{j-1}\otimes\nu_{j-1} + u_j^{\otimes 2}$$

where u_j is a polynomial in $\{\nu_1,\ldots,\nu_{j-1}\}$.

<u>Proof</u>: In terms of the splitting of §9.6(b), $J_4 \simeq B\Sigma_2\times J_1$, put

$$\tilde{\theta}_j = 1\otimes\nu_j + e_1\otimes w_j \quad \text{where } 0 \neq e_1\in\tilde{K}_o(B\Sigma_2;Z/_2).$$

The result follows from the formula (7.2) and Lemma 9.7.

<u>Theorem 9.9</u>: If, for $i = 1$ or 2, there is a homotopy equivalence

$$(SG)_2 \simeq (J_i)_2\times(\text{Coker } J)_2$$

then $\tilde{K}_*(\text{Coker } J;Z/_2) = 0$.

<u>Proof</u>: By §9.6 it suffices to treat J_1.

We have a split monomorphism

$$(J_1)_2 \longrightarrow (SG)_2 = [(QS^o)_o]_2$$

and hence a monomorphism

$$\pi: Z/_2[\nu_1,\nu_2,\ldots] \subset K_o(J_1;Z/_2) \longrightarrow K_o((QS^o)_o;Z/_2) \cong Z/_2[\tilde\theta_1,\tilde\theta_2,\ldots].$$

This monomorphism is not necessarily a map of algebras. However π preserves primitives and the skeletal filtration. The primitives are generated by $\{\nu_1^{2^j}\}$ and $\{\tilde\theta_1^{2^j}\}$, by §7.3. Since $\nu_1^{2^j}$ and $\tilde\theta_1^{2^j}$ are carried by the 2^{j+1}-skeleton and by no lower skeleton then π is an isomorphism on the primitives, by a dimension-counting argument using induction up the skeletal filtration. This induction yields the formula

$$\pi(\nu_1^k) = \tilde\theta_1^k + \sum_{j<k} \lambda_j \tilde\theta_1^j, (\lambda_j \in Z/_2),$$

when $k = 2^n$. The map π commutes with Δ_* and, arguing as in §7.9

$$\pi(\nu^{\underline{\beta}}) = \tilde\theta^{\underline{\beta}} \quad \text{(modulo lower filtration)}$$

where the monomials in the $\{\tilde\theta_i\}$ are ordered as in §7.9. Hence π is an epimorphism and, from the Künneth formula, $K_*(\text{Coker } J;Z/_2)= 0$.

<u>Remark 9.9.1</u>: The method of Theorem 9.9 could also have been used to prove $0 = \tilde K_*(\text{Coker } J;Z/_p)$ $(p \neq 2)$ in Theorem 9.3.

<u>Hypothesis 9.10</u>: There exists a splitting $(SG)_2 \simeq (J_i)_2 \times (\text{Coker } J)_2$ for $i = 1$ or 2.

Remark 9.10.1: Despite the claim in [C-S] that a splitting exists (§9.10) will continue to be given the status of a hyppthesis, in these notes, because I have not seen a proof and no proof will be given in these notes.

Corollary 9.11: If (§9.10) is true then

$$K_*(J_1;Z/_2) \cong Z/_2[\nu_1,\nu_2,\ldots] \cong K_o((OS^o)_o;Z/_2)$$

and no splitting for J_1, as in Theorem 9.9, is a splitting H-space.

Proof: By §§9.9 and 7.10, $\pi(\nu_1)^2 = \pi_1(\nu_1)$ in $K_o(SG;Z/_2)$.

Lemma 9.12: In $K_o(J_3;Z/_2)$ there are non-zero elements $\{\lambda_i\}$ $(i \geq$ such that

(a) if §9.10 is true then

$$K_*(J_3;Z/_2) \cong Z/_2[\lambda_1,\lambda_2,\ldots]$$

(b) the element λ_1 is primitive and $(1 < j)$

$$\Delta_*(\lambda_j) = \lambda_j \otimes 1 + 1 \otimes \lambda_j + w_j^{\otimes 2}$$

where w_j is a polynomial in $\{\lambda_1,\ldots,\lambda_{j-1}\}$.

Proof: Consider the homotopy commutative diagram

$$\begin{array}{ccc}
X_o \times J_3 & \longrightarrow & X_o \times BSO \\
\downarrow & & \downarrow \\
B\Sigma_2 \times J_3 & \longrightarrow & B\Sigma_2 \times BSO \\
\cong \downarrow & & \cong \downarrow \\
J_1 & \longrightarrow & BO
\end{array}$$

$K_*(J_1;Z/_2) \cong Z/_2[\nu_1,\nu_2,\ldots]$, by §9.11. Put

$(j \geq 2) \quad 1\otimes\lambda_j + e_1\otimes S_j = \nu_j \in K_o(B\Sigma_2;Z/_2)\otimes K_o(J_3;Z/_2)$.

Hence, as in §9.8, (b) is true for $j > 2$. Also, by §8.7,

$$K_o(BO;Z/_2) \cong \frac{Z/_2[\tilde{x},\tilde{y}_1,\tilde{y}_2,\ldots]}{(\tilde{x}^2 + \tilde{x} + \tilde{y}_1)}$$

where $u_o\tilde{x} = x$ and $u_o^2.\tilde{y}_j = y_j$. Since $K_o(J_1;Z/_2) \to K_o(BO;Z/_2)$ is monic and ν_1 is the only non-zero primitive in $K_o(J_1;Z/_2)$ carried by the 2-skeleton, the image of ν_1 is $\tilde{x}$. Put $\lambda_1 = \nu_1^2 + \nu_1$ in $K_o(BSO;Z/_2) \cap K_o(J_1;Z/_2) = K_o(J_3;Z/_2)$. If $\Delta_1(t) = \Delta_*t - t\otimes 1 - 1\otimes t$ then $\Delta_1(\nu_2) = \nu_1\otimes\nu_1 + \mu\nu_1^2\otimes\nu_1^2$ $(\mu\in Z/_2)$. However $\tilde{\Theta}_2$ in §9.7 is defined modulo multiples of ν_1^2 so μ is independent of the choice of $\tilde{\Theta}_2$ and ν_2. Hence if $\varepsilon:K_o(B\Sigma_2;Z/_2) \to Z/_2$ is the augmentation we have, in

$$K_o(B\Sigma_2;Z/_2)^{\otimes 2} \otimes K_o(BSO;Z/_2)^{\otimes 2}$$

$$\Delta_1(\nu_2) = e_1\otimes e_1\otimes 1\otimes 1 + \mu[1\otimes\tilde{y}_1 + e_1\otimes 1]^{\otimes 2}$$

and

$$\begin{aligned}\Delta_1(\lambda_2) &= (\varepsilon\otimes\varepsilon)\Delta_1(\nu_2)\\ &= \mu(\tilde{y}_1\otimes\tilde{y}_1)\\ &= \mu(\lambda_1\otimes\lambda_1)\ .\end{aligned}$$

Hence if μ is zero then λ_2 is primitive. However λ_2 cannot be primitive since, by Theorem 9.9, the image of the vector space generated by $\{\lambda_2^{2^i}, \lambda_1^{2^i}\}$ would be primitive and would give too large a dimension for the vector space of primitives in $K_o(SG;Z/_2)$

carried by the n-skeleton, for some n.

Hence $\mu \equiv 1 (\mathrm{mod}\ 2)$ and the proof of part (b) is complete.

To show $Z/_2[\lambda_1,\lambda_2,\ldots] \cong K_0(J_3;Z/_2)$ it suffices to make the following observations.

By §9.11, $$K_*(J_1;Z/_2) \cong \frac{Z/_2[\nu_1,\lambda_1,\lambda_2,\ldots]}{(\nu_1^2+\nu_1+\lambda_1)}$$

and in terms of these generators it is clear from §8.7 that

$$Z/_2[\lambda_1,\lambda_2,\ldots] = K_0(BSO;Z/_2) \cap K_0(J_1;Z/_2) ,$$

by counting dimensions of the vector subspaces of elements carried by the finite skeletons.

<u>Corollary 9.13</u>: Suppose §9.10 is true. In terms of the splitting of §9.6(c) and the elements of §9.12

$$\frac{Z/_2[e,\lambda_1,\lambda_2,\ldots]}{(e^2+e)} \cong K_*(J_2;Z/_2)$$

where e is the image of $0 \neq e_1 \in \tilde{K}_0(B\Sigma_2;Z/_2)$ under $B\Sigma_2 \to SO \to J_2$.

<u>Proof</u>: The construction of the splitting

$$B\Sigma_2 \times J_3 \longrightarrow J_2 \times J_2 \longrightarrow J_2$$

implies that the algebra structure must be as stated for products of polynomials in the $\{\lambda_i\}$ and for products of such polynomials with the element, e. It remains to determine e^2.

There is a multiplicative fibration

$$Spin \longrightarrow SO \xrightarrow{f} B\Sigma_2$$

satisfying $0 \neq f_*(e_1)$ and $f_*(e_1)^2 = f_*(e_1)$. Hence, by §8.11, $(e_1 \otimes 1)^2 = e_1 \otimes 1 + 1 \otimes a + e_1 \otimes b$ in $K_0(SO;Z/_2) \cong K_0(B\Sigma_2;Z/_2) \otimes K_0(Spin;Z/_2)$ where $a,b \in \tilde{K}_0(Spin;Z/_2)$. Since $SO \to J_2$ is an H-map, $K_*(Spin;Z/_2)$ is generated in odd dimensions and $K_1(J_2;Z/_2) = 0$ we obtain $e^2 = e \in K_0(J_2;Z/_2)$.

__Lemma 9.14__:

$$K_*(SG/_{SO};Z/_2) \cong \frac{K_*(SG;Z/_2)}{(\psi_1)} \otimes K_*(BSO;Z/_2)$$

and

$$K_*(SG/_{Spin};Z/_2) \cong K_*(SG;Z/_2) \otimes K_*(BSpin;Z/_2)$$

with algebra structures such that (H = Spin or SO)

$$K_0(SG;Z/_2) \longrightarrow K_0(SG/_H;Z/_2) \longrightarrow K_0(BH;Z/_2)$$

are maps of algebras.

__Proof__: Consider the Rothenberg-Steenrod dual to that used in [Sn 1]

$$\mathrm{Tor}^{K_*(H;Z/_2)}(K_*(SG;Z/_2),Z/_2) \Longrightarrow K_*(SG/_H;Z/_2).$$

For $H = SO$ the action factors through

$$K_*(SO;Z/_2) \otimes K_*(SG;Z/_2) \longrightarrow K_*(SO;Z/_2)/\!/_{K_*(Spin;Z/_2)} \otimes K(SG;Z/_2)$$

and $e_1 \otimes t$ goes to $\psi_1 . t \in K_0(SG;Z/_2)$

so $$\mathrm{Tor}^{K_*(SO;Z/_2)}(K_*(SG;Z/_2),Z/_2)$$

is isomorphic to $\dfrac{K_*(SG;Z/_2)}{(\psi_1)} \otimes \mathrm{Tor}^{K_*(SO;Z/_2)}(Z/_2,Z/_2)$

and the spectral sequence collapses as does the spectral sequence

$$\mathrm{Tor}^{K_*(SO;Z/_2)}(Z/_2,Z/_2) \Longrightarrow K_*(BSO;Z/_2),$$

which gives the result for SO. For Spin the module action is trivial and the result follows, as for SO.

§9.15

In this section we study maps $\gamma : J_2 \to SG$ and

$$\gamma_* : Z/_2\frac{[e,\lambda_1,\lambda_2,\ldots]}{(e^2+e)} \cong K_*(J_2;Z/_2) \to K_*(SG;Z/_2) \cong Z/_2\frac{[\psi_1,\nu,\psi_2,\ldots]}{(\psi_1^2+\psi_1)}$$

Suppose henceforth

(9.15.1)
$$\gamma_*(e) = \psi_1$$
$$\gamma_*(\lambda_1) = \nu + a\psi_1 \qquad (a \in Z/_2),$$

as will be the case, for example, if γ_* is a monomorphism. This is seen by looking at the dimensions of the skeletons carrying these elements. By §9.12 (proof) consideration of $(\gamma_*\otimes\gamma_*)\Delta_1 = \Delta_1\gamma_*$ shows

(9.15.2) $$\gamma_*(\lambda_2) = \psi_2 + \nu\psi_1 + \text{primitives}.$$

Now the image of λ_1 in $K_o(BSO;Z/_2)$ is

$$\tilde{y}_1 \in Z/_2[\tilde{x},\tilde{y}_1,\tilde{y}_2,\ldots] = K_o(BO;Z/_2).$$

The map $J_2 \to BSO$ is a map of infinite loopspaces which induces a monomorphism in $K_*(-;Z/_2)$.

Using §8.8 to compute $Q(\tilde{y}_1,u_o^2) \subset K_o(BO\times Z;Z/_2)$ and applying the Cartan formula we obtain $(b\in Z/_2)$

(9.15.3) $$\lambda_2 + \lambda_1^2 + b\lambda_1 \in Q(\lambda_1) \subset K_o(J_2;Z/_2),$$

since $$(y_2 + y_1).u_o^2 + y_1^2 \in Q(y_1) \subset K_o(BO\times Z;Z/_2).$$

We now investigate the action of the Q-operation on $\nu \in K_o(SG;Z/_2) = K_o(SF;Z/_2)$. Since we have to distinguish between the algebra structures and Q-structures coming from the composition product on SF and those from the loopspace addition on $(\Omega S^o) = \lim\limits_{\rightarrow} \Omega^n S^n$, we will write $a \circ b$ and $\hat{Q}(x)$ for the algebra multiplication and Q-operation coming from the composition or smash product. From [Mad §1.8] we have the following commutative diagram in which $F = QS^o$ as a monoid under smash product.

$$\begin{array}{ccccc}
K_*^{\Sigma_2}((F^2)^2;Z/_2) & & \xrightarrow{\Delta_*} & & K_*^{\Sigma_2}((F^2;Z/_2)^{\otimes 3} \\
(\mu^2)_* \downarrow & & & & \downarrow \hat{\Theta}_1 \otimes \hat{\Theta}_2 \otimes \hat{\Theta}_3 \\
K_*^{\Sigma_2}(F^2;Z/_2) & \xrightarrow[\hat{\Theta}]{} & K_*(F;Z/_2) & \xleftarrow[\mu_*(\mu_* \otimes 1)]{} & K_*(F;Z/_2)^{\otimes 3}
\end{array} \quad (9.15.4)$$

In this diagram $\hat{\Theta}$ is the H^∞-structure map for $\hat{Q}$ and $\mu: F \times F \to F$ is loop addition.

Also in (9.15.4)

$$\hat{\Theta}_i[i_*(a_1 \otimes a_2 \otimes b_1 \otimes b_2)] = \begin{cases} \varepsilon(a_2)\varepsilon(b_2)(a_1 \circ b_1), & i = 1, \\ \varepsilon(a_1)\varepsilon(b_1)(a_2 \circ b_2), & i = 2, \\ (a_1 \circ b_2)(a_2 \circ b_1), & i = 3, \end{cases}$$

where $\varepsilon: K_*)-;Z/_2) \to Z/_2$ is augmentation.

Furthermore in (9.15.4)

$$\hat{\Theta}_i q(a \otimes b) \equiv \begin{cases} \hat{Q}(\varepsilon(b)a), & i = 1, \\ \hat{Q}(\varepsilon(a)b), & i = 2, \end{cases}$$

while

$$\hat{\Theta}_3(a \otimes \Theta_1^j) \equiv Q(a \circ (\Theta_1^j))$$

(modulo elements of the form $\hat{\Theta}_3 i_*[x\otimes\Theta_1^j]^{\otimes 2}$).

Chasing $q(\nu\otimes\Theta_1^3)=q(\Theta_2.\Theta_2.\Theta_1^{-3}\otimes\Theta_1^3)$ both ways round (9.15.4), using the formula for $q(x+y+z)$ and the Cartan formula we obtain

(9.15.5)
$$\begin{aligned}
&\hat{Q}(\Theta_2.\Theta_2) \\
&\equiv \hat{Q}(\nu)\Theta_1^{15} + (\nu\circ\nu)\{\hat{Q}(\Theta_1^3).\Theta_1^6 + \Theta_1^9\, Q(\Theta_1^3)\} \\
&+ \hat{Q}[\Theta_1].\Theta_2^4\Theta_1^7 \\
&+ \hat{Q}[\Theta_3]\Theta_1^{10} + Q(\Theta_2^2\Theta_1^{-1})\Theta_1^{10} \\
&+ \Theta_2^4.\Theta_1^8
\end{aligned}$$

where $\Theta = \Theta_3 \in K_o((QS^o)_3;Z/_2)$.

However we may also evaluate $\hat{Q}(\Theta_2.\Theta_2)$ using diagram (7.1) for the group homomorphisms

$$(\Sigma_2\times\Sigma_2)\int\Sigma_2 \longrightarrow \Sigma_4\int\Sigma_2 \longrightarrow \Sigma_{16}.$$

For this diagram the splitting of the operation into $\hat{\Theta}_1, \hat{\Theta}_2$ and $\hat{\Theta}_3$ can be accomplished at the level of groups. Let $\mathbb{R}^4 = \mathbb{R}_1^2 \oplus \mathbb{R}_2^2$ where e_1, e_2 is a basis of $\mathbb{R}_1^2$ and

e_3, e_4 is a basis of $\mathbb{R}_2^2$.

Thus $\Sigma_4\int\Sigma_2$ acts on the sixteen basis elements

$$\{e_i\otimes e_j;\ 1 \le i,j \le 4\} \quad\text{and}\quad (\Sigma_2\times\Sigma_2)\int\Sigma_2$$

is generated by involutions

$$\tau_1\otimes 1 = (e_1,e_2)\otimes 1$$
$$\tau_2\otimes 1 = (e_3,e_4)\otimes 1$$

and τ defined by $\tau(e_i \otimes e_j) = e_j \otimes e_i$.

However $\mathbb{R}^4 \otimes \mathbb{R}^4$ decomposes as the direct sum of $(\Sigma_2 \times \Sigma_2) \int\int \Sigma_2$ modules as follows:

(a) $\{e_1 \otimes e_2, e_2 \otimes e_1, e_1 \otimes e_1, e_2 \otimes e_2\}$,

(b) replace $\{e_1, e_2\}$ by $\{e_3, e_4\}$ in (a),

(c) {the orbit of $e_3 \otimes e_1$}.

On (a) $(\Sigma_2 \times \Sigma_2 \int\int \Sigma_2$ acts through the homomorphism

$$\Theta_1 = (\pi_1 \int\int 1) : (\Sigma_2 \times \Sigma_2) \int\int \Sigma_2 \longrightarrow (\Sigma_2 \times 1) \int\int \Sigma_2 = \Sigma_2 \int\int \Sigma_2$$

since τ_2 acts trivially in (a). Similarly on (b) the action factors through $\Theta_2 = (\pi_2 \int\int 1)$. Put

$$\Sigma_2 \otimes \Sigma_2 = \{(1,2)(3,4);\ (1,3)(2,4)\}, \text{ the four-group in } \Sigma_4.$$

On (c) the action factors through a homomorphism

$$(\Sigma_2 \times \Sigma_2) \int\int \Sigma_2 \xrightarrow{\Theta_3} (\Sigma_2 \otimes \Sigma_2) \int \Sigma_2 \subset \Sigma_8$$

defined by

$$\Theta_3(\tau_1 \otimes 1) = (1,3)(2,4)$$

$$\Theta_3(\tau) = \tau$$

$$\Theta_3(\tau_2 \otimes 1) = \tau(1,2)(3,4)\tau.$$

Hence we have a group theoretic analogue of (9.15.4)

(9.15.6)
$$\begin{array}{ccc} (\Sigma_2 \times \Sigma_2) \int\int \Sigma_2 & \xrightarrow{\Delta} & [(\Sigma_2 \times \Sigma_2) \int\int \Sigma_2]^3 \\ \cap \big\downarrow & & \big\downarrow \Theta_1 \times \Theta_2 \times \Theta_3 \\ \Sigma_{16} & \xleftarrow{\supset} & \Sigma_4 \times \Sigma_4 \times \Sigma_8 \end{array} .$$

Also Θ_1, Θ_2 and Θ_3 induce homomorphisms between the $K_o(-;Z/_2)$ groups of the classifying spaces which satisfy the same formula as $\hat{\Theta}_1$, $\hat{\Theta}_2$ and $\hat{\Theta}_3$ and in addition

$$\Theta_3 q(a\otimes b) = q(a\otimes\tau_*(b)) = q(a\otimes b) \subset K_o(B\Sigma_8;Z/_2)$$

$(a,b \in K_o(B\Sigma_2;Z/_2))$.

To calculate $\hat{\Omega}(\Theta_2^2)$ we may use (9.15.6) and evaluate

$(\Theta_1\otimes\Theta_2\otimes\Theta_3)\Delta_* q(\Theta_2\otimes\Theta_2)$.

Now Θ_2 is primitive so $\Delta_*(\Theta_2\otimes\Theta_2)$ is the sum of nine monomials in Θ_2 and Θ_1 and $\Delta_* q(\Theta_2\otimes\Theta_2)$ is evaluated using the formula for $q(x+y)$ and the Cartan formula. It is easy to check that the only term in this evaluation which is not a decomposable in algebra structure derived from loop addition is the term $\hat{\Omega}(O_2).\Theta_1^{12}$, $(O = O_2 \in K_o(B\Sigma_2;Z/_2))$, which is a square in the o-algebra.

<u>Proposition 9.16</u>: Let $\gamma:J_2 \to SG$ be a map satisfying

$$\gamma_*(e) = \psi_1 \quad \text{and} \quad \gamma_*(\lambda_1) = \nu + a\psi_1 \quad (a \in Z/_2)$$

in $K_*(SG;Z/_2)$, then

$$\gamma_*\Omega(\lambda_1) \nsubseteq \hat{\Omega}(\gamma_*\lambda_1).$$

<u>Proof</u>: From the preceding discussion we have that $\gamma_*(\lambda_2) \equiv \psi_2 + \nu.\psi_1 + c\psi_1$ (modulo primitive squares) and

$$\lambda_2 + \lambda_1^2 + b\lambda_1 \in \Omega(\lambda_1) , \qquad (b,c \in Z/_2).$$

Hence there is an element of the form

$$\psi_2 + \nu\psi_1 + \text{(a primitive polynomial in } \psi_1)$$

in $\gamma_*\Omega(\lambda_1)$.

However $\hat{Q}(\nu + a\psi_1) \equiv \hat{Q}(\nu) + a\nu\psi_1 + \hat{Q}(a\psi_1)$ modulo sqaures in the o-lagebra. Hence if $\gamma_*\Omega(\lambda_1) \subset \hat{Q}(\gamma_*\lambda_1)$ there will occur an indecomposable monomial, $\psi_2.\theta_1^{15}$, in $\hat{Q}(\nu).\theta_1^{15}$ and this cannot happen for the following reason. $\hat{Q}(\nu).\theta_1^{15}$ is evaluated by equating the expressions for $\hat{Q}(\theta_2^2)$ obtained from (9.15.4) and (9.15.6), modulo o-algebra squares. No monomial $\psi_2.\theta_1^{15} = \theta_4.\theta_1^8$ occurs either in the expression (9.15.5) or that obtained from (9.15.6), by the previous discussion. Hence the monomial $\psi_2.\theta_1^{15}$ must come from a o-algebra square. However, in the sense of Lemma 7.9, the "leading term" of a o-algebra monomial when written in the loop-addition product is given by replacing one product formally by the other and if $\psi_2.\theta_1^{15}$ were a monomial in a o-algebra square this would imply the existence of loop-addition-square monomials involving the $\{\psi_i; i \geq 2\}$ in the expression for $\hat{Q}(\theta_2^2)$ and this is impossible by (9.15.5) and the previous discussion.

<u>Corollary 9.17</u>: There exists no splitting $(SG)_2 \simeq (J_2)_2 \times (\text{Coker } J)_2$ as threefold loopspaces.

<u>Proof</u>: From the exact sequence obtained by applying $K_*^{\Sigma_2}(-;Z/_2)$ to the pair of Σ_2-spaces $(D^4 \times X^2, S^3 \times X^2)$, with the antipodal action on the first factor and the switching action on the last two factors, it is clear that the Q-operation originates in $K_*^{\Sigma_2}(S^3 \times X^2; Z/_2)$ and is therefore natural with respect to maps

of four-fold loopspaces. However in [Sn4] it is shown that $K_*^{\Sigma_2}(S^2 \times X^2; Z/_2) \to K_*^{\Sigma_2}(S^3 \times X^2; Z/_2)$ is onto from which it is clear that the operation Q(mod decomposables) originates in $K_*^{\Sigma_2}(S^2 \times X^2; Z/_2)$ and hence is natural for maps of three-fold loopspaces. Since the computation of §9.16 shows that γ_* does not commute with Q(mod decomposables) the result follows from the fact that any splitting map, γ, would satisfy the hypothesis of §9.16.

<u>Remark 9.18</u>: I hope to examine more closely elsewhere what the Q and $\hat{Q}$ operations dictate for loopspace maps $\gamma : J_i \to SG$ (i = 1,2). For example (§§9.16,9,17) imply that neither α_1 nor α_2 in the diagram (2-localised)

$$\begin{array}{ccccc} SG & \longrightarrow & SG/_{SO} & \longrightarrow & BSO \\ \uparrow{\scriptstyle \alpha_1} & & \uparrow{\scriptstyle \alpha_2} & & \uparrow{\scriptstyle 1} \\ J_2 & \longrightarrow & BSO & \xrightarrow[\psi^3 - 1]{} & BSO \end{array}$$

can be of the form $\Omega^3 f$. However for maps fitting into this particular diagram Madsen and Priddy have proofs using singular homology to show that α_1 and α_2 are not even H-maps. The advantage of applying $K_*(-;Z/_2)$ to study γ is that γ_* is very restricted from the start, because of the "small number" of primitives in $K_o(SG; Z/_p)$.

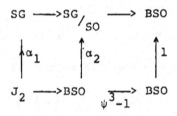

Appendix I

§A.1:

We use the notation of §3.3. In particular p is an odd prime.

Proposition A1.1:

Given $x \in K^{\alpha}(U,V;Z/_p)$ there exists $z(x) \in K^{\alpha+1}(U,V;Z/_p)$ such that $(\iota \otimes x^{\otimes p} + \xi \otimes z(x)^{\otimes p}) \otimes e_o \in E_2^{o,\alpha}(S^1_{\pi} \times (U,V)^p;\pi_p;Z/_p)$ is a permanent cycle. Also $z(x)$ is unique and $z(-)$ is additive.

Proof:

(i) By [A-T,II § 7;An, §2.2] the multiplication pairing on the representing spectrum (c.f.§1.1) of $\tilde{K}^*(-;Z/_p)$ is homotopy commutative. Hence if $X = BU^{M_p}$ or U^{M_p} there is a homotopy, H, between

$$(-)^p \text{ and } (-)^p \circ \sigma\colon X^p \to X$$

where $(-)^p$ is the p-fold product and $\sigma \in \pi_p$ is the canonical p-cycle $\sigma = (1,2,\dots,p) \in \Sigma_p$. Putting $H, H\circ\sigma,\dots,H\circ\sigma^{p-1}$ end to end yields a π_p-map

$$G : S^1_{\pi} \times X^p \to X$$

such that $s_o \times X^p \subset S^1_{\pi} \times X^p \to X$ is just $(-)^p$. By considering the universal example we see that there exists

$$u \in K^{\alpha}_{\pi_p}(S^1_{\pi} \times (U,V)^p;Z/_p)$$

such that $i_1^*(u) = \iota \otimes x^{\otimes p} + \xi \otimes w \in K^{\alpha}(S^1_{\pi} \times (U,V)^p;Z/_p)$.

However w must be invariant under the action of π_p on $K^*(U,V;Z/_p)^{\otimes p}$, since π_p acts trivially on $K^*(S^1_{\pi};Z/_p)$.

Since $(i_1)^*(i_1)_!(w_1 \otimes \dots \otimes w_p) = \sum_{g \in \pi_p} g^*(w_1 \otimes \dots \otimes w_p)$ the element

$w \in K^{\alpha+1}((U,V)^p;Z/_p)$ can be reduced to the form $z(x)^{\otimes p}$.

(ii) Since $z_1^{\otimes p} + z_2^{\otimes p} \equiv (z_1 + z_2)^{\otimes p} (\mathrm{mod}\ \mathrm{im}(i_1^*) \circ (i_1)_!)$ it is clear that if $z(x)$ is unique then $z(-)$ is additive. Also the above congruence shows it is sufficient to show that if

$$\xi \otimes z^{\otimes p} \otimes e_o \in E_2^{o,\alpha}(S_\pi^1 \times (U,V)^p;\pi_p;Z/_p)$$

is a permanent cycle then z is zero. In Proposition A1.4 it is shown that

$$d_2(\xi \otimes e_o) = \iota \otimes e_2 \in E_2^{2,o}(S_\pi^1;\ \pi_p;\ Z/_p)\ .$$

Also there exists z_1 such that $(\iota \otimes z^{\otimes p} + \xi \otimes z_1^{\otimes p}) \otimes e_o$ is a permanent cycle. Hence

$$\begin{aligned} 0 &= d_2(\xi \otimes z^{\otimes p} \otimes e_o) \\ &= d_2((\xi \otimes e_o)(\iota \otimes z^{\otimes p} \otimes e_o + \xi \otimes z_1^{\otimes p} \otimes e_o)) \\ &\qquad\qquad , \text{ since } \xi^2 = 0, \\ &= \iota \otimes z^{\otimes p} \otimes e_2 + \xi \otimes z_1^{\otimes p} \otimes e_2 \end{aligned}$$

so $z = 0 = z_1$.

Remark A1.2:

Modulo subsequent results on the spectral sequences $\{E_r(S_\pi^{2t+1},\ \pi_p;Z/_p)\}$ we have proved a strong form of Proposition $3.3(i)_o$. The proof of Proposition 3.3 will follow the inductive pattern $\{3.3(i)\}_t$ implies $\{3.3(i)_{t+1}$ and $3.3(ii)_{t-1}\}$ and finally $\{3.3(i)_t$ for $t > p\}$ implies $\{3.3(iii)\}$. However it is in Proposition A1.1 that the difference between odd primes and the case $p = 2$ appears. This time the difference is homotopy theoretic, for $p = 2$ $(-)^2 : [(BU)^{M_2}]^2 \to BU^{M_2}$ is not homotopy commutative.

§A1.3.1:

We will need the following exact sequences. Let D^{2t}_{π} be the 2t-disc with π_p-action given by $D^{2t}_{\pi} = CS^{2t-1}_{\pi}$. We have Thom isomorphisms

$$K^{*}_{\pi_p}((D^{2t}_{\pi}, S^{2t-1}_{\pi}) \times (X,Y);Z/_p) \cong K^{*}_{\pi_p}(X,Y;Z/_p)$$

and

$$K^{*}_{\pi_p}((D^{2t},S^{2t-1}) \times S^{1}_{\pi} \times (X,Y);Z/_p) \cong K^{*}_{\pi_p}(S^{1}_{\pi} \times (X,Y);Z/_p).$$

Also there are relative homeomorphisms

$$\begin{aligned} &(S^{2t-1} \times D^{2}_{\pi} \cup D^{2t}_{\pi} \times S^{1}_{\pi}, S^{2t-1}_{\pi} \times D^{2}_{\pi}) \\ \cong\ &(D^{2t}_{\pi} \times S^{1}_{\pi}, S^{2t-1}_{\pi} \times S^{1}_{\pi}) \\ \cong\ &(D^{2t} \times S^{1}_{\pi}, S^{2t-1} \times S^{1}_{\pi}) , \end{aligned}$$

the first being an excision and the second being given by $m(z,z') = (zz',z')$ $(z \in D^{2t}, z' \in S^{1}_{\pi})$. Hence the commutative diagram of inclusions

$$\begin{array}{ccc} S^{2t-1} \times D^{2}_{\pi} & \longrightarrow & D^{2t}_{\pi} \times D^{2}_{\pi} \\ & \searrow & \uparrow \\ & & S^{2t-1}_{\pi} \times D^{2}_{\pi} \cup D^{2t}_{\pi} \times S^{1}_{\pi} = S^{2t+1}_{\pi} \end{array}$$

yields a commutative diagram with exact rows.

$$
\begin{array}{ccc}
\vdots & & \vdots \\
\downarrow & & \downarrow \\
K^{\beta}_{\pi_p}(X,Y;Z/_p) & \xrightarrow{\ \alpha\ } & K^{\beta}_{\pi_p}(S^1_{\pi}\times(X,Y);Z/_p) \\
\Big\downarrow{\scriptstyle (-.\sigma^t)} & & \Big\downarrow{\scriptstyle \beta_t} \\
K^{\beta}_{\pi_p}(X,Y;Z/_p) & \longrightarrow & K^{\beta}_{\pi_p}(S^{2t+1}_{\pi}\times(X,Y);Z/_p) \\
\Big\downarrow & & \Big\downarrow \\
K^{\beta}_{\pi_p}(S^{2t-1}_{\pi}\times(X,Y);Z/_p) & \xrightarrow[\ 1\ ]{} & K^{\beta}_{\pi_p}(S^{2t-1}_{\pi}\times(X,Y);Z/_p) \\
\Big\downarrow{\scriptstyle \delta} & & \Big\downarrow{\scriptstyle \bar{\delta}} \\
K^{\beta+1}_{\pi_p}(X,Y;Z/_p) & \xrightarrow[\ \alpha\ ]{} & K^{\beta+1}_{\pi_p}(S^1_{\pi}\times(X,Y);Z/_p) \\
\downarrow & & \downarrow \\
\vdots & & \vdots
\end{array}
\qquad \text{(A1.3.2)}
$$

where $\sigma = (1-y) \in R(\pi_p) \otimes Z/_p \cong Z/_p[\sigma]/(\sigma^p)$.

Proposition A1.3.3:

In the diagram (A1.3.2) $\alpha = j^*$ where j^* is induced by $j\colon S^1_{\pi} \to$ (point).

Proof:

Let $\Lambda^*(t\,\mathbb{C}_{\pi})$ be the Koszul complex of the representation $ty \in R(\pi_p)$, [Se, §3.1 et sequ.]

For $x \in K^{\beta}_{\pi_p}(X,Y;Z/_p)$ also denote by x a representing complex of vector bundles over $(X,Y) \times (M_p,m_o) \times (D^{\beta},S^{\beta-1})$. In terms of complexes of vector bundles the homomorphism, α, is given by sending x to the element represented by the complex

$$m^*[\Lambda^*(t\,\mathbb{C}_{\pi}) \otimes (S^1_{\pi} \times x)] \quad \text{over}$$

$(D^{2t}, S^{2t-1}) \times S^1_{\pi} \times (X,Y) \times (M_p,m_o)$. However $\Lambda^*(t\,\mathbb{C}_{\pi})$ is the external tensor product of t copies of the complex

$$0 \longrightarrow \mathbb{C} \times D_\pi^2 \xrightarrow{\ d\ } \mathbb{C}_\pi \times D_\pi^2 \longrightarrow 0$$

given by $d(z,w) = (zw,w)$. Thus the commutative diagram

$$\begin{array}{ccc} \mathbb{C} \times D_\pi^2 \times S_\pi^1 & \xrightarrow{\ d\times 1\ } & \mathbb{C}_\pi \times D_\pi^2 \times S_\pi^1 \\ f_1 \Big\uparrow & & \Big\uparrow f_2 \\ \mathbb{C} \times D^2 \times S_\pi^1 & \xrightarrow{\ \bar{d}\ } & \mathbb{C} \times D^2 \times S_\pi^1 \end{array}$$

given by

$$\bar{d}(z,w_1,w_2) = (zw_1,w_1,w_2) \quad ,$$

$$f_1(z,w_1,w_2) = (z,w_1w_2,w_2) \quad ,$$

$$f_2(z,w_1,w_2) = (zw_2,w_1w_2,w_2)$$

shows that $m^*(\Lambda(t\mathbb{C}_\pi)) = \Lambda^*(t\mathbb{C})$. Hence

$m^*(\Lambda^*(t\,\mathbb{C}_\pi) \otimes (S^1 \times x))$ represents $j^*(x)$.

<u>§A1.4.1</u>:

We now examine the spectral sequence $\{E_r(S_\pi^{2t+1} \times (U,V)^p;\ \pi_p;Z/_p)\}$, $(t \geq o)$, in the case $(U,V) = (\text{point}, \phi)$. Firstly we observe that as $g \in \pi_p$ acts trivially on $K^*(S_\pi^1;Z/_p)$ the E_2-terms are given by isomorphisms

(A1.4.2) $$E_2^{q,\alpha}(S_\pi^{2t+1} \times (U,V)^p;\pi_p;Z/_p)$$

$$\cong \iota \otimes E_2^{q,\alpha}((U,V)^p;\pi_p;Z/_p) \oplus \xi \otimes E_2^{q,\alpha+1}((U,V)^p;\pi_p;Z/_p).$$

<u>Proposition A1.4.3</u>:

(i) For $\alpha = 0$ or 1 there exist isomorphisms

$$K^{\alpha}_{\pi_p}(S^{2t-1}_{\pi};Z/_p) = \begin{cases} Z/_p[\sigma]/_{(\sigma^t)} & , \quad (t \leq p-1), \\ Z/_p[\sigma]/_{(\sigma^p)} & , \quad (t \geq p) , \end{cases}$$

where the generators are $\iota_t \in K^{o}_{\pi_p}(S^{2t-1}_{\pi};Z/_p)$ and $x_t \in K^{1}_{\pi_p}(S^{2t-1}_{\pi};Z/_p)$ satisfying $(i_1^*)(\iota_t) = \iota$ and $(i_1^*)(x_t) = 0$.

(ii) In $E_2((S^{2t-1}_{\pi},\phi);\ \pi_p;Z/_p)$ the elements ι_t and x_t are respectively represented by $\iota \otimes e_o$ and $\iota \otimes e_1$.

(iii) $\{E_r((S^{2t-1}_{\pi},\phi);\ \pi_p;Z/_p)\}$ $(r \geq 2)$ is determined by its only non-zero differential being

$$d_{\mu(t)}(\xi \otimes e_q) = \lambda(\iota \otimes e_{q+\mu(t)}) , \quad (0 \neq \lambda \in Z/_p),$$

$$d_{\mu(t)}(\iota \otimes e_q) = 0$$

where $\mu(t) = \begin{cases} 2t & (t \leq p-1) \\ 2p & (t \geq p) . \end{cases}$

<u>Proof:</u>

(i) The isomorphisms follow from the left column of (A1.3.2). Choosing ι_t corresponding to $1 \in R(\pi_p) \otimes Z/_p$ guarantees $(i^*)(\iota_t) = \iota$. Choosing x_t $(t \leq p-1)$ so that $\delta(x_t) = \sigma^{p-t}$ + (higher powers of σ) implies that $0 = \alpha(\delta(x_t)) = \bar{\delta}(x_t)$ so we may choose x_{t+1} to restrict to x_t and to satisfy $\delta(x_{t+1}) = \sigma^{p-t-1}$ + (higher power of σ).

But the inclusion $S^{2t-1}_{\pi} \subset S^{2t+1}_{\pi}$ is non-equivariantly nullhomotopic so $(i_1^*)(x_t) = 0$ for $t \leq p$. That $(i_1^*)(x_t) = 0$ for $t \geq p$ follows from the fact that if $t \geq p$ the inclusion

$S_\pi^{2t-1} \subset S_\pi^{2t+1}$ induces an isomorphism

$K^*_{\pi_p}(S_\pi^{2t+1};Z/_p) \xrightarrow{\cong} K^*_{\pi_p}(S_\pi^{2t-1};Z/_p)$.

(ii) The representation of ι_t is a restatement of $(i_1^*)(\iota_t) = \iota$. If A1.4.3(iii) is true for $t = 1$ then the only non-zero groups in E_∞ are $E_\infty^{0,0}$ and $E_\infty^{1,0}$ so x_1 must be represented by $\iota \otimes e_1$. Hence the fact that x_{t+1} restricts to x_t and $(i_1^*)(x_t) = 0$ imply that x_t is represented by $\iota \otimes e_1$ for $t \geq 1$.

<u>(iii)(a) t = 1</u>:

By (i) some differential is non-zero on $\xi \otimes e_0$, and $0 = \sigma\iota$ implies that $\iota \otimes e_2$ cannot survive. Hence the only possibility is that $0 \neq d_2(\xi \otimes e_0) = \lambda(\iota \otimes e_2)$.

<u>(b) t > 1</u>:

We may now use the results of (ii), without producing a circular argument. Hence all the $\iota \otimes e_q$ are permanent cycles. The elements $\iota\otimes e_{2q}$ and $\iota\otimes e_{2q+1}$ respectively represent $\sigma^q \iota_t$ and $\sigma^q x_t$. Therefore $\iota \otimes e_j$ survives for $j \leq \mu(t) - 1$ and does not survive otherwise.

<u>§A1.5</u>:

We now determine the spectral sequence $\{E_r(S_\pi^1 \times (U,V)^p;\pi_p;Z/_p);\ r \geq 2\}$. We identify the E_2-term by the isomorphisms of (§A1.4.2).

<u>Proposition A1.5.1</u>:

There is an additive operation

$$z(-):K^\alpha(-;Z/_p) \longrightarrow K^{\alpha+1}(-;Z/_p) ,$$

stable in the sense of Proposition 3.2 (proof), which satisfies the following properties. For $w \in K^{\alpha}(U,V;Z/_p)$

(i) $(\iota \otimes w^{\otimes p} + \xi \otimes z(w)^{\otimes p}) \otimes e_o \in E_2^{o,\alpha}(S^1_\pi \times (U,V)^p;\pi_p;Z/_p)$ is a permanent cycle.

(ii) There is $0 \neq \lambda \in Z/_p$ such that

$$d_2(\xi \otimes w^{\otimes p} \otimes e_q) = \lambda(\iota \otimes w^{\otimes p} + \xi \otimes z(w)^{\otimes p} \otimes e_{q+2} ,$$

$$d_2(\iota \otimes w^{\otimes p} \otimes e_q) = \lambda[- \iota \otimes z(w)^{\otimes p} \otimes e_{q+2}] .$$

Also for $w_i \in K^{\beta_i}(U,V;Z/_p)$, $(1 \leq i \leq p)$,

$$\sum_{g \in \pi_p} (a\iota + b\xi) \otimes g^*(w_1 \otimes \ldots \otimes w_p) \in E_2^{o,\beta}, \quad (\beta = \Sigma \beta_i) ,$$

is a permanent cycle, $(a,b \in Z/_p)$.

Finally $E_3^{o,*} = E_\infty^{o,*}$ is generated by the permanent cycles described above and there is a natural isomorphism

$$\gamma : K^*(U,V;Z/_p) \longrightarrow E_3^{1,*} = E_\infty^{1,*}(S^1_\pi \times (U,V)^p;\pi_p;Z/_p)$$

given by $\gamma(w) = (\iota \otimes w^{\otimes p} + \xi \otimes z(w)^{\otimes p}) \otimes e_1$,

$$E_\infty^{j,*} = 0 \quad \text{for} \quad j > 1 .$$

<u>Proof:</u>

Let $w \in K^{\beta}(U,V;Z/_p)$. The operation $z(-)$ is that of Proposition A1.1 and is stable by the remarks in the proof of Proposition 3.2(iii). Hence Proposition A1.4.3 implies

$$d_2(\xi \otimes w^{\otimes p} \otimes e_q)$$

$$= \quad d_2((\xi \otimes e_o)\ulcorner \iota \otimes w^{\otimes p} + \xi \otimes z(w)^{\otimes p} \urcorner \otimes e_q)$$

$$= \quad \lambda(\iota \otimes w^{\otimes p} + \xi \otimes z(w)^{\otimes p}) \otimes e_{q+2} .$$

Also $d_2(\iota \otimes w^{\otimes p} \otimes e_q)$

$$= \quad - d_2(\xi \otimes z(w)^{\otimes p} \otimes e_q)$$

$= \quad \lambda[-\iota \otimes z(w)^{\otimes p} - \xi \otimes z(z(w))^{\otimes p}] \otimes e_{q+2}$

$= \quad (-\lambda\iota) \otimes z(w)^{\otimes p} \otimes e_{q+2}$

since $z(-)$ [c.f. Proposition 3.2(iii)(proof)] is a linear combination of the $\{\psi^i \circ \beta_p\}$ and $\beta_p^2 = 0$. The element $(a\iota + b\xi) \otimes \sum_g g^*(w_1 \otimes \ldots \otimes w_p) \otimes e_o$ is a permanent cycle representing $(i_1)_!([a\iota + b\xi] \otimes w_1 \otimes \ldots, \otimes w_p)$, by Proposition 2.2. From the differentials the computation of the spectral sequence is straightforward, d_2 being the only differential.

§A1.6:

We may now conclude the proof of Proposition 3.3, by proving the following strong form of (§3.3(i)).

Proposition A1.6.1(t):

There is an additive stable operation

$$z_t(-): K^{\alpha}(-;Z/_p) \longrightarrow K^{\alpha+1}(-;Z/_p)$$

such that $(\iota \otimes w^{\otimes p} + \xi \otimes z_t(w)^{\otimes p} \otimes e_o$ is a permanent cycle in $E_2(S_\pi^{2t-1} \times (U,V)^p;\pi_p;Z/_p)$.

Proof:

Proposition A1.5.1 starts the induction. Consider the homomorphism $\bar{\delta}$ in the right column of (A1.3.2). This induces a homomorphism of bidegree (0,1)

$$E_2^{o,*}(S_\pi^{2t-1} \times (U,V)^p;\pi_p;Z/_p) \longrightarrow E_2^{o,*}(S_\pi^1 \times (U,V)^p;\pi_p;Z/_p)$$

which sends permanent cycles to permanent cycles and satisfies

$\bar{\delta}([\iota \otimes w^{\otimes p} + \xi \otimes z_t(w)^{\otimes p}] \otimes e_o)$

$= \quad \bar{\delta}(\xi) \otimes z_t(w)^{\otimes p} \otimes e_o$

$= \quad \iota \otimes z_t(w)^{\otimes p} \otimes e_o.$

As in Proposition 3.2 (proof), since $z_t(-)$ satisfies $z_t(xy) = z_t(x)y$ if y is an integral class, $z_t(-)$ is zero if it is zero on (M_p, m_o). Now $\tilde{K}^{\alpha}(M_p; Z/_p) \cong Z/_p$ with generators, $u \in \tilde{K}^1$ and $\beta_p u \in \tilde{K}^o$, so $z_t(u) = \lambda\beta_p(u)$, $(\lambda \in Z/_p)$. Suppose $\lambda \neq 0$. Since $\beta_p(u)$ is an integral class $(\iota \otimes \beta_p u^{\otimes p} \otimes e_o)$ represents $\alpha(\beta_p u^{\boxtimes p}) \in K^1_{\pi_p}(S^1_\pi \times (M_p, m_o)^p; Z/_p)$ modulo the lower filtration, which is killed by β_t, then

$$0 = \sigma^t(\beta_p(u))^{\boxtimes p} \in K^1_{\pi_p}(S^{2t+1} \times (M_p, m_o)^p; Z/_p) .$$

However, from the $R(\pi_p) \otimes Z/_p$-module structure of $\{E_r((U,V)^p; \pi_p; Z/_p)\}$ $\sigma^t(\beta_p u)^{\boxtimes p}$ is not a multiple of σ^{t+1} so $\sigma^t(\beta_p u)^{\boxtimes p} \neq 0$. Hence $z_t(-)$ is zero. We have now all we need for Proposition 3.3. However, to continue, the permanent cycle represented by

$(\iota \otimes w^{\otimes p} + \xi \otimes z_t(w)^{\otimes p}) \otimes e_o = (\iota \otimes w^{\otimes p} \otimes e_o)$ must map under $\bar{\delta}$ into the smaller filtration

$$K^{\beta}(U,V; Z/_p) \cong E^{1,\beta}_{\infty} \subset K^*_{\pi_p}(S^1_\pi \times (U,V)^p; \pi_p; Z/_p) .$$

Since $\bar{\delta}$ induces a map of spectral sequences which kills the smaller filtration of $K^*_{\pi_p}(S^{2t-1} \times (U,V)^p; Z/_p)$ which correspond to $E^{j,*}_{\infty}$ $(j \geq 1)$ we may use $\bar{\delta}$ to induce an operation $D: K^{\beta}(-; Z/_p) \to K^{\beta}(-; Z/_p)$. D is defined by sending $w \in K^{\beta}(U,V; Z/_p)$ to $(\gamma^{-1})\, \bar{\delta}$ (element represented by $[\iota \otimes w^{\otimes p} + \xi \otimes z_t(w)^{\otimes p}] \otimes e_o$). As in Proposition 3.2(iii),

D will be additive and stable. The results on integral K-theory of [Sn 2, §2.1] show that $D(w)$ is zero if w is an integral class, and hence $D = 0$ in general (c.f. Proposition 3.2(iii)(proof)). Thus if $y \in K^*_{\pi_p}(S^{2t-1} \times (U,V)^p; Z/_p)$ is represented by $[\iota \otimes w^{\otimes p} + \xi \otimes z_t(w)^{\otimes p}] \otimes e_o = \iota \otimes w^{\otimes p} \otimes e_o$ then $\bar{\delta}(y) = 0$. So there is an element in $K^*_{\pi_p}(S^{2t+1} \times (U,V)^p; Z/_p)$ represented by $(\iota \otimes w^{\otimes p} + \xi \otimes z_{t+1}(w)^{\otimes p} \otimes e_o$ and (arguing as in Proposition A1.1 using §A1.4.3) $z_{t+1}(-)$ is an additive operation which is stable in the sense of Proposition 3.2(iii) (proof).

Incidently, the fact that the restriction map kills $\xi \in K^1(S^{2t+1}; Z/_p)$ shows that the existence of $z_{t+1}(-)$ implies $z_t(-)$ is zero, which was shown by an alternative method in the course of the above discussion.

Proof of Proposition 3.3:

$3.3(i)_t$:

This is part of Proposition A1.6.1(t).

$3.3(ii)_t$:

The fact that $z_t(-) = 0$ was proved in the proof of Proposition A1.6.1(t).

3.3(iii):

Take $t \geq p$ in the exact sequences of (§A1.3.2), then $\sigma^t = 0$ and the exact sequence reduces to

$$0 \to K^*_{\pi_p}((U,V)^p; Z/_p) \xrightarrow{f} K^*_{\pi_p}(S^{2t-1}_{\pi_p} \times (U,V)^p; Z/_p) \xrightarrow{\delta} K^*_{\pi_p}((U,V)^p; Z/_p) \to 0$$

Also from Proposition A1.6.1(proof) the only non-zero differen-

tial in $\{E_r(S_\pi^{2t-1} \times (U,V)^p;\pi_p;Z/_p);\ r \geq 2\}$ is given by
$d_{2p}(\xi \otimes w^{\otimes p} \otimes e_j) = \iota \otimes w^{\otimes p} \otimes e_{j+2p}$.
Hence $\ker(-.\sigma) \subset K^*_{\pi_p}(S_\pi^{2t-1} \times (U,V)^p;Z/_p)$ sits in an exact sequence

$$0 \to E_\infty^{2p-1,*} \cong K^*(U,V;Z/_p) \xrightarrow{g} \ker(-.\sigma) \xrightarrow{h} E_\infty^{2p-2,*} \cong K^*(U,V;Z/_p) \to 0$$

obtained from the spectral sequence ($\deg h \equiv 1$, $\deg g \equiv 0 \pmod 2$).
Now if $w \in K^\beta(U,V;Z/_p)$ has $g(w) \in \mathrm{im}(f)$ then there exists an element $z \in \ker(-.\sigma)$ represented by $\iota \otimes w^{\otimes p} \otimes e_{2p-1} \in E_\infty^{2p-1,\beta+1}$ such that $z = f(z')$. If $x_t \in K^1_{\pi_p}(S_\pi^{2t-1};\ Z/_p)$ is the element of Proposition 1.4.3 then $\delta(x_t z) = \delta(x_t)z' = z' \in K^\beta_{\pi_p}((U,V)^p;Z/_p)$. However the multiplicative structure of the spectral sequence implies $x_t z = 0$ so $z' = 0$ and $w = 0$. Hence all non-zero elements of $\ker(-.\sigma) \cap \mathrm{im}(f)$ are represented by non-zero elements of $E_\infty^{2p-2,*}$ and by the $R(\pi_p) \otimes Z/_p$-module structure all elements of $\mathrm{im}(f)$ are represented by classes in some $E_\infty^{2q,*}(S_\pi^{2t-1} \times (U,V)^p;\pi_p;Z/_p)$, $0 \leq q \leq p-1$. Finally, by counting dimensions, all elements of each $E_\infty^{2q,*}$ represent elements of $\mathrm{im}(f)$. Hence $\iota \otimes w^{\otimes p} \otimes e_o$ is a permanent cycle representing an element of $\mathrm{im}\, f$ and $w^{\otimes p} \otimes e_o \in E_2^{o,\beta}((U,V)^p;\pi_p;Z/_p)$ is a permanent cycle.

Appendix II

§A.2:

This appendix contains several K-theoretic proofs, involving direct constructions with vector bundles. The proofs are straightforward, but are notationally cumbersome and for this reason each construction is illustrated. The illustrations show what differentials are used on a specified family of vector bundles to make it into a complex of vector bundles.

Let Y be a compact space with a closed subspace, B.

Proposition A.2.1:

Let $y \in K(Y,B)$ satisfy $2.y = 0$ and let $0 \neq a \in K(M_2,m_o) \cong Z/_2$. If $x \in \langle y,2,a\rangle$ there exist elements $y_1 \in \langle y,2,y\rangle$ and $0 \neq a_1 = \langle a,2,a\rangle \in K^{-1}((M_2,m_o)^2) \cong \tilde{K}^{-1}(M_2;Z/_2) \cong Z/_2$, such that $x^{\otimes 2} \in \langle y_1,2,a^{\otimes 2}\rangle + \langle y^{\otimes 2},2,a_1\rangle \subset K((Y,B)^2 \times (I,\partial I)^2 \times (M_2,m_o)^2)$. More generally if $y_i \in K(Y_i,B_i)$, $n_i \in K(W_i,A_i)$ $p.y_i = 0 = p.n_i$ $(p \in Z; i = 1,2)$ and $x_i \in \langle y_i,p,n_i\rangle$ there exist elements $n' \in \langle n_1,p,n_2\rangle$ and $y' \in \langle y_1,p,y_2\rangle$ such that

$$(-x_1 \otimes x_2) \in \langle y_1 \otimes y_2,p,n'\rangle + \langle y',p,n_1 \otimes n_2\rangle .$$

Proof: We will prove only the mod 2 statement, since the proof of the general statement differs only in requiring more elaborate notation. Choose representing complexes of vector bundles

(Y,d_Y) over (Y,B) for y,

(A,d_A) over (M_2,m_o) for a

and $(\mathbb{C}^2,d_2 = 0)$ for $2 \in Z = K(pt,\phi)$

which satisfy the following properties. There exist homotopies of differentials

H_t on $Y \otimes \mathbb{C}^2$ over (Y,B) , $(t \in I)$,

and G_t on $\mathbb{C}^2 \otimes A$ over (M_2,m_o)

such that (i) $H_o = d_Y \otimes d_2$, $G_o = d_2 \otimes d_A$ (ii) H_1 and G_1 are exact (iii) $(H_t|B) = d_Y \otimes d_2$, $(G_t|m_o) = d_2 \otimes d_A$.

By [Sn 3,I §3.4 & II §2; At,§2.6.12] the conditions may be fulfilled by some defining system of complexes and homotopies for the Massey product element, x, and in future such conditions will be assumed possible without any further remark.

The element $x^{\otimes 2}$ is then represented in the following way [see fig(i)]. Represent I^2 as a polygon with vertices $\{i|\ 1 \leq i \leq 8\}$. Take the underlying family of vector bundles of $(Y \otimes \mathbb{C}^2 \otimes A \otimes Y \otimes \mathbb{C}^2 \otimes A) \times I^2$ with the following differentials to make a complex of vector bundles over $(Y,B)^2 \times (I,\partial I)^2 \times (M_2,m_o)^2$. We remark once for all that the differential need only be specified where it is required to be exact. Over the point (y_1,y_2,z,m_1,m_2) $(y_i \in Y, m_i \in M_2, z \in I^2)$ the differential is given by

$$d_Y(y_1) \otimes d_2 \otimes d_A(m_1) \otimes d_Y(y_2) \otimes d_2 \otimes d_A(m_2) \quad \text{if some}$$

$y_i \in B$ of $m_i = m_o$ and over ∂I^2 the differential is given by

$d_Y \otimes G_t \otimes d_Y \otimes G_1$ on the edge (2,3),
$d_Y \otimes G_1 \otimes d_Y \otimes G_t$ on (4,3), $d_Y \otimes G_1 \otimes H_t \otimes d_A$ on (4,5),
$d_Y \otimes G_t \otimes H_1 \otimes d_A$ on (6,5), $H_t \otimes d_A \otimes H_1 \otimes d_A$ on (6,7),
$H_1 \otimes d_A \otimes H_t \otimes d_A$ on (8,7) and $H_1 \otimes d_A \otimes d_Y \otimes G_t$ on (8,1)

in the senses indicated.

To simplify notation we will sometimes omit obvious isomorphisms of tensor products in definitions of representing complexes. For example, H_t on $\mathbb{C}^2 \otimes Y$ will mean the transpose of H_t on $Y \otimes \mathbb{C}^2$.

fig(i) $x^{\otimes 2} \epsilon K(Y,B)^2 \times (I,\partial I)^2 \times (M_2,m_o)^2)$, differentials on $(Y \otimes \mathbb{C}^2 \otimes A \otimes Y \otimes \mathbb{C}^2 \otimes A) \times I^2$

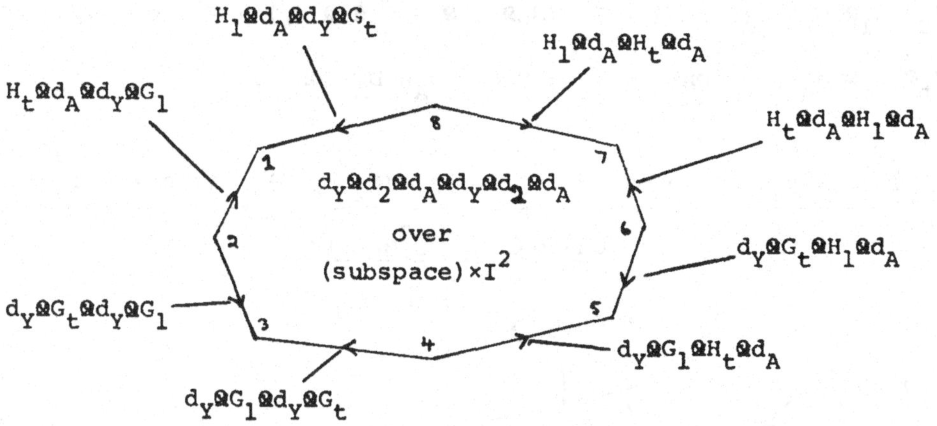

We now construct representatives for elements of the iterated Massey products, $\langle y_1,2,a^{\otimes 2})$ and $\langle y^{\otimes 2},2,a_1\rangle$ whose sum is $x^{\otimes 2}$. The elements $y_1 \epsilon \langle y,2,y\rangle$ and $a_1 \epsilon \langle a,2,a\rangle$ will each have a symmetric representation using a defining system with the same homotopy of differentials used twice. For example y_1 is represented by the Massey product construction on the underlying vector bundles of $Y \otimes \mathbb{C}^2 \otimes Y$ using $d_Y \otimes H_t$ and $H_t \otimes d_Y$.

An element $z_1 \epsilon \langle y_1,2,a^{\otimes 2}\rangle$ is represented in the following way. Take the underlying vector bundles of $(Y \otimes \mathbb{C}^2 \otimes Y \otimes \mathbb{C}^2 \otimes A \otimes A) \times I^2$ where I^2 is represented as a polygon with vertices, $\{i \mid 1 \leq i \leq 7\}$. Let $F_t \epsilon \mathrm{Aut}_{\mathbb{C}}(\mathbb{C}^2 \otimes \mathbb{C}^2)$ be a homotopy from the identity to T, the switching map. The following differentials [see fig.(ii)] define a complex of vector bundles over $(Y,B)^2 \times (I,\partial I)^2 \times (M_2,m_o)^2$.

Over the point (y_1, y_2, z, m_1, m_2) the differential is given by $d_Y(y_1) \otimes d_2 \otimes d_Y(y_2) \otimes d_2 \otimes d_A(m_1) \otimes d_A(m_2)$ is some $y_i \in B$ or $m_i = m_o$, along ∂I^2 the differential is given by

$H_1 \otimes d_Y \otimes G_t \otimes d_A$ on the edge (1,2),

$H_t \otimes d_Y \otimes G_1 \otimes d_A$ on (3,2), $d_Y \otimes H_t \otimes G_1 \otimes d_A$ on (3,4),

$d_Y \otimes H_1 \otimes G_t \otimes d_A$ on (5,4), $(d_Y \otimes d_2 \otimes H_1 \otimes d_A \otimes d_A) \circ F_t$ on (6,5),

$H_t \otimes H_1 \otimes d_A \otimes d_A$ on (6,7), $H_1 \otimes H_t \otimes d_A \otimes D_A$ on (1,7).

fig(ii) $z_1 \in \langle y_1, 2, a^{\otimes 2} \rangle \subset K((Y,B)^2 \times (I,\partial I)^2 \times (M_2,m_o)^2)$, differentials on $(Y \otimes \mathbb{C}^2 \otimes Y \otimes \mathbb{C}^2 \otimes A \otimes A) \times I^2$

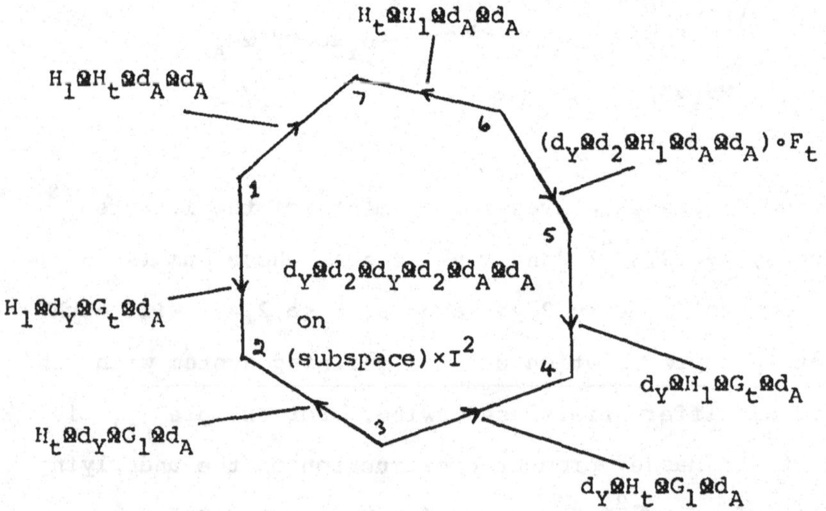

An element $z_2 \in \langle y^{\otimes 2}, 2, a_1 \rangle$ is represented in the following way. Take the underlying vector bundle of $(Y \otimes Y \otimes \mathbb{C}^2 \otimes A \otimes \mathbb{C}^2 \otimes A) \times I^2$ where I^2 is represented as a polygon with vertices $\{i \mid 1 \leq i \leq 7\}$. The following differentials [see fig(iii)] define a complex of vector bundles over $(Y,B)^2 \times (I,\partial I)^2 \times (M_2,m_o)^2$. Over the point, (y_1, y_2, z, m_1, m_2), the differential is given by

$d_Y(y_1) \otimes d_Y(y_2) \otimes d_2 \otimes d_A(m_1) \otimes d_2 \otimes d_A(m_2)$ if some $y_i \in B$ or $m_i = m_o$, along ∂I^2 the differential is given by

$H_t \otimes d_Y \otimes G_1 \otimes d_A$ on the edge (1,2),

$H_1 \otimes d_Y \otimes G_t \otimes d_A$ on (3,2), $H_1 \otimes d_Y \otimes d_A \otimes G_t$ on (3,4),

$H_t \otimes d_Y \otimes d_A \otimes G_1$ on (5,4), $d_Y \otimes d_Y \otimes G_t \otimes G_1$ on (5,6);

$d_Y \otimes d_Y \otimes G_1 \otimes G_t$ on (7,6), $(d_Y \otimes d_Y \otimes G_1 \otimes d_2 \otimes d_A) \circ F_t$ on (7,1).

fig(iii) $z_2 \in \langle y^{\otimes 2}, 2, a_1 \rangle \subset K((Y,B)^2 \times (I,\partial I)^2 \times (M_2, m_o)^2)$, differentials on $(Y \otimes Y \otimes \mathbb{C}^2 \otimes A \otimes \mathbb{C}^2 \otimes A) \times I^2$

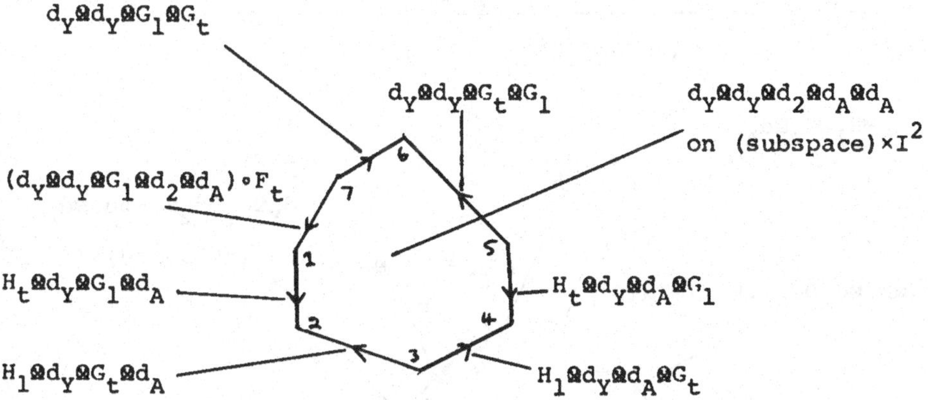

The element z_1+z_2 is represented in the following way. Identify $Y \otimes \mathbb{C}^2 \otimes Y \otimes \mathbb{C}^2 \otimes A \otimes A$ and $Y \otimes Y \otimes \mathbb{C}^2 \otimes A \otimes \mathbb{C}^2 \otimes A$ by the switching map $(1 \otimes T \otimes T \otimes 1)$. Add the representing complexes for z_1 and z_2 by juxtaposing along the common edges and cancelling these edges. Hence on the family of vector bundles of $(Y \otimes Y \otimes \mathbb{C}^2 \otimes A \otimes \mathbb{C}^2 \otimes A) \times I^2$, where I^2 is represented as a polygon with vertices $\{i \mid 1 \leq i \leq 10\}$, the differentials given as follows [see fig(iv)]. Over (y_1, y_2, z, m_1, m_2) the differential is

$d_Y(y_1) \otimes d_Y(y_2) \otimes d_2 \otimes d_A(m_1) \otimes d_2 \otimes d_A(m_2)$ if some $y_i \in B$ or $m_i = m_o$, over ∂I^2 the differential is given by

$(H_t \otimes H_1 \otimes d_A \otimes d_A)(1 \otimes T \otimes T \otimes 1)$ on (1,2), $(H_1 \otimes H_t \otimes d_A \otimes d_A)(1 \otimes T \otimes T \otimes 1)$ on (3,2)

$H_1 \otimes d_Y \otimes d_A \otimes G_t$ on (3,4), $H_t \otimes d_Y \otimes d_A \otimes G_1$ on (5,4)

$d_Y \otimes d_Y \otimes G_t \otimes G_1$ on (5,6), $d_Y \otimes d_Y \otimes G_1 \otimes G_t$ on (7,6)

$(d_Y \otimes d_Y \otimes G_1 \otimes d_2 \otimes d_A) \circ F_t$ on (7,8), $d_Y \otimes H_t \otimes G_1 \otimes d_A$ on (8,9)

$d_Y \otimes H_1 \otimes G_t \otimes d_A$ on (10,9), $(d_Y \otimes d_2 \otimes H_1 \otimes d_A \otimes d_A) \circ F_t \circ (1 \otimes T \otimes T \otimes 1)$ on (1,10)

fig(iv) $z_1+z_2 \in K((Y,B)^2 \times (I,\partial I)^2 \times (M_2,m_o)^2)$, differentials on $(Y \otimes Y \otimes \mathbb{C}^2 \otimes A \otimes \mathbb{C}^2 \otimes A) \times I^2$

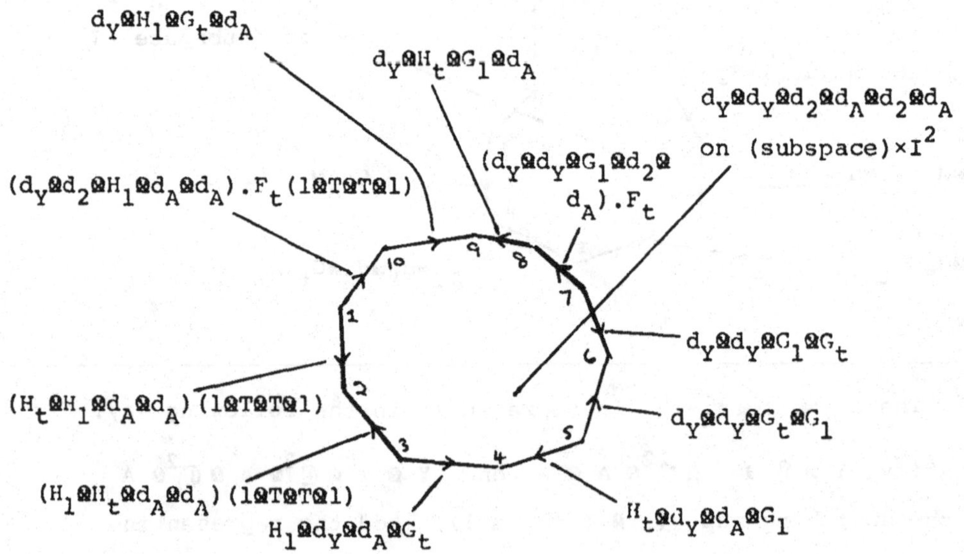

The element $x^{\otimes 2} + z_1+z_2$ is now represented in the following way [see fig(v)] on the underlying family of vector bundles of $(Y \otimes \mathbb{C}^2 \otimes A \otimes Y \otimes \mathbb{C}^2 \otimes A) \times I^2$ when I^2 is a polygon with vertices $\{i \mid 1 \leq i \leq 6\}$. Over the point (y_1, y_2, z, m_1, m_2) the differential is

$d_Y(y_1)\otimes d_2\otimes d_A(m_1)\otimes d_Y(y_2)\otimes d_2\otimes d_A(m_2)$ if some $y_i \in B$ or $m_i = m_o$, over ∂I^2 the differential is given by

$(d_Y\otimes G_t\otimes H_1\otimes d_A)\circ F_1$ on $(1,2)$, $(d_Y\otimes G_1\otimes H_t\otimes d_A)\circ F_1$ on $(3,2)$
$(d_Y\otimes G_1\otimes d_Y\otimes d_2\otimes d_A)\circ F_t$ on $(4,3)$, $d_Y\otimes G_1\otimes H_t\otimes d_A$ on $(4,5)$
$(d_Y\otimes G_t\otimes H_1\otimes d_A)$ on $(6,5)$, $(d_Y\otimes d_2\otimes d_A\otimes H_1\otimes d_A)\circ F_t$ on $(6,1)$.

fig(v) $x^{\otimes 2}+z_1+z_2\in K((Y,B)^2\times(I,\partial I)^2\times(M_2,m_o)^2)$, differentials on $(Y\otimes\mathbb{C}^2\otimes A\otimes Y\otimes\mathbb{C}^2\otimes A)\times I^2$

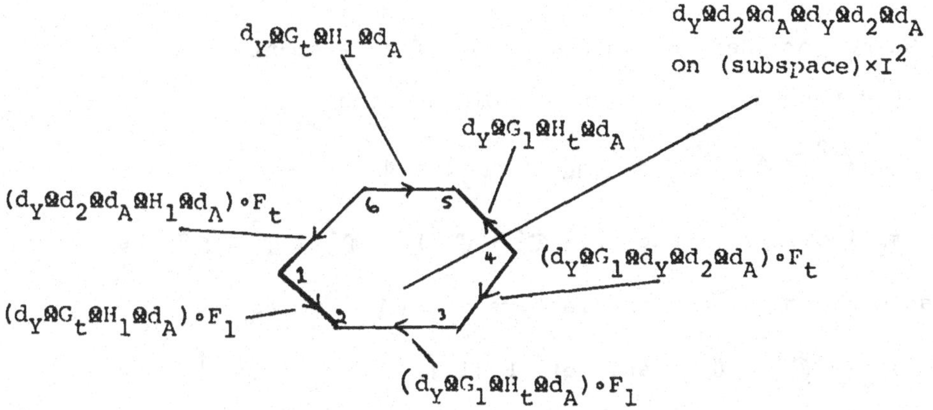

Finally this representation of $x^{\otimes 2}+z_1+z_2$ clearly represents zero since the differential on $(4,5,6)$ can be extended over I^2 by composing it with the homotopy $F_t \in \mathrm{Aut}\ (\mathbb{C}^2\otimes\mathbb{C}^2)$, and the "proof by pictures" of Proposition A.2.1 is complete.

Now suppose that the element, y, of Proposition A.2.1 has the form $y = 2.w$ and that the representing complexes in the proof of Proposition A.2.1 have the form $(Y,d_Y) = (W\otimes\mathbb{C}^2, d_W\otimes d_2)$, when (W,d_W) is a complex of vector bundles over (Y,B). Since the symmetric square, $(\mathbb{C}^2)^{\otimes 2} \in R(\pi_2)$, is isomorphic to $(\mathbb{C}^5 + \mathbb{C}^3_Y) \otimes \mathbb{C}^2$ as a π_2-representation we have that

$(4.w)^{\boxtimes 2} = [(\mathbb{C}^5 + \mathbb{C}^3_Y).w^{\boxtimes 2}].2 = 0$ and

$<(\mathbb{C}^5 + \mathbb{C}^3_Y).w^{\boxtimes 2},2,a> \subset K_{\pi_2}((Y,B)^2 \times (I,\partial I) \times (M_2,m_o))$

is defined. Now that the element, y, has become 2w we may revert to denoting by $y \in R(\pi_2)$ the one-dimensional involution representation. Let $v \in <5+3y).w^{\boxtimes 2},2,a>$ be the element represented by the defining system using the underlying vector bundles of $(W \otimes \mathbb{C}^2) \otimes \mathbb{C}^2 \otimes (W \otimes \mathbb{C}^2) \otimes \mathbb{C}^2 \otimes A$, and homotopies, $H_t^{\boxtimes 2}$ and G_t.

We now proceed to a description of an element $v_1 \in \text{quad}(v)$ but first we need some notation.

Let $\phi: (\mathbb{C}^2 \otimes \mathbb{C}^2)^{\boxtimes 2} \overset{\simeq}{\to} (\mathbb{C}^5 + \mathbb{C}^3_y) \otimes \mathbb{C}^2$

be a π_2-isomorphism and $\mu: (\mathbb{C}^5 + \mathbb{C}^3_y) \overset{\equiv}{\to} \mathbb{C}^2 \otimes \mathbb{C}^2 \otimes \mathbb{C}^2$ be an isomorphism. Index the factors of $(\mathbb{C}^2)^{\otimes 4}$ as $\mathbb{C}^2_1 \otimes \mathbb{C}^2_2 \otimes \mathbb{C}^2_3 \otimes \mathbb{C}^2_4$ and let $F_t(i,j) \in \text{Aut}_{\mathbb{C}}((\mathbb{C}^2)^{\otimes 4}$ be the homotopy induced by F_t from the identity to $T(i,j)$,

the map which switches the i-th and j-th coordinate. Let $J_j \in \text{Aut}((\mathbb{C}^2)^{\otimes 4})$ be a homotopy between the identity and $J_1 = (\mu \otimes 1) \circ \phi^{-1}$. Take the underlying family of vector bundles of $(W \otimes \mathbb{C}^2_1) \otimes \mathbb{C}^2_2 \otimes (W \otimes \mathbb{C}^2_3) \otimes \mathbb{C}^2_4 \otimes A \times I^2$ when I^2 is represented as a polygon with vertices $\{i \mid 1 \leq i \leq 12\}$.
The following differential defines a complex of vector bundles, [see fig(vi)] over $(Y,B)^2 \times (I^2,\partial I^2) \times (M_2,m_o)$, where the π_2-action used in the quadratic construction switches the $(Y,B)^2$ factors and interchanges the $(W \otimes \mathbb{C}^2_1 \otimes \mathbb{C}^2_2)$ and $(W \otimes \mathbb{C}^2_3 \otimes \mathbb{C}^2_4)$

factors, which represents $v_1 \in \mathrm{quad}(v)$. Over (y_1, y_2, z, m) the differential is $d_W \otimes d_2 \otimes d_2 \otimes d_W \otimes d_2 \otimes d_2 \otimes d_A$ if some $y_i \in B$ or $m = m_o$., on ∂I^2 the differential is given by

$H_1 \otimes H_t \otimes d_A$ on (2,1), $H_t \otimes H_1 \otimes d_A$ on (12,1)

$(H_1 \otimes d_W \otimes d_2 \otimes d_2 \otimes d_A) \circ F_t(2,4)$ on (2,3), $(d_W \otimes d_2 \otimes d_2 \otimes H_1 \otimes d_A) \circ F_t(24)$ on (12,11)

$(H_1 \otimes d_W \otimes d_2 \otimes G_t) \circ T(2,4)$ on (3,4) and its conjugate by τ on (11,10),

$(H_1 \otimes d_W \otimes d_2 \otimes G_1) \circ F_t(2,4)$ on (5,4) and its conjugate by τ on (9,10)

$J_t^{-1} \circ (H_1 \otimes d_W \otimes d_2 \otimes G_1) \circ J_t$ on (5,6) and its conjugate by τ on (9,8)

$J_1^{-1} \circ (H_t \otimes d_W \otimes d_2 \otimes G_1) \circ J_1$ on (7,6) and its conjugate τ on (7,8).

Notice that at the vertex, (7), the differential is π_2-equivariant since it is induced by J_1 form $d_W^{\boxtimes 2} \otimes d_{5+3y} \otimes G_1$ on $W^{\boxtimes 2} \otimes (\mathbb{C}^5 + \mathbb{C}_y^3) \otimes \mathbb{C}^2 \otimes A$.

fig(vi) $v_1 \in \mathrm{quad}(v) \subset K((Y,B)^2 \times (I^2, \partial I^2) \times (M_2, m_o))$, differentials on $([W \otimes \mathbb{C}_1^2] \otimes \mathbb{C}_2^2) \otimes ([W \otimes \mathbb{C}_3^2] \otimes \mathbb{C}_4^2) \otimes A \times I^2$

$[(\ )^\tau \equiv$ conjugate by $\tau]$

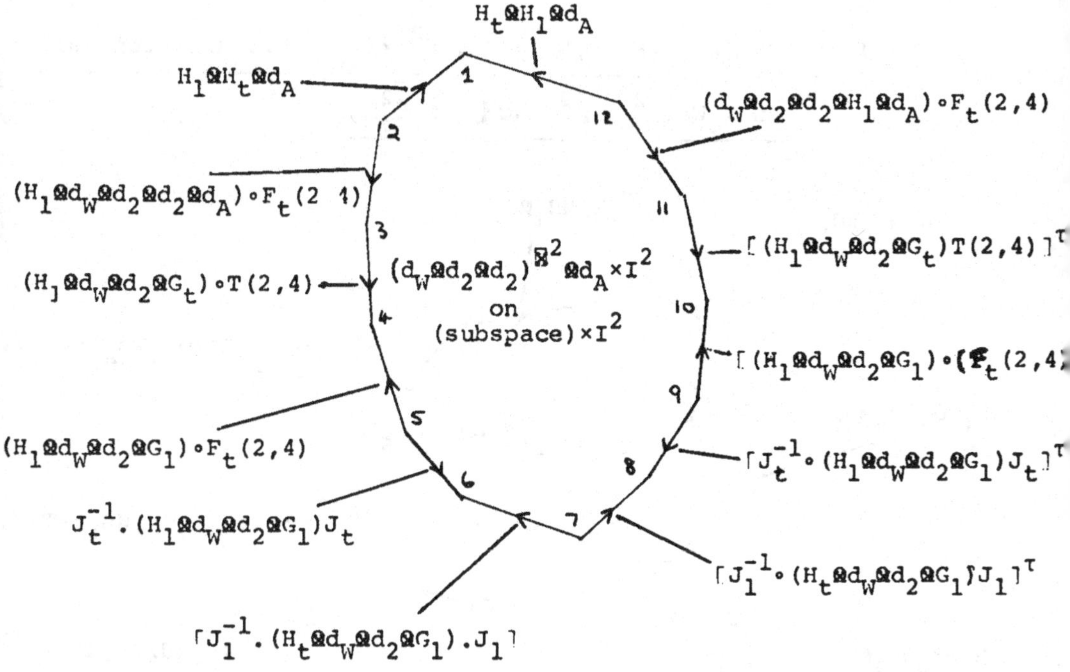

We now describe a representative of an element $z_3 \in \langle y_1, 2, a\rangle \subset K((Y,B)^2 \times (I,\partial I)^2 \times (M_2, m_o))$ which is designed to be the restriction of the element $z_1 \in \langle y_1, 2, a^{\otimes 2}\rangle$ of Proposition A.2.1 in the case when $y = 2w$. [c.f.fig(ii)]. Take the underlying family of vector bundles of $(W \otimes \mathbb{C}_1^2) \otimes \mathbb{C}_2^2 \otimes (W \otimes \mathbb{C}_3^2) \otimes \mathbb{C}_4^2 \times I^2$ when I^2 is represented as a polygon with vertices $\{i \mid 1 \leq i \leq 7\}$. The following differential defines a complex of vector bundles [see fig(vii)] over $(Y,B)^2 \times (I^2, \partial I^2) \times (M_2, m_o)$ representing z_3. Over (y_1, y_2, z, m) the differential is $d_W \otimes d_2 \otimes d_2 \otimes d_W \otimes d_2 \otimes d_2 \otimes d_A$ if some $y_i \in B$ or $m = m_o$, on ∂I^2 the differential is given by

$H_1 \otimes d_W \otimes d_2 \otimes G_t$ on $(1,2)$, $H_t \otimes d_W \otimes d_2 \otimes G_1$ on $(3,2)$,

$(d_W \otimes d_2 \otimes H_t \otimes G_1) \circ T(2,3)$ on $(3,4)$, $(d_W \otimes d_2 \otimes H_1 \otimes G_t) \circ T(2,3)$ on $(5,4)$,

$(d_W \otimes d_2 \otimes d_2 \otimes H_1 \otimes d_A) \circ F_t(2,4)$ on $(6,5)$, $(H_t \otimes H_1 \otimes d_A)$ on $(6,7)$,

$H_1 \otimes H_t \otimes d_A$ on $(1,7)$.

fig(vii) $z_3 \in \langle y_1, 2, a\rangle \subset K((Y,B)^2 \times (I^2, \partial I^2) \times (M_2, m_o))$, differentials on $([W \otimes \mathbb{C}_1^2] \otimes \mathbb{C}_2^2 \otimes [W \otimes \mathbb{C}_3^2] \otimes \mathbb{C}_4^2 \otimes A) \times I^2$

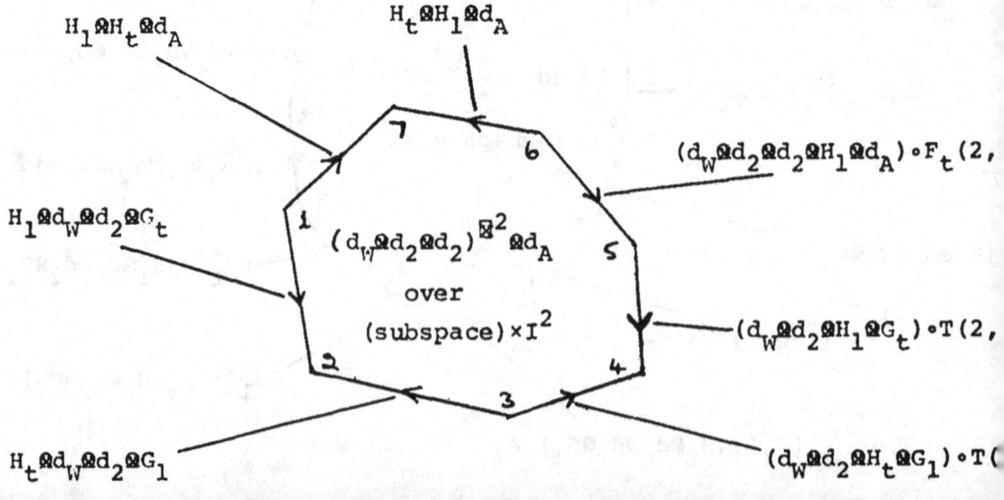

We now examine the representative of $z_3 + v_1$ obtained by adding the representatives constructed so far. We wish to show $0 = z_3 + v_1$ by applying a series of homotopies to the representative we have constructed. This will be done in simple stages. To obtain the sum of representatives of z_3 and v_1 already constructed take the underlying family of vector bundles of $(W \otimes \mathbb{C}_1^2 \otimes \mathbb{C}_2^2 \otimes W \otimes \mathbb{C}_3^2 \otimes \mathbb{C}_4^2) \times I^2$ when I^2 is represented as a polygon with vertices $\{i \mid 1 \leq i \leq 13\}$. The following differential defines a complex over $(Y,B)^2 \times (I^2.,\partial I^2) \times (M_2,m_o)$ [see fig(viii)] representing $z_3 + v_1$. Over the point (y_1,y_2,z,m) the differential is $d_W \otimes d_2 \otimes d_2 \otimes d_W \otimes d_2 \otimes d_2 \otimes d_A$ if some $y_i \in B$ or $m = m_o$, on ∂I^2 the differential is given by

$H_1 \otimes d_W \otimes d_2 \otimes G_t$ on $(2,1)$, $(H_1 \otimes d_W \otimes d_2 \otimes d_2 \otimes d_A) \circ F_t(2,4)$ on $(2,3)$,

$(H_1 \otimes d_W \otimes d_2 \otimes G_t) \circ T(2,4)$ on $(3,4)$ and its conjugate by τ on $(11,10)$

$(H_1 \otimes d_W \otimes d_2 \otimes G_1) \circ F_t(2,4)$ on $(5,4)$ and its conjugate by τ on $(9,10)$,

$J_t^{-1} \circ (H_1 \otimes d_W \otimes d_2 \otimes G_1) \circ J_t$ on $(5,6)$ and its conjugate by τ on $(9,8)$,

$J_1^{-1} \circ (H_t \otimes d_W \otimes d_2 \otimes G_1) \circ J_1$ on $(7,6)$ and its conjugate by τ on $(7,8)$,

$(d_W \otimes d_2 \otimes H_1 \otimes G_t) \circ T(2,3)$ on $(11,12)$, $(d_W \otimes d_2 \otimes H_t \otimes G_1) \circ T(2,3)$ on $(13,12)$,

$H_t \otimes d_W \otimes d_2 \otimes G_1$ on $(13,1)$.

<u>fig(viii)</u> $z_3+v_1\in K((Y,B)^2\times(I^2,\partial I^2)\times(M_2,m_o))$, differentials on $(W\otimes\mathbb{C}_1^2)\otimes\mathbb{C}_2^2\otimes(W\otimes\mathbb{C}_3^2)\otimes\mathbb{C}_4^2\otimes A\times I^2$

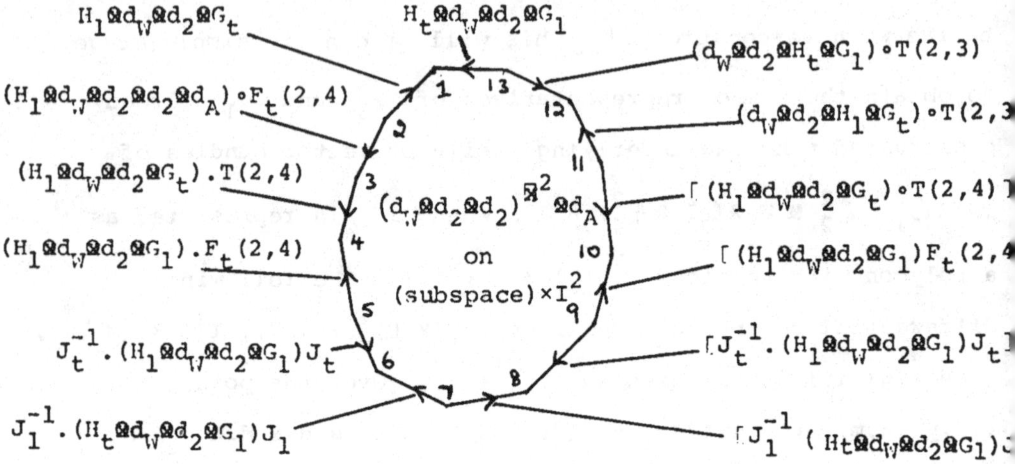

The differentials along (11,12) and (11,10) are equal, since they use the same homotopies and the differentials are in the same places relative to the $\{\mathbb{C}_i^2\}$, H_1 uses $\mathbb{C}_2^2\otimes\mathbb{C}_3^2$ with factors reversed and G_t uses $\mathbb{C}_4^2$. Also the homotopy of differentials along (3,2,1) is homotopic, relative to the end points, to $(H_1\otimes d_W\otimes G_t)\circ F_{1-t}(2,4)$ and so is the homotopy of differentials along (3,4,5). Using these observations we may cancel some differentials on ∂I^2 to obtain the following representation of $z_3 + v_1$ on $W\otimes\mathbb{C}_1^2\otimes\mathbb{C}_2^2\otimes W\otimes\mathbb{C}_3^2\otimes\mathbb{C}_4^2\otimes A\times I^2$ where I^2 is represented as a polygon with vertices $\{i|\ 5\le i\le 11\}$, whose differentials on ∂I^2 are those of the previous representative on each edge with the exception of edges (11,10) and (11,5) on which they are respectively the former differentials on sides (13,12) and (13,1), [see fig(ix)].

fig(ix) reduced form of (z_3+v_1), differentials on $W\otimes\mathbb{C}_1^2\otimes\mathbb{C}_2^2\otimes W\otimes\mathbb{C}_3^2\otimes\mathbb{C}_4^2\otimes A\times I^2$

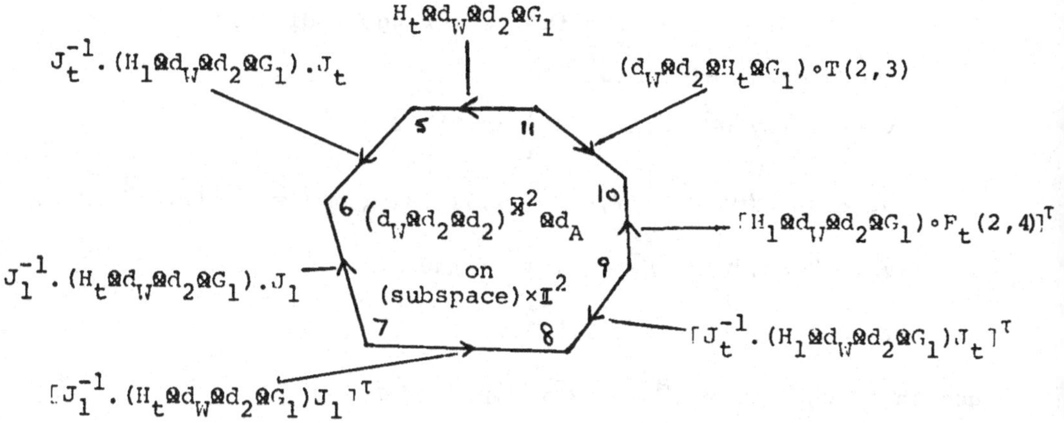

Now, in the representative of fig(ix) at every point of ∂I^2 the differential involves the exact differential, G_1. Thus the homotopy, H_t, can be used to homotopy this representation of (z_3+v_1) to one on the same underlying family of vector bundles having the following differentials. Over (y_1,y_2,z,m) the differential is $d_W\otimes d_2\otimes d_2\otimes d_W\otimes d_2\otimes d_2\otimes d_A$ if some $y_i \in B$ or $m = m_0$ and on ∂I^2 the differentials are obtained at each point by replacing H_t or H_1 by $H_0 = d_W\otimes d_2\otimes d_2$. Now the representative has a factor $(W\otimes W, d_W\otimes d_W)$ which may be extracted. Hence there is an element $b \in K((I^2,\partial I^2)\times(M_2,m_0)) \cong Z/_2$ such that $z_3 + v_1 = w^{\otimes 2}.b \in K((Y,B)^2\times(I,\partial I)^2\times(M_2,m_0))$. The element, b, is represented by the complex of vector bundles shown in fig(ix) when $(Y,B) = (\text{point},\phi)$ and (W,d_W) is $\mathbb{C}$ with zero differential. Hence b is independent of all the other variables since all Massey products $\langle y,2,a\rangle$ can be formed using the same homotopy, G_t, and the representative of b depends only on J_t, $F_t(2,4)$ and G_t.

Proposition A.2.2:

Let $w \in K(Y,B)$ satisfy $4w = 0$.

(i) There is an element $b \in K(M_2,m_o) \cong Z/_2$, independent of w, Y and B, satisfying the following conditions.

There exist elements

$v \in \langle(5+3y)w^{\otimes 2},2,a\rangle \subset K_{\pi_2}^{-1}((Y,B)^2;Z/_2)$,

$v_1 \in \text{quad}(v) \subset K((Y,B)^2\times(I^2,\partial I^2)\times(M_2,m_o)) \cong K^o((Y,B)^2;Z/_2)$,

$y_1 \in 2w,2,2w\rangle \subset K^{-1}((Y,B)^2)$ and

$z_3 \in \langle y_1,2,a\rangle \subset K^o((Y,B)^2:Z/_2)$

such that $z_3 + v_1 = w^{\otimes 2}b \in K^o((Y,B)^2;Z/_2)$.

(ii) In fact b is zero.

Proof:

(i) This follows by using the elements constructed in the foregoing discussion.

(ii) Let $j:RP^3 \to RP^4$ be the canonical inclusion of projective spaces. We show, by looking at this map, that $0 \neq b \in Z/_2$ is impossible for $(Y,B) = (RP^4$, point). The integral K-groups of RP^3 and RP^4 are [At 2,§2.7]

$K^1(RP^4) = 0$, $K^1(RP^3) = Z$ and

$j^*: \tilde{K}^o(RP^4) \cong Z/_4 \to \tilde{K}^o(RP^3) \cong Z/_2$ is onto.

Let $w \in \tilde{K}^o(RP^4)$ be a generator. Arguing as in the derivation of Proposition 3.7 from Proposition 3.6 the fact that $z_3 + v_1 = w^{\otimes 2}$ implies the following behaviour in the spectral sequence $\{E_r((RP^4,p_o)^2;\pi_2;Z/_2)\}$. If $x \in \langle 2w,2,a\rangle$ (this Massey product is the non-zero <u>element</u> in $K^1(RP^4;Z/_2) \cong Z/_2$ then $(x^{\otimes 2} + w^{\otimes 2}) \otimes e_1 \in E_2^{1,o}((RP^4,p_o)^2;\pi_2;Z/_2)$ is a permanent cycle. Since $w \in \tilde{K}^{o}(RP^{4})$ is an integral class the results of [Sn 2,§2] imply

$$w^{\otimes 2} \otimes e_4 = d_3(w^{\otimes 2} \otimes e_1) = d_3(x^{\otimes 2} \otimes e_1).$$

However $j^*(x) = j^*(\langle 2w,2,a\rangle) \subset \langle j^*(2w),2,a\rangle = \langle 0,2,a\rangle$ is in $\mathrm{im}\{K^1(RP^3) \to K^1(RP^3;Z/_2) \cong Z/_2 \oplus Z/_2\}$ so $j^*(x)$ is a non-zero integral class which is not in the image of β_2.

$$\begin{aligned} \text{Also}\quad j^*(w^{\otimes 2} \otimes e_4) &= j^*(d_3(x^2 \otimes e_1)) \\ &= d_3(j^*(x)^{\otimes 2} \otimes e_1) \\ &= j^*(x)^{\otimes 2} \otimes e_4 \in E_2^{4,o} \cong \ker \beta_2/_{\mathrm{im}\,\beta_2} \end{aligned}$$

so $0 \neq j^*(x) = j^*(w) \in \ker \beta_2/_{\mathrm{im}\,\beta_2}$ which is impossible, for they have different degrees.

Hence $b = 0$ in general.

<u>Proof of Proposition 3.6</u>:

Putting (U,V) for (Y,B) in Proposition A2.1 we have

$$x^{\otimes 2} \in \langle w_1,2,a^{\otimes 2}\rangle + \langle (2w)^{\otimes 2}, 2,a_1\rangle$$

in $K^{-2}((Y,B)^2 \times (M_2,m_o)^2) \cong K^o((Y/_B) \wedge M_2 \wedge M_2)$

where $w_1 \in \langle 2w,2,2w\rangle$ and $a_1 \in \langle a,2,a\rangle$. Also $4w = 0$ so $\langle 2w)^{\otimes 2},2,a_1\rangle = \langle 0,2,a_1\rangle \subset K^o((Y,B)^2).a_1$.

However the homomorphism induced by $\alpha\colon N_2 \to M_2 \wedge M_2$ [c.f.§1] is given by

$$0 = \alpha^*(a_1) \in K^*(RP^2) \oplus K^*(S^2)$$

and $\alpha^*(a^{\otimes 2}) = (a,0)$. Hence the mod 2 product $x^{\otimes 2} \in \langle w_1,2,\alpha^*(a^{\otimes 2})\rangle = \langle w_1,2,a\rangle$. This proves the first half of Proposition 3.6. The second part follows from Proposition

A2.2 once it is observed that the explicit representative of $\langle w_1, 2, a^{\otimes 2}\rangle$ constructed for Proposition A2.1 maps to the explicit representative of

$$\langle w_1, 2, (a,0)\rangle \subset K((Y,B)^2 \times (I^2, \partial I^2) \times (N_2, n_o)) \text{ under } \alpha.$$

For α may be considered as being induced by a stable map $N_2 \to M_2 \times M_2$ which therefore affects only the N_2 and $M_2 \times M_2$ coordinates and there is no indeterminacy associated to varying representing complexes for $a \in \tilde{K}^o(RP^2)$ or homotopies corresponding to the relation $2a = 0$.

Appendix III

§A3.1:

Let F be a compact space and $CF = [o,1]\times F/_{\approx}$ where $(o,f) \approx O$ and $F = \{1\}\times F \subset CF$. In this section we consider the homomorphisms

$$\begin{array}{ccc} K_G^{\alpha}(F^p,*;Z/_p) & \overset{\delta}{\underset{\cong}{\to}} & K_G^{\alpha+1}((CF)^p,F^p;Z/_p) \\ & & \uparrow p^* \\ & & K_G^{\alpha+1}((CF,F)^p;Z/_p) \end{array}$$

when $G = \Sigma_p$ or π_p. For $x \in K^*(CF,F;Z/_p)$ such that, in the notation of §4, $x^{\otimes p}\otimes e_o \in K^*_{\Sigma_p}((CF,F)^p;Z/_p)$ is defined we determine $(\delta^{-1})p^*(x^{\otimes p}\otimes e_o)$ in terms of the homomorphism $\delta':K^{\beta}(F,*;Z/_p) \overset{\cong}{\to} K^{\beta+1}(CF,F;Z/_p)$.

We begin by discussing the various cases when $p = 2$.

Case (i): $\delta:K^1_{\Sigma_2} \to K^o_{\Sigma_2}$ integral classes

Let $x \in K^{-1}(F)$ be represented as a complex over $(CF, o \cup F)$ in the following manner. Let $H_t:F \to \mathrm{End}(\mathbb{C}^k)$, $(t \in [o,1])$, be a homotopy such that H_o is constant and $H_o(F)$, $H_1(F)$ lie in $\mathrm{Aut}(\mathbb{C}^k)$. Let (X,d_X) be the complex over $(CF, o \cup F)$ given over $[t,f]$ by $O \longrightarrow \mathbb{C}^k \xrightarrow[H_t(f)]{} \mathbb{C}^k \longrightarrow O$.

The map

$$\begin{array}{ccc} \delta:\ K^{\beta}_{\Sigma_2}(C(F)^2,\ o \cup F^2) & \longrightarrow & K^{\beta}_{\Sigma_2}((CF)^2 \cup [1,2]\times F^2,\{2\}\times F^2) \\ \downarrow\cong & & \downarrow\cong \\ K^{\beta-1}_{\Sigma_2}(F^2) & & K^{\beta}((CF)^2,F^2) \end{array}$$

is induced by the map

$$\delta:\ (CF)^2 \cup [1,2]\times F^2 \longrightarrow C(F^2)$$

which is given by

$$\delta(t_1,f_1,t_2,f_2) = [0,f_1,f_2]$$

$$\delta(s,f_1,f_2) = [s-1,f_1,f_2],\ (t_i\in[0\ 1];\ f_i\in F; s\in[1,2]).$$

There is a homeomorphism of Σ_2-spaces, $I_o\times I_\pi \cong I^2$, where the respective actions are given by

$$\tau(x,y) = (x,1-y) \text{ and } \tau(x,y) = (y,x),\ (1 \neq \tau\in\Sigma_2; x,y\in[0,1]).$$

The element $x'\in K_{\Sigma_2}(C(F^2),\ o\cup F^2)$ determined by $(X,d_X)^{\boxtimes 2}|\ (I_o,\partial I_o)\times F^2/_{(\approx)}$ reduces mod 2 to the element, $x^{\otimes 2}\otimes e_1 \in K^1_{\Sigma_2}(F^2;Z/_2)$. Take the underlying family of Σ_2-vector bundles of $(0\to \mathbb{C}^k \to \mathbb{C}^k \to 0)^{\boxtimes 2}\times\{(CF)^2\cup[1,2]\times F^2\}$ and on this form the following homotopy, $\underline{\phi}_t$, of complexes of Σ_2-vector bundles over $((CF)^2\cup[1,2]\times F^2,\{2\}\times F^2)$.

Choose functions

$$a:\ [1,2]\times[01] \to [01]$$

and

$$b:\ [01]\times[01] \to [01]$$

satisfying

$$a(s,1) = s-1,\ \ a(s,o) = 1,\ \ a(2,t) = 1,$$

$$b(t,1) = 0,\ b(t,o) = t,\ b(o,t) = 0 \text{ and}$$

$a(1,t) = b(1,t)$. Define the differential, Φ_t, by

$$\underline{\phi}(t,s,f_1,f_2) = d_X(a(s,t),f_1)\otimes d_X(a(s,t),f_2)$$

and

$$\underline{\phi}(t,[t_1,f_1,t_2,f_2]) = d_X(b(t_1,t),f_1)\otimes d_X(b(t_2,t),f_2).$$

When $t = 1$ this complex represents

$\delta[X,d_X)^{\boxtimes 2}| \ (I_o,\partial I_o) \times F^2/_{\sim}]$

and when $t = 0$ it represents $p^*(\delta'(x)^{\boxtimes 2})$.

Case (ii): $\delta : K^1_{\Sigma_2} \to K^o_{\Sigma_2}$, non-integral classes.

In the notation of Appendix II let $x \in K^1(F;Z/_2)$, $w \in K^o(F)$ and $x \in \langle 2w,2,a\rangle$. Let x be represented by the complex $W \equiv (o \to W_o \xrightarrow{o} W_1 \longrightarrow o)$ of vector bundles over F. Let $b \in K(D^2,S^1)$ be the Thom class represented by the complex $(o \longrightarrow \mathbb{C} \xrightarrow{(-.z)} \mathbb{C} \longrightarrow o)$, $(z \in D^2 \subset \mathbb{C})$.
Hence $w.b \in K(F \times (I^2,\partial I^2))$ is represented by any complex of the form

$$0 \longrightarrow W_o \oplus E_o \xrightarrow[o]{\{(-.z)\oplus 1\}} (W_o \oplus E_o) \oplus (W_1 \oplus E_1) \xrightarrow{(o,(-.z)\oplus 1)} (W_1 \oplus E_1) \longrightarrow 0.$$

Taking E_o and E_1 so that $W_i \oplus E_i \cong \mathbb{C}^k \times F$, $w.b$ is representable by a complex of trivial bundles
$(W',d') \equiv (0 \to \mathbb{C}^k \to \mathbb{C}^{2k} \to \mathbb{C}^k \to 0)$ over $F \times (I^2,\partial I^2)$
which satisfies the condition that $d'(f,t_1,t_2)$ is independent of $f \in F$ if $t_i = 0$ for $i = 1$ or 2.
A homotopy, H_t, of differentials on $W \otimes \mathbb{C}^4$ gives one, $H_t \otimes (-.z) = H_t'$, on $W \otimes (o \to \mathbb{C} \to \mathbb{C} \to o) \otimes {}^4$ and hence one, $H_t'' = H_t' \otimes 1_{E_o \otimes \mathbb{C}^4} \otimes 1_{E_1 \otimes \mathbb{C}^4}$ on $W' \otimes \mathbb{C}^4$. If H_t is the homotopy used in the construction of an element of $\langle 2w,2,a\rangle$ in Appendix II then $H_t''(f,t_1,t_2)$ is independent of $f \in F$ if $t_i = o$ for $i = 1$ or 2, since d' is exact at such points and so H_t'' is $d' \otimes 1_{\mathbb{C}^4}$ there. Now form a representative of an element of $\langle (5+3y).W^{\boxtimes 2}.b^{\boxtimes 2},2,a\rangle$ as

a complex over $F^2 \times (I,\partial I)^2 \times (I,\partial I)^2 \times (I,\partial I) \times (M_2,m_o)$ on the underlying family of vector bundles of $W' \otimes W' \otimes \mathbb{C}^4 \otimes \mathbb{C}^4 \otimes A$ as in Appendix II, using the homotopy, H''_t. Let this complex be (Z,d_Z). By the choice of H''_t this element of $\langle (5+3y)\, W^{\otimes 2}.b^{\otimes 2},2,a\rangle$ will be equal to the product of $b^{\otimes 2}$ with the element of $\langle (5+3y).W^{\otimes 2},2,a\rangle$ constructed using the homotopy, H_t. However, because of the independence of the differentials on $f \in F$ when t_1 or t_2 is zero, the representative defines a complex over

$(C(F^2),o \cup F^2) \times (I_\pi,\partial I_\pi) \times [(I,\partial I)^2] \times (I,\partial I) \times (M_2,m_o)$

(when $I_o \subset I^2$ is the cone-coordinate) representing the element of

$\langle (5+3y)\, W^{\otimes 2}.b^{\otimes 2},2,a\rangle$ considered as a subset of

$K_{\Sigma_2}((C(F^2),o\cup F^2)\times(I_\pi,\partial I_\pi)\times[(I_o,\partial I_o)\times(I_\pi,\partial I_\pi)]\times(I,\partial I)\times(M_2,m_o))$

$\cong K^1_{\Sigma_2}(F^2;Z/_2)$.

It is now simple to describe a representative of the restriction to

$[(CF)^2\cup[1,2]\times F^2,\{2\}\times F^2]\times[(I,\partial I)^2]\times(I,\partial I)\times(M_2,m_o)$

of the image under δ of this element, by means of the map, δ, used in case(i). Take the underlying family of vector bundles of the complex

$\{\{o\to \mathbb{C}^k\to \mathbb{C}^{2k}\to \mathbb{C}^k\to o\}^{\otimes 2}\otimes(\mathbb{C}^4)^{\otimes 2}\times[(CF)^2\cup[1,2]\times F^2]\times[I^2]\times I\}\otimes A$.

Define a homotopy, Φ_t, of complexes on the family of vector bundles over

$[(CF)^2\cup[1,2]\times F^2,\{2\}\times F^2]\times[(I,\partial I)^2]\times(I,\partial I)\times(M_2,m_o)$ by

$\Phi(t,s,f_1,f_2,u_1,u_2,v,m) = d_Z(f_1,f_2,a(s,t),a(s,t),u_1,u_2,v,m)$

and

$$\Phi(t,[t_1,f_1,t_2,f_2],u_1,u_2,v,m) = d_Z(f_1,f_2,b(t_1,t),b(t_2,t),u_1,u_2,v,m)$$

$$(s\in[1,2];t_1,t_2,t,u_1,u_2,v \in [01];m\in M_2;f_i\in F).$$

If $x^{\otimes 2} \otimes e_1 \in \langle(5+3y).W^{\otimes 2},2,a\rangle$ and $q \in \langle 5+3y).\delta'(W)^{\otimes 2},2,a\rangle$ are the canonical elements constructed in Appendix II then the complex at $t = 1$ represents

$$\delta[(x^{\otimes 2}\otimes e_1).b^{\otimes 2}]|[(CF)^2\cup[1,2]\times F^2,\{2\}\times F^2]\times(I,\partial I)^2\times(I,\partial I)\times(M_2,m_o)$$

and at $t = 0$ represents $p^*(q)$.

Now $b^{\otimes 2} \in K_{\Sigma_2}((I^2,\partial I^2)^2) \cong K_{\Sigma_2}(\mathbb{C}\oplus\mathbb{C}_y)$ is the Thom class and therefore $b^{\otimes 2}|(I_o^2,\partial I_o^2) = \sigma.b$.

Thus $q^*(q) = $ (restriction of $\delta((x^{\otimes 2}\otimes e_1).b^{\otimes 2}$))

$$= \sigma.b\delta(x^{\otimes 2}\otimes e_1) \in K_{\Sigma_2}^{-2}((CF)^2,F^2;Z/_2).$$

However $\delta'(x)^{\otimes 2}\otimes e_2 = \delta'(x)^{\otimes 2} \in$ quad(q) and quad(-), [c.f. Sn 2,§2] is the operation $i^*.j^{-1}$ obtained from the homomorphisms

$$K_{\Sigma_2}^{-2}([(CF)^2,F^2]\times(I_\pi,\partial I_\pi);Z/_2) \xleftarrow{j} K_{\Sigma_2}^{-2}([(CF)^2,F^2]\times(I_\pi,\partial I_\pi)^2;Z/_2)$$

$$\swarrow i^*$$

$$K^{-4}((CF)^2,F^2;Z/_2) \cong E_1^{2,o}((CF)^2,F^2;\Sigma_2;Z/_2).$$

Thus, in the notation of §4,

$[\delta(x^{\otimes 2}\otimes e_1) - (\delta'(x)^{\otimes 2}\otimes e_o)] \in \ker(\sigma.-) = \mathrm{im}(i_!)$ and

$\delta(x^{\otimes 2}\otimes e_1) = \delta'(x)^{\otimes 2}\otimes e_o + i_!(z)$ for some $z \in K^o((CF)^2,F^2;Z/_2)$.

<u>Case(iii): $\delta:K_{\Sigma_2}^o \to K_{\Sigma_2}^1$, integral classes.</u>

Let $w \in K^o(F)$ be represented by the complex $0 \to W_o \to W_1 \to 0$ as in case(ii) and form the complex (W',d')

over $F \times (I^2,\partial I^2)$ representing $w.b \in K^o(F \times (I^2,\partial I^2))$.

The complex $(W',d')^{\boxtimes 2}$ restricted to $[F^2 \times (I_o,\partial I_o)] \times (I_o,\partial I_o)$

represents $\sigma.b.w^{\boxtimes 2} \in K^{-2}_{\Sigma_2}(F^2)$. Also this complex in fact

defines a complex over $(CF^2, o \cup F^2) \times (I_o,\partial I_o)$ from which

a representative of $\delta(\sigma.w^{\boxtimes 2}.b) = \delta(i_!(w^{\otimes 2}).b) = \delta(i_!(w^{\otimes 2})].b$

may be constructed. Take the underlying family of vector bundles of the complex

$$(0 \to \mathbb{C}^k \to \mathbb{C}^{2k} \to \mathbb{C}^k \to 0)^{\boxtimes 2} \times [(CF)^2 \cup [12] \times F^2] \times I_o$$

and construct a homotopy, Φ_t, of complexes over

$((CF)^2 \cup [1,2] \times F^2, \{2\} \times F^2) \times (I_o,\partial I_o)$. The differential,

Φ_t, is given by

$$\Phi(t,s,f_1,f_2,u) = (d')^{\boxtimes 2}(f_1,f_2,a(s,t),a(s,t),u,u)$$

and

$$\Phi(t,[t_1,f_1,t_2,f_2],u) = (d')^{\boxtimes 2}(f_1,f_2,b(t_1,t),b(t_2,t),u,u),$$

$$(t,t_1,t_2,u \in [0,1];\ s \in [1,2];\ f_i \in F).$$

At $t = 1$ this represents $\delta(i_!(w^{\otimes 2})).b$ and at $t = 0$ it

represents $p^*(\delta'(w)^{\boxtimes 2} | ((CF)^2,F^2) \times (I_o,\partial I_o))$.

Hence, from [Sn 2, §2],

$$\delta(i_!(w^{\otimes 2})) = p^*(\delta'(w)^{\otimes 2} \otimes e_1) \in K^1_{\Sigma_2}((CF)^2,F^2).$$

<u>Case(iv)</u>: <u>$\delta : K^o_{\Sigma_2} \to K^1_{\Sigma_2}$, non-integral classes.</u>

Let $x \in K^1(F)$ be represented by a complex

$(X,d_X) \equiv (0 \longrightarrow \mathbb{C}^k \xrightarrow{H_t} \mathbb{C}^k \to 0)$ as in case(i). Suppose

$y \in \langle 2x,2,a \rangle \subset K(F \times (I^2,\partial I^2);Z/_2)$ is represented by a Massey

product complex over $F \times (I^2,\partial I^2) \times (M_2,m_o)$ on the underlying family of vector bundles of $X \otimes \mathbb{C}^4 \otimes A$. The differentials will be independent of the F-coordinate when $t_i = 0$ for $i = 1$ or 2. As in Appendix II form a complex (Y, d_Y) over $F^2 \times [(I,\partial I)^2] \times (I,\partial I) \times (M_2, m_o)$ on the family of vector bundles of $X^{\otimes 2} \otimes (\mathbb{C}^4)^{\otimes 2} \otimes A$ representing an element $z \in \langle (5+3y).x^{\otimes 2}, 2, a\rangle$. The restriction of (Y, d_Y) to $F^2 \times [I_o, \partial I_o] \times (I,\partial I) \times (M_2, m_o)$ in fact defines a complex over $(C(F^2), o \cup F^2) \times (I,\partial I) \times (M_2, m_o)$. By Appendix II this complex represents

$$b.[i_!(y^{\otimes 2})] \in K_{\Sigma_2}((CF^2, o \cup F^2) \times (I,\partial I) \times (M_2, m_o))$$

$$\cong K_{\Sigma_2}^{-1}(CF^2, o \cup F^2; Z/_2) \cong K_{\Sigma_2}^{-2}(F^2; Z/_2).$$

From this complex it is simple to describe a representative of $\delta(b.i_!(y^{\otimes 2})) = b.\delta(i_!(y^{\otimes 2}))$. Take the underlying family of vector bundles of the complex

$$\{(0 \to \mathbb{C}^k \to \mathbb{C}^k \to 0)^{\otimes 2} \otimes (\mathbb{C}^4)^{\otimes 2} \times [(CF)^2 \cup [1,2] \times F^2]\} \otimes A$$

and construct a homotopy, Φ_t, of complexes over $((CF)^2 \cup [1,2] \times F^2, \{2\} \times F^2) \times (I,\partial I) \times (M_2, m_o)$. The differential, Φ_t, is given by

$$\Phi(t,s,f_1,f_2,u,m) = d_Y(f_1,f_2,a(s,t),a(s,t),u,m)$$

and

$$\Phi(t,[t_1,f_1,t_2,f_2]u,m) = d_Y(f_1,f_2,b(t_1,t),b(t_2,t)u,m),$$

$$(s \in [1,2];\ t,t_1,t_2,u \in [0,1];\ m \in M_2; f_i \in F).$$

At $t = 1$ this represents $b.\delta(i_!(y^{\otimes 2}))$ and at $t = 0$ it is the restriction, under p, of the representative constructed in Appendix II for the element

$$\delta'(y)^{\otimes 2} \otimes e_1 \in \langle (5+3y).\delta'(x)^{\otimes 2}, 2, a\rangle \subset K^1_{\Sigma_2}((CF,F)^2;Z/_2).$$

Thus $\delta(i_!(y^{\otimes 2})) = p^*(\delta'(y)^{\otimes 2} \otimes e_1) \in K^1((CF)^2,F^2;Z/_2)$.

In §A3.1 we have shown the following result.

Proposition A3.2:

Let $\Delta_{\Sigma_2} = (\delta^{-1})\circ p^*: K^\beta_{\Sigma_2}((CF,F)^2;Z/_2) \to K^{\beta-1}_{\Sigma_2}(F^2,*;Z/_2)$ and $\Delta_1 = \delta^{-1}: K^\beta(CF,F;Z/_2) \to K^{\beta-1}(F,*;Z/_2)$.

(i) If $x \in \ker \beta_2 \subset K^0(CF,F;Z/_2)$ then

$\Delta_{\Sigma_2}(x^{\otimes 2} \otimes e_0) = \Delta_1(x)^{\otimes 2} \otimes e_1 + i_!(z)$ for some

$z \in K^*(F^2;Z/_2)$.

(ii) If $x \in \ker \beta_2 \subset K^1(CF,F;Z/_2)$ then

$\Delta_{\Sigma_2}(x^{\otimes 2} \otimes e_1) = i_!(\Delta_1(x)^{\otimes 2})$.

§A3.3:

We now examine the homomorphism, δ, when $p \neq 2$. Let $(D,E) = (M_{p},m_o)$ or (point,ϕ). We determine the behaviour of the homomorphisms

$$\begin{array}{ccc} K^\alpha_{\pi_p}((F^p,*)\times(D,E)^p) & \xrightarrow[\cong]{\delta} & K^{\alpha+1}_{\pi_p}([(CF)^p,F^p]\times(D,E)^p) \\ & & \uparrow p^* \\ & & K^{\alpha+1}_{\pi_p}((CF,F)^p\times(D,E)^p) \end{array}$$

on suitable integral classes.

Case(a): $\delta: K^1_{\pi_p} \to K^0_{\pi_p}$.

Let $x \in K^1(F\times(D,E)) \cong K(F\times(I,\partial I)\times(D,E))$ be represented by a complex of trivial bundles

$(X,d_X) \equiv (0 \to \mathbb{C}^k \to \mathbb{C}^k \to 0)$ such that d_X is independent of $f \in F$ at the point, $(f,0,d)$. If $(D,E) = (\text{point},\phi)$ this representation is achieved as in §A3.1 by a difference element [c.f.Se;Sn 1,§3] representation if $(D,E) = (M_p,m_o)$ this representation is possible for integral classes in $K(F \times (I,\partial I) \times (M_p,m_o)) \cong K^1(F;Z/_p)$. However the other classes are represented by Massey products of the form $\langle z,p,a\rangle \subset K^1(F;Z/_p)$ and at one end of the interval, I, the Massey product differential is $d_Z \otimes G_1$ where (Z,d_Z) represents $z \in K^o(F)$ and so d_Z may be taken as zero.

Now form $(X,d_X)^{\otimes p}$ and restrict this to $F^p \times (I_o,\partial I_o) \times (D,E)^p$ where $I_o \subset I^p$ is the diagonal. From [Sn 2,§2], this complex represents an element

$$\left[\prod_{j=1}^{(p-1)/2} (1 - y^j)\right] x' \in K^{-1}_{\pi_p}(F^p \times (D,E)^p)$$

where x' is an element of

$$x^{\otimes p} \otimes e_o = (i_1^*)^{-1}(x^{\otimes p}) \subset K^{-1}_{\pi_p}(F^p \times (D,E)^p).$$

Now $(X,d_X)^{\otimes p}$ restricted in this manner in fact defines a complex on

$$(C(F^p),\ 0 \cup F^p) \times (D,E)^p$$

from which we may construct a representative of $\delta([\prod_j (1-y^j)]x')$ since δ is induced by the map $\delta: (CF)^p \cup [1,2] \times F^p \to C(F^p)$ given by

$$\delta(s,f_1,\dots,f_p) = [s-1,f_1,\dots,f_p] \text{ and}$$

$$\delta([t_1,f_1,\dots,t_p,f_p]) = 0\ ,$$

$$(s \in [1,2],\ t_i \in [0,1],\ f_i \in F).$$

Take the underlying family of π_p-vector bundles of

$$(0 \to \mathbb{C}^k \to \mathbb{C}^k \to 0)^{\boxtimes p} \times [(CF)^p \cup [1,2] \times F^p] \times D^p$$

and on this form the following homotopy, Φ_t, of complexes of π_p-vector bundles over

$$((CF)^p \cup [1,2] \times F^p, \{2\} \times F^p) \times (D,E)^p.$$

The differential, Φ_t, is given by

$$\Phi(t,s,f_1,\ldots,f_p,d_1,\ldots,d_p) = \mathop{\boxtimes}_{i=1}^{p} d_X(f_i,a(s,t),d_i)$$

and

$$\Phi(t,[t_1,f_1,\ldots,t_p,f_p]d_1,\ldots,d_p) = \mathop{\boxtimes}_{i=1}^{p} d_X(f_i,b(t_i,t),d_i),$$

($s \in [1,2]$, $t_i \in [0,1]$, $d_i \in D$, $f_i \in F$) where a and b are the functions used in §A3.1. At $t = 1$ this complex represents $\delta([\prod_j (1-y^j)]x') = [\prod_j(1-y^j)]\,\delta(x')$ and at $t = 0$ it represents $p^*(\delta'(x)^{\boxtimes p})$.

<u>Case(b): $\delta: K^0_{\pi_p} \to K^1_{\pi_p}$</u>.

Let $z \in K^0(F \times (D,E))$ and let $b.z \in K(F \times (I,\partial I) \times (I,\partial I) \times (D,E))$ be represented by a complex of trivial bundles $(Z,d_Z) \equiv (0 \to \mathbb{C}^k \to \mathbb{C}^k \to 0)$ such that d_Z is independent of $f \in F$ at the point (f,t_1,t_2,d) when $t_i = 0$ for $i = 1$ or 2. As in case(a) any element (to within an irrelevant sign which appears because triple Massey products have to be taken "upside down") can be represented in this manner. Now form $(X,d_X)^{\boxtimes p}$ and restrict it to $F^p \times (I_0,\partial I_0) \times (I,\partial I)^p \times (D,E)^p$. Now I^p is π_p-homeomorphic to the product of I_0 with the unit disc in the complex representation $(\sum_{j=1}^{(p-1)/2} y^j)$ of π_p, [c.f.Sn 2,§2]. If $\Lambda_{-1}(\sum_{1}^{(p-1)/2} y^j)$ is the Thom class of this representation the complex constructed above represents

$$b.z^{\otimes p}[\prod_{j=1}^{(p-1)/2}(1-y^{p-j})]\,\Lambda_{-1}(\sum_{1}^{(p-1)/2} y^j) \text{ in}$$

$$K_{\pi_p}(F^p \times (I_o,\partial I_o) \times (I,\partial I)^p \times (D,E)^p) \cong K_{\pi_p}^{-2}(F^p \times (D,E)^p),$$

since $b^{\otimes p} = b\,\Lambda_{-1}(\sum_{1}^{(p-1)} y^j) \in K_{\pi_p}((I^2,\partial I^2)^p) \cong K_{\pi_p}^{-2}(\sum_{j=1}^{(p-1)} y^j)$.

However $(Z,d_Z)^{\otimes p}$ restricted in this manner defines a complex on $(C(F^p),\, o \cup F^p) \times (I,\partial I)^p \times (D,E)^p$ from which, via the map δ, we may construct a representative of

$$\delta(b.\Lambda_{-1}.z^{\otimes p}.\,[\prod_{j}(1-y^{p-j})]) = b.\Lambda_{-1}.\prod_{j=1}^{(p-1)/2}(1-y^{p-j})\,\delta(z^{\otimes p}).$$

Take the underlying family of π_p-vector bundles of $(O \to \mathbb{C}^k \to \mathbb{C}^k \to O)^{\otimes p} \times [(CF)^p \cup [12] \times F^2] \times I^p \times D^p$ and on this form the following homotopy, Φ_t, of complexes of π_p-vector bundles over

$[(CF)^p \cup [1,2] \times F^p, \{2\} \times F^p] \times (I,\partial I)^p \times (D,E)^p$. The differential, Φ_t, is given by

$$\Phi(t,s,f_1,\ldots,f_p,v_1,\ldots,v_p,d_1,\ldots,d_p) = \bigotimes_{i=1}^{p} d_Z(f_i,a(s,t),v_i,d_i)$$

and

$$\Phi(t,[t_1,f_1,\ldots,t_p,f_p]\,v_1,\ldots,v_p,d_1,\ldots,d_p) = \bigotimes_{i=1}^{p} d_Z(f_i,b(t_i,t),v_i,d_i),$$

$(s \in [1,2];\ t_i,v_i \in [0,1];\ f_i \in F;\ d_i \in D)$.

At $t = 1$ this complex represents

$b.\Lambda_{-1}(\Sigma y^j).\,[\prod^{(p-1)/2}(1-y^{p-j})]\,\delta(z^{\otimes p})$ and at $t = 0$ it represents $p^*(\delta'(z))^{\otimes p}$ which is [Sn 2,§2] $\Lambda_{-1}(\sum_1^{(p-1/2}y^j)p^*(z')$ where z' is an element of

$\delta'(z)^{\otimes p} \otimes e_o = (i_1^*)^{-1}(\delta'(z)^{\otimes p}) \subset K^1_{\pi_p}(F^p \times (D,E)^p).$

From the discussion of §A.3.3 we have the following results. In $K^*_G((U,V)^p;Z/_p)$ $(G = \pi_p$ or $\Sigma_p)$ let $x^{\otimes p}\otimes e_{2_j}$ denote the subset of elements which are represented by the element $x^{\otimes p} \otimes e_{2_j} \in E^{2j',*}((U,V)^p;G;Z/_p)$. Hence $x^{\otimes p}\otimes e_{2p-2}$ is an element and $x^{\otimes p} \otimes e_o = (i^*)^{-1}(x^{\otimes p})$. For $G = \pi_p$ or Σ_p let $\Delta_G = (\delta^{-1})\circ p^*:K^\beta_G((CF,F)^p;Z/_p) \to K^{\beta-1}_G(F^p,*;Z/_p)$ and let $\Delta_1 = (\delta^{-1}): K^\beta_G(CF,F;Z/_p) \to K^{\beta-1}_G(F,*;Z/_p)$.

<u>Lemma A3.4</u>:

Let $p \neq 2$ be a prime. If $x \in K^\alpha(CF,F;Z/_p)$ then

(i) $(i_1)^* \Delta_{\pi_p}(x^{\otimes p} \otimes e_o) = 0 \in K^{\alpha-1}(F^p,*;Z/_p)$

(ii) $(i)^* \Delta_{\Sigma_p}(x^{\otimes p} \otimes e_o) = 0 \in K^{\alpha-1}(F^p,*;Z/_p)$.

<u>Proof</u>:

(ii) follows from (i) by naturality and the fact that there exist elements in $(x^{\otimes p} \otimes e_o) \cap \mathrm{im}(i_2^*)$.

(i) For integral classes, x, the discussion of §A3.3 shows

$$\Delta_{\pi_p}(x^{\otimes p}\otimes e_o) = \begin{cases} [\prod_{j=1}^{(p-1)/2}(1-y^j)](\Delta_1(x)^{\otimes p}\otimes e_o), \text{ if } \\ \qquad \alpha \equiv 0 \pmod 2 \\ [\prod^{(p-1)/2}(1-y^{p-j})](\Delta_1(x)^{\otimes p}\otimes e_o), \text{ if } \\ \qquad \alpha \equiv 1 \pmod 2 \end{cases}$$

which is in $\ker(i_1^*)$. For non-integral classes we show that $i_1^*\,\Delta_{\pi_p}(x^{\otimes p}\otimes e_o)$ is in the kernel of the monomorphism

$$S: K^*(F^p;Z/_p) \to K^*(F^p\times S^1_\pi;Z/_p).$$

By Appendix I there is a natural homomorphism, induced by a (stable) π_p-equivariant map,

$t: K^*_{\pi_p}((-)^p\times(M_p,m_o)^p) \to K^*_{\pi_p}((-)^p\times S^1_\pi;Z/_p)$ such that if

$$x\in K^*((CF,F)\times(M_p,m_o)) \cong K^*(CF,F;Z/_p)$$

$$t(x^{\boxtimes p}) = S(x^{\otimes p}\otimes e_o)\in K^*_{\pi_p}((CF,F)^p\times S^1_\pi;Z/_p).$$

Using the results of §A3.3 on the homomorphism $(\delta^{-1})\circ p^*$ described there (when $(D,E) = (M_p,m_o)$) we have $i_1^*\,\Delta\,(x^{\boxtimes p}) = 0$ and hence

$$\begin{aligned} 0 &= t\,i_1^*\,\Delta(x^{\boxtimes p}) \\ &= i_1^*(\delta^{-1})p^*t(x^{\boxtimes p}) \\ &= i_1^*(\delta^{-1})p^*\,t(x^{\boxtimes p}) \\ &= i_1^*(\delta^{-1})p^*\,S(x^{\otimes p}\otimes e_o) \\ &= S\,i_1^*\,\Delta_{\pi_p}(x^{\otimes p}\otimes e_o). \end{aligned}$$

Proposition A3.5:

Let $p \neq 2$ be a prime. Let $n_o = (\prod_{j=1}^{(p-1/_2)} j) \in Z/_p$.

For $x \in K^\alpha(CF,F;Z/_p)$

$$\Delta_{\Sigma_p}(x^{\otimes P} \otimes e_o) = n_o \, \Delta_1(x)^{\otimes P} \otimes e_{p-1}, \quad \text{if} \quad \alpha \equiv 0 (\text{mod } 2)$$

and

$$\Delta_{\Sigma_p}(x^{\otimes P} \otimes e_{p-1}) = (- {}^1/_{n_o}) \, \Delta_1(x)^{\otimes P} \otimes e_{2p-2}, \quad \text{if} \quad \alpha \equiv 1 (\text{mod } 2)$$

Proof:

The map i_2^* embeds $\{E_s^{**}((U,V)^P;\Sigma_p;Z/_p)\}$ in $\{E_s^{**}((U,V)^P;\, \pi_p;Z/_p)\}$ as a direct summand, sending $z \otimes e_j$ to itself. Hence the results of §A3.3 show that if $x \in K^\alpha(CF,F;Z/_p)$ is an integral class the following results are true. If $\alpha \equiv 0$(mod then

$$i_2^*(\Delta_{\Sigma_p}(x^{\otimes P} \otimes e_o)) = [\prod_1^{(p-1)/_2} (1-y^j)](\Delta_1(x)^{\otimes P} \otimes e_o)$$

$$= n_o (d^{p-1/2} + \text{ higher terms})(\Delta_1(x)^{\otimes P} \otimes e_o)$$

$$= n_o(\Delta_1(x)^{\otimes P} \otimes e_{p-1})$$

and if $\alpha \equiv 1$(mod 2)

$$i_2^*(\Delta_{\Sigma_p}(x^{\otimes P})) = \Delta_{\pi_p}(\prod_1^{(p-1)/_2}(1-y^j)\, x^{\otimes P} \otimes e_o) = \Delta_{\pi_p}(n_o x^{\otimes P} \otimes e_{p-1})$$

$$= \prod_1^{p-1} (1-y^j)(x^{\otimes P} \otimes e_o) = (p-1)!(x^{\otimes P} \otimes e_{2p-2}).$$

Thus the result is true for integral classes. In general we have for $x \in K^\alpha(CF,F;Z/_p)$ that

$$\Delta_{\Sigma_p}(x^{\otimes P} \otimes e_{p-1}) \in E_\infty^{2p-2,o} \cong K^o(F) \quad \text{if} \quad \alpha \equiv 1 (\text{mod } 2)$$

and

$$\Delta_{\Sigma_p}(x^{\otimes p} \otimes e_o) \in E_\infty^{p-1,1} \cong K^1(F) \qquad \text{if} \quad \alpha \equiv 0 (\mathrm{mod}\ 2).$$

Using these natural isomorphisms we may construct a natural transform $\Delta_\beta\colon \tilde{K}^\beta(F;Z/_p) \to \tilde{K}^\beta(F;Z/_p)$ defined by

$\Delta_\beta(F)(z) = \Delta_{\Sigma_p}(\delta'(z)^{\otimes p} \otimes e_j)$, ($j = 0$ or $p-1$, as appropriate).

This additive operation is defined for the Z-graded theory and commutes with multiplication by the Thom class $b \in K(D^2,S^1)$. Thus for each $\beta \in Z/_2$ $\Delta_\beta(-)$ is stable in the sense of Proposition 3.2 (proof). The calculations of [M] show, as in Proposition 3.2 (proof), that $\Delta_\beta(-)$ is determined by its behaviour on integral class, and this is just multiplication by n_o or by $(-^1/n_o)$.

REFERENCES

[An] D.W. Anderson: Universal Coefficient Theorems for K-theory (to appear).

[An-H] D.W. Anderson & L. Hodgkin: The K-theory of Eilenberg-Maclane complexes. (Topology 7 (1968) 317-29).

[A-T] S. Araki & H. Toda: Multiplicative structures in mod q cohomology theories I,II. (Osaka J.Math. 2 (1965) 71-115 & 3(1966) 81-120).

[A-Y] S.Araki & Z. Yosimura: Differential Hopf Algebras modelled on K-theory mod p. I. (Osaka J.Math. 8(1971) 151-206.

[At 1] M.F. Atiyah: Characters and cohomology of finite groups. (Pub.Math. No.9 IHES Paris).

[At 2] M.F. Atiyah: K-theory (Benjamin Press (1968)).

[At 3] M.F. Atiyah: Power operations in K-theory. (Quart.J.Math. 17(1966) 165-193 - reprinted in "K-theory"

[At-Se] M.F. Atiyah & G.B. Segal: Equivariant K-theory and completion. (J.Diff.Geom.3 (1969) 1-18).

[B] M.G. Barratt: Homotopy ringoids and homotopy groups. (Quart.J.Math. 5 (1954) 271-290).

[D-L] E. Dyer & R.K. Lashof: Homology of iterated loopspaces. (Am.J. Math. 84 (1962) 35-88).

[E-Mo] S. Eilenberg & J.C. Moore: Limits and spectral sequences. (Topology 1 (1961) 1-23).

[H1] L. Hodgkin: Dyer-Lashof operations in K-theory. (to appear in Proc.Oxford Symposium in Algebraic Topology (1972)).

[H2] L. Hodgkin: The K-theory of some well-known spaces - I.QS^o (Topology 11(1972) 371-375).

[K] S.O. Kochman: The homology of the classical groups over the Dyer-Lashof algebra. (Bull.A.M.Soc. 77(1) (1971) 142-147).

[K-P] D.S. Kahn & S.B. Priddy: Applications of the transfer to stable homotopy. (Bull.A.M.Soc. 78(6) (1972) 981-987).

[L] P.S. Landweber: On symmetric maps between spheres and equivariant K-theory. (Topology 9(1970) 55-61).

[M] C.R.F. Maunder: Stable operations in mod p K-theory. (Proc.Cambs.Phil.Soc. 63 (1967) 631-646).

[Ma] J.P. May: Geometry of Iterated Loopspaces. (Springer-Verlag, Lecture Notes in Maths. 271).

[P] S.B. Priddy: Dyer-Lashof operations for the classifying spaces of certain matrix groups. (preprint).

[R-S] M. Rothenberg & N.E. Steenrod: The cohomology of classifying spaces of H-spaces. (Bull.A.M.Soc. 71 (1961) 872-5 and preprint).

[Se] G.B. Segal: Equivariant K-theory. (Pub.Math. No.34 (1968) IHES Paris 129-151).

[Sn 1] V.P. Snaith: On the K-theory of homogeneous spaces and conjugate bundles of Lie groups. (Proc.L.M.Soc.(3) 22 (1971) 562-84.

[Sn 2] V.P. Snaith: On cyclic maps. (Proc.Cambs.Phil.Soc. (1972) 71, 449-456).

[Sn 3] V.P. Snaith: Massey Products in K-theory I,II. (Proc.Cambs.Phil.Soc. (1970) 68, 303-320 & (1971) 69, 259-289).

[W] G.W. Whitehead: Generalised homology theories. (Trans.A.M.Soc. 102 (1962) 227-283).

[Mi] R.J. Milgram: The mod 2 spherical characteristic classes (Annals of Maths (2)(1970)92, pp238-261).

[A] J.F. Adams: J(X) I-IV (Topology (2,3 & 5)(1963-1966)).

[C-S] R.R. Clough & J.D. Stasheff: BSJ does not map correctly into BSF mod 2. (manuscripta math.7, 205-214 (1972)).

[Su] D. Sullivan: Geometric Topology Part I (mimeo notes: MIT (1970)).

[Mad] I. Madsen: On the action of the Dyer-Lashof action in $H_*(G)$ and $H_*(G/_{Top})$. (Ph.D. dissertation U. Chicago (1970).

[Q] D. Quillen: On the cohomology and K-theory of the general linear groups over a finite field. (Annals of Math.96(1972)552-586).

[Sn4] V.P. Snaith: On $K_*(\Omega^2 X;Z/_2)$. (to appear).

Vol. 399: Functional Analysis and its Applications. Proceedings 1973. Edited by H. G. Garnir, K. R. Unni and J. H. Williamson. II, 584 pages. 1974.

Vol. 400: A Crash Course on Kleinian Groups. Proceedings 1974. Edited by L. Bers and I. Kra. VII, 130 pages. 1974.

Vol. 401: M. F. Atiyah, Elliptic Operators and Compact Groups. V, 93 pages. 1974.

Vol. 402: M. Waldschmidt, Nombres Transcendants. VIII, 277 pages. 1974.

Vol. 403: Combinatorial Mathematics. Proceedings 1972. Edited by D. A. Holton. VIII, 148 pages. 1974.

Vol. 404: Théorie du Potentiel et Analyse Harmonique. Edité par J. Faraut. V, 245 pages. 1974.

Vol. 405: K. J. Devlin and H. Johnsbråten, The Souslin Problem. VIII, 132 pages. 1974.

Vol. 406: Graphs and Combinatorics. Proceedings 1973. Edited by R. A. Bari and F. Harary. VIII, 355 pages. 1974.

Vol. 407: P. Berthelot, Cohomologie Cristalline des Schémas de Caracteristique $p > o$. II, 604 pages. 1974.

Vol. 408: J. Wermer, Potential Theory. VIII, 146 pages. 1974.

Vol. 409: Fonctions de Plusieurs Variables Complexes, Séminaire François Norguet 1970–1973. XIII, 612 pages. 1974.

Vol. 410: Séminaire Pierre Lelong (Analyse) Année 1972–1973. VI, 181 pages. 1974.

Vol. 411: Hypergraph Seminar. Ohio State University, 1972. Edited by C. Berge and D. Ray-Chaudhuri. IX, 287 pages. 1974.

Vol. 412: Classification of Algebraic Varieties and Compact Complex Manifolds. Proceedings 1974. Edited by H. Popp. V. 333 pages. 1974.

Vol. 413: M. Bruneau, Variation Totale d'une Fonction. XIV, 332 pages. 1974.

Vol. 414: T. Kambayashi, M. Miyanishi and M. Takeuchi, Unipotent Algebraic Groups. VI, 165 pages. 1974.

Vol. 415: Ordinary and Partial Differential Equations. Proceedings 1974. XVII, 447 pages. 1974.

Vol. 416: M. E. Taylor, Pseudo Differential Operators. IV, 155 pages. 1974.

Vol. 417: H. H. Keller, Differential Calculus in Locally Convex Spaces. XVI, 131 pages. 1974.

Vol. 418: Localization in Group Theory and Homotopy Theory and Related Topics. Battelle Seattle 1974 Seminar. Edited by P. J. Hilton. VI, 172 pages 1974.

Vol. 419: Topics in Analysis. Proceedings 1970. Edited by O. E. Lehto, I. S. Louhivaara, and R. H. Nevanlinna. XIII, 392 pages. 1974.

Vol. 420: Category Seminar. Proceedings 1972/73. Edited by G. M. Kelly. VI, 375 pages. 1974.

Vol. 421: V. Poénaru, Groupes Discrets. VI, 216 pages. 1974.

Vol. 422: J.-M. Lemaire, Algèbres Connexes et Homologie des Espaces de Lacets. XIV, 133 pages. 1974.

Vol. 423: S. S. Abhyankar and A. M. Sathaye, Geometric Theory of Algebraic Space Curves. XIV, 302 pages. 1974.

Vol. 424: L. Weiss and J. Wolfowitz, Maximum Probability Estimators and Related Topics. V, 106 pages. 1974.

Vol. 425: P. R. Chernoff and J. E. Marsden, Properties of Infinite Dimensional Hamiltonian Systems. IV, 160 pages. 1974.

Vol. 426: M. L. Silverstein, Symmetric Markov Processes. X, 287 pages. 1974.

Vol. 427: H. Omori, Infinite Dimensional Lie Transformation Groups. XII, 149 pages. 1974.

Vol. 428: Algebraic and Geometrical Methods in Topology, Proceedings 1973. Edited by L. F. McAuley. XI, 280 pages. 1974.

Vol. 429: L. Cohn, Analytic Theory of the Harish-Chandra C-Function. III, 154 pages. 1974.

Vol. 430: Constructive and Computational Methods for Differential and Integral Equations. Proceedings 1974. Edited by D. L. Colton and R. P. Gilbert. VII, 476 pages. 1974.

Vol. 431: Séminaire Bourbaki – vol. 1973/74. Exposés 436–452 IV, 347 pages. 1975.

Vol. 432: R. P. Pflug, Holomorphiegebiete, pseudokonvexe Gebiete und das Levi-Problem. VI, 210 Seiten. 1975.

Vol. 433: W. G. Faris, Self-Adjoint Operators. VII, 115 pages. 1975.

Vol. 434: P. Brenner, V. Thomée, and L. B. Wahlbin, Besov Spaces and Applications to Difference Methods for Initial Value Problems. II, 154 pages. 1975.

Vol. 435: C. F. Dunkl and D. E. Ramirez, Representations of Commutative Semitopological Semigroups. VI, 181 pages. 1975

Vol. 436: L. Auslander and R. Tolimieri, Abelian Harmonic Analysis Theta Functions and Function Algebras on a Nilmanifold. V, 99 pages. 1975.

Vol. 437: D. W. Masser, Elliptic Functions and Transcendence XIV, 143 pages. 1975.

Vol. 438: Geometric Topology. Proceedings 1974. Edited b L. C. Glaser and T. B. Rushing. X, 459 pages. 1975.

Vol. 439: K. Ueno, Classification Theory of Algebraic Varietie and Compact Complex Spaces. XIX, 278 pages. 1975

Vol. 440: R. K. Getoor, Markov Processes: Ray Processes an Right Processes. V, 118 pages. 1975.

Vol. 441: N. Jacobson, PI-Algebras. An Introduction. V, 115 pages 1975.

Vol. 442: C. H. Wilcox, Scattering Theory for the d'Alember Equation in Exterior Domains. III, 184 pages. 1975.

Vol. 443: M. Lazard, Commutative Formal Groups. II, 236 pages 1975.

Vol. 444: F. van Oystaeyen, Prime Spectra in Non-Commutativ Algebra. V, 128 pages. 1975.

Vol. 445: Model Theory and Topoi. Edited by F. W. Lawvere C. Maurer, and G. C. Wraith. III, 354 pages. 1975.

Vol. 446: Partial Differential Equations and Related Topics Proceedings 1974. Edited by J. A. Goldstein. IV, 389 pages 1975.

Vol. 447: S. Toledo, Tableau Systems for First Order Numbe Theory and Certain Higher Order Theories. III, 339 pages. 1975

Vol. 448: Spectral Theory and Differential Equations. Proceeding 1974. Edited by W. N. Everitt. XII, 321 pages. 1975.

Vol. 449: Hyperfunctions and Theoretical Physics. Proceeding 1973. Edited by F. Pham. IV, 218 pages. 1975.

Vol. 450: Algebra and Logic. Proceedings 1974. Edited by J. N Crossley. VIII, 307 pages. 1975.

Vol. 451: Probabilistic Methods in Differential Equations. Proceedings 1974. Edited by M. A. Pinsky. VII, 162 pages. 1975.

Vol. 452: Combinatorial Mathematics III. Proceedings 197 Edited by Anne Penfold Street and W. D. Wallis. IX, 233 page 1975.

Vol. 453: Logic Colloquium. Symposium on Logic Held at Bosto 1972–73. Edited by R. Parikh. IV, 251 pages. 1975.

Vol. 454: J. Hirschfeld and W. H. Wheeler, Forcing, Arithmeti Division Rings. VII, 266 pages. 1975.

Vol. 455: H. Kraft, Kommutative algebraische Gruppen und Ring III, 163 Seiten. 1975.

Vol. 456: R. M. Fossum, P. A. Griffith, and I. Reiten, Trivial E tensions of Abelian Categories. Homological Algebra of Trivi Extensions of Abelian Categories with Applications to Rir Theory. XI, 122 pages. 1975.